AAPS Advances in the Pharmaceutical Sciences Series

Volume 51

The AAPS Advances in the Pharmaceutical Sciences Series, published in partnership with the American Association of Pharmaceutical Scientists, is designed to deliver volumes authored by opinion leaders and authorities from around the globe, addressing innovations in drug research and development, and best practice for scientists and industry professionals in the pharma and biotech industries. Indexed in
Reaxys
SCOPUS Chemical Abstracts Service (CAS) SCImago
EMBASE

Mine Orlu • Fang Liu
Editors

Pharmaceutical Formulations for Older Patients

Springer

aaps®

Editors
Mine Orlu
UCL School of Pharmacy
University College London
London, UK

Fang Liu
Department of Clinical Pharmaceutical and Biological Sciences
School of Life and Medical Sciences
University of Hertfordshire
Hatfield, UK

ISSN 2210-7371 ISSN 2210-738X (electronic)
AAPS Advances in the Pharmaceutical Sciences Series
ISBN 978-3-031-35810-4 ISBN 978-3-031-35811-1 (eBook)
https://doi.org/10.1007/978-3-031-35811-1

This Springer imprint is published by the registered company Springer Nature Switzerland AG
The registered company address is: Gewerbestrasse 11, 6330 Cham, Switzerland

Paper in this product is recyclable.

Preface

In the last decade, the pharmaceutical community responded positively to the demographic changes of the society with increasing proportion of older people (65+ years). Patient-centric pharmaceutical design becomes a topic of discussion in regulatory reflections, conference proceedings, review articles and book chapters. The complexity of the subject is recognised; many factors are at play and intertwined affecting the suitability of a product to the needs of older patients: the heterogeneity of the aged population, the multi-dimensional product features, the requirements of health authorities and the involvement of healthcare professionals and caregivers. Pharmaceutical formulations play a unique and critical role in linking the puzzling pieces. Our understanding has improved on how the attributes of those "classic" dosage forms, such as tablets and capsules, affect older patients taking their medicines. New, innovative technologies emerge adding to the "toolbox" of formulators to enhance product performance and patient experience. This book provides a critical review of the contributions of formulations, old and new, in the pursuit of better therapies for older adults.

The book begins with an overview of the important aspects of medicinal treatments in older people, the medication optimisation, safety and acceptability (Chaps. 1–3). The topics cover the social and clinical backgrounds of medicine use, the interactions of patient characteristics and pharmaceutical products, the roles of healthcare professionals and the unique design of a standardised, multivariate approach to evaluate acceptability of medicines in older adults and identify drivers for improvement. The following two chapters (Chaps. 4 and 5) discuss the design of patient-centric dosage forms from industrial perspectives, from regulatory framework, target product profiles to human factor studies. The second part of the book (Chaps. 6–9) explore emerging technologies and their applications in therapies relevant to older patients, including advancement in oral sustained release formulations, 3D printing, the ageing microbiome and tailored vaccines for older individuals. We are truly grateful to all of the authors for their knowledge and expertise reflected to their chapters and their commitment to this book. These chapters provide an understanding of the relevance of ageing physiology on specific formulation design, a comprehensive review of the available technologies and prospects of their use in personalised, precision medicines.

There is still a long way to go in achieving better medicines for later life. Pharmaceutical development is largely driven by regulatory guidelines; however, keeping the end user in mind in product design is both rewarding and market differentiating. This book will be an interesting read for everyone engaged in the research, development, prescribing and assisting the use of medicines in older adults.

London, UK Mine Orlu
Hatfield, UK Fang Liu

Contents

About the Editors

Mine Orlu is Professor of Pharmaceutics at the UCL School of Pharmacy. Professor Orlu's research focuses on designing pharmaceutical formulations tailored considering the physiological alterations and medicine administration related needs occurring at advanced age. Professor Orlu has published over 100 papers in leading journals of pharmaceutics. She is one of the World Economic Forum (WEF)'s Young Scientists (Class of 2020).

Fang Liu is Professor of Pharmaceutics at the Department of Clinical Pharmaceutical and Biological Science, University of Hertfordshire. Fang's research interest is in the development of age-appropriate medicines for children and older patients especially using oral solid dosage forms. She is the Founder Director of Fluid Pharma Ltd, a University of Hertfordshire spin out to advance paediatric and geriatric medicine development. Fang is the Co-Chair of Age-related Medicines Focus Group at the Academy of Pharmaceutical Sciences of Great Britain and a Member of the European Paediatric Formulation Initiative.

Medication Optimisation in Older People

Emma L. Smith and Ian Maidment

Contents

Abstract The population of the UK and many countries worldwide is ageing. Older people tend to suffer from more diseases and therefore take more medication. This is in addition to other factors, such as physical and mental frailty, which can make medication optimisation more challenging in older people.

This chapter introduces some of the key challenges associated with optimising medications for older people. It explores key social and clinical areas in older people and how this relates to medication optimisation including areas such as evidence-based prescribing, frailty, dementia, polypharmacy, deprescribing and swallowing difficulties. Ultimately, it aims to encourage person-centred care when optimising medications in the older person.

E. L. Smith (✉) · I. Maidment
Aston Pharmacy School, College of Health and Life Sciences, Aston University, Birmingham, UK
e-mail: e.smith9@aston.ac.uk; i.maidment@aston.ac.uk

M. Orlu, F. Liu (eds.), *Pharmaceutical Formulations for Older Patients*, AAPS Advances in the Pharmaceutical Sciences Series 51,
https://doi.org/10.1007/978-3-031-35811-1_1

Keywords Medicines optimisation · Person-centred care · Frailty · Dementia · Caregivers

1 Introduction

We are all living longer – this is something to be celebrated. As a consequence of this, the UK, in common with many countries across the world, has an ageing population. In 2019, Age UK reports that there are approaching 12 million people aged 65 plus; of these, 5.4 million people are aged 75 or more, and 1.6 million are aged 85 plus [1]. Looking to the future, in 50 years, it is estimated that there will be an extra 8.6 million people aged 65 years plus, approximately the current population of Scotland and Wales combined. The 85 plus population is the fastest-growing group and estimated to treble by 2066 to 5.1 million, making it to be around 7% of the total UK population. There are differences in health and social care systems across the world. However, although this chapter uses the UK to illustrate medication optimisation in older people, many of the issues discussed will likely be applicable outside of the UK particularly in countries with similar health need and healthcare systems.

Multimorbidity, which is defined as two or more long-term conditions, is common in older people [2]. Around 65% of people aged 65 plus and 82% of people aged 85 plus have two or more long-term conditions [3]. This multimorbidity often results in polypharmacy; there is no consistent definition of polypharmacy, but it is often considered to be taking five or more medicines [4]. Alternatively, problematic polypharmacy is considered the prescribing of multiple medications inappropriately, or where the intended benefit of the medications is not realised [5].

Over the last few years, polypharmacy in older people has rapidly increased. The number of older people taking five or more different medications has quadrupled from 12% to 49%, whilst the proportion of older people not on any medication has gone from 1 in 5 to 1 in 13 [6]. In older people, polypharmacy, or taking five plus medications, "is the new normal" [7].

There are a number of other key definitions related to medication taking that are used throughout this chapter; these are discussed below.

Medication management is the entire way that medicines are selected, procured, delivered, prescribed, administered and reviewed to optimise the contribution that medicines make to producing informed and desired outcomes of patient care [8].

Medication optimisation is a person-centred approach to safe and effective medicine use, to ensure people obtain the best possible outcomes from their medicines [5].

Adherence to medicines is the extent to which the patient's action matches the agreed recommendations with the prescriber [9].

The term 'older person' not only covers a wide range of ages but also a wide spectrum of levels of physical and mental health and ultimately frailty. We cannot assume that an 85-year-old is more frail than a 65-year-old, for example. There will also be intersectionality with other characteristics such as gender and ethnicity to consider.

Person-centred care is something that should always be at the forefront of clinical practice, focusing on the needs and wants of each individual person. This concept is not new, although it has evolved over the years. Balint [10] described patient-centred medicine and proposed that the patient should be "understood as a unique human being" [10]. The term has evolved to person centred to capture that a patient is foremost a person, not their symptoms or diagnosis, and to enable true holistic care [11, 12]. Clinical decisions and care should be a partnership between the person, any family carer or other caregiver (including social aspects) and the healthcare team, ensuring that preferences, needs and values produce decisions that are responsive and respectful for each individual [12]. There is also the need to link with any social care provider.

Medication optimisation is multifaceted, and whilst this chapter is split into sections, there are many connections between each, and they should not be considered in isolation. Similarly, of course, every person is different, and not each section will apply to everyone or to the same extent. This chapter will elaborate on some key social, clinical and medication formulation areas relating to medicines optimisation in older people, exploring areas such as evidence-based prescribing, frailty, dementia, polypharmacy, deprescribing and swallowing difficulties. The ultimate aim is to encourage person-centred care when optimising medications in the older person.

2 Key Clinical and Social Issues in Older People and Related Medication Optimisation Challenges

This section focuses on some key clinical and social issues that can pose a challenge in medication optimisation in the older person. They should not be considered in isolation but rather how they connect and may perpetuate each other.

2.1 Frailty

Frailty is often considered a condition of ageing, whereby an individual is more vulnerable to external stressors due to there being a poor resolution of homeostasis when a stressor event occurs. Essentially, a person cannot recover as easily to their baseline functioning [13, 14]. This external stressor could be one of many events and may appear quite minor, such as a new medication being started or a nonserious infection [15]. However, this may result in marked change in health status, for example, a change from someone being independent to becoming dependent [16].

Frail older people are among those at the greatest risk of adverse events such as falls, hospital admission, the need for long-term residential care and ultimately mortality. Although frailty is related to ageing, as well as other factors such as disability and comorbidity, it is a different concept. Whilst frailty prevalence increases with age, it is not linearly related to chronological age [16]. It has been estimated that around a quarter to half of people over the age of 85 years are frail, which means between half and three quarters of people older than 85 are not defined as frail [15].

There are several criterion or scales that can be used to assess and define frailty. One commonly used is known as either the frailty index score or the Clinical Frailty Scale. This was proposed by Rockwood and Mitnitski and assesses the level of dependence on care providers [13]. There are 9 categories, ranging from 1 which is 'very fit' and 9 which is 'terminally ill'. People move from 4, which is 'vulnerable', to 5, which is 'mildly frail', when they start to need help with activities of daily living such as shopping and housework [13, 14]. This means that even a small incident could cause an irreversible decline in functioning.

In England, screening for frailty identification is routine for people who are 65 years or older. The National Health Service (NHS) reports that it was the first health system in the world to use a 'population-based stratification approach' to systematically identify those 65 years old or older who have moderate or severe frailty [17].

As mentioned, medication can worsen frailty [18]. Polypharmacy and medication with sedative or anticholinergic effects are particularly concerning. A cohort study found the regular usage of sedative medication as assessed by the total sedative load was associated with frailty [19]. Conversely, some medication may help, for example, supplementing with vitamin D with calcium may reduce the rate of falls among older people who are housebound [20]. The British Geriatric Society recommends a structured medication review (SMR) as a key clinical intervention to limit frailty in older people [21], and NHS England requires Primary Care Networks to identify and prioritise people who would benefit from a SMR, which includes those living in care homes and those with severe frailty [22].

The British Geriatric Society in May 2020 published guidance on end-of-life care in frailty and the importance of medication management in older people [23].

The guidance covers:

- Medication review.
- Who should do a medication review.
- Triggers for a medication review.
- The role of the patient and carer (specifically informal, also called family, carer).
- Making every contact count (e.g. the pharmacy delivery driver could ask the older person and family carer how they are getting on with their medication) [7, 24, 25].
- Tools to support structured medication reviews.

2.2 Falls

A fall in an older person, whether they are frail or not, can be life-threatening and devastate quality of life. Many never regain the same functioning after a fall. Many different medications are potentially associated with falls and subsequent fractures including medications with anticholinergic activity [26, 27], high-dose 'z-drugs'

[28], antidepressants [29], benzodiazepines [30] and treatments for hypertension [31]. Acetyl cholinesterase inhibitors used to treat the symptoms of dementia can also lead to syncope and falls [32].

The risk of a fall is generally believed to increase at higher doses, for example, a recent cohort study, funded by the NIHR, found that z-drugs (e.g. zopiclone) are associated with a dose-dependent increase in fracture risk [33]. Decisions to prescribe any medication associated with falls should balance the potential benefits against the risks. Falls prevention strategies may be required. Again, a structured medication review should be done on any medication that might increase the risk of falls.

2.3 Evidence-Based Medicine in Older People

It is important that the use of medicines is evidence based and guidelines are utilised in practice to support evidence-based prescribing. However, fixed guidelines for each condition can lead to people who have multiple conditions being prescribed numerous medications, leading to polypharmacy. Older people are more likely to fit into this category. This puts them at greater risk of adverse drug reactions [34]. Guidelines may also not take into consideration certain personal characteristics that could impact management, and this could include increasing age and frailty. This is likely linked to underrepresentation in clinical trials, discussed in this section.

Some guidelines do make recommendations on age, but frailty is also a consideration. A 60-year-old could be considered frail, whereas an 80-year-old may not be frail. Of course, guidelines cannot factor in all variables and are not 'one size fits all'; therefore, it is important health and social care professionals ensure the care they give is person centred – 'treat the person and not the guideline'. However, there is an argument for consideration of both age and frailty in guidelines.

As we age, how the body handles active pharmaceutical compounds changes, and so we expect to see some variability in pharmacodynamics and pharmacokinetics. Older people may also experience different adverse drug reactions to those younger or due to multimorbidity have conditions or be on combinations of medications not studied. In general, older people are not adequately represented in the trials that are used to make the evidence-based guidelines. Strict inclusion criteria may exclude them from drug trials. There is a particular under-representation of the 'oldest' of older people due to the age limits imposed, although this does seem to be improving [35]. Although some medications may have recommendations for a different dose for 'the elderly', more research needs to include older people with different levels of frailty [36].

The European Medicines Agency (EMA) has been working on increasing the number of older people in clinical trials focusing on investigational medicines. Cherubini et al. in 2011 reported that there was an arbitrary upper age limit in clinical trials for heart failure and that over 40% of potential participants had one or more exclusion criteria that was felt to be unjustified [37]. To be able to understand why this is, the PREDICT study [38, 39] investigated the views of professionals, patients and carers on the under-representation of older people in clinical trials. Older people do want to see themselves represented in trials and for the outcomes of trials to be meaningful for them.

Further research is now being done into managing long-term conditions in older people, taking into consideration frailty where possible. Type 2 diabetes is one condition where there is currently published literature. In England, those 65 years old and older constitute around 52% of all adults with type 2 diabetes [40]. Older people are more susceptible to hypoglycaemia and the consequences of this, which includes falls, fractures, admission to hospital and increased risk of cardiovascular events [41, 42]. Some older people with type 2 diabetes are over-treated. This can be for many reasons including their medications and targets not being reviewed as their age and frailty advance [42].

Frailty has been recognised as a key determinant of the management of diabetes and should be taken into consideration routinely as part of reviewing older people with type 2 diabetes, including consideration of blood glucose targets and prescribing of medication [42]. When considering prescribing of medication for type 2 diabetes, there are now recommendations to include the consideration of age and also frailty [41].

2.4 Polypharmacy and Overprescribing

The King's Fund published a report in 2013 focusing on polypharmacy [43]. In this, they discussed that polypharmacy could be 'appropriate' and 'problematic'. Polypharmacy can bring many risks to the older person, especially if they are already frail. There is an observed relationship between polypharmacy and adverse outcomes including adverse drug reactions, impact on cognitive and physical functioning, hospital admission and ultimately mortality [44]. It is also thought that polypharmacy increases the incidence of poor adherence to medication, although reported levels of non-adherence in those who are living with multimorbidity varies [45]. A study by Paisansirikul et al. [46], based in Thailand, did find that in older people, polypharmacy was one of the factors significantly associated with non-adherence.

However, it is important that people are not automatically denied medication that would be classed as adding to 'appropriate polypharmacy' as this can extend life expectancy and quality of life [43]. The example of prescribing vitamin D was discussed earlier in the chapter.

A UK report published in 2021 estimated that around 10% of medicines may be overprescribed, although they acknowledged that it not possible to know the full extent [47]. Overprescribing can take several forms: someone may be prescribed a medication when non-pharmacological management would have been more appropriate; a medication may be considered generally appropriate to treat a condition but is not appropriate for the individual person it is prescribed to; or a person's condition may change, and the medication is no longer appropriate for them, but it is continued without review, or the medication is no longer needed or benefits the person, but they continue to be prescribed it without review [47].

2.5 *Anticholinergic Burden*

Medicines with anticholinergic activity inhibit the activity of acetylcholine at its receptor and are used to treatment of many conditions including Parkinson's disease, urinary incontinence, psychosis, depression and epilepsy [48, 49]. Medicines with anticholinergic activity are used by between 10% and 50% of older and middle-aged people at any one time [48, 49]. Anticholinergic effects are common and include peripheral effects, frequently dry mouth, blurred vision and constipation, and central effects such as confusion, delirium and cognitive impairment [50].

Many medicines have anticholinergic activity. So-called anticholinergics can be classified into two groups:

1. Medicines that exert their therapeutic activity via anticholinergic activity, for example, bladder medications such as oxybutynin.
2. Medicines whose therapeutic activity is not believed to be dependent on anticholinergic activity (e.g. antipsychotics, H2-antagonists, antidepressants).

There are numerous scales (20 plus) that have been developed to rate anticholinergic activity. Essentially, the vast majority are consensus scales (developed by experts) that rate anticholinergic burden at 0, 1, 2, or 3 (with 3 being more potent) [51]. Medicines in group one, which exert therapeutic activity via the cholinergic system, tend to be more potent anticholinergics (e.g. score 3 on the scales). The total score for each medication, that someone is taking, is summed to obtain the overall anticholinergic burden score for that an individual.

All these scales should be used to guide prescribing with the overall focus being to reduce the total medication burden rather than focusing on reducing the score.

2.6 *Deprescribing*

Deprescribing is something that can be implemented in response to polypharmacy as part of a medication review and seeks to stop any medications that are now inappropriate or now not indicated for a person [52–54]. Reducing inappropriate polypharmacy can have many advantages. It can help minimise side effect burden, risk of adverse drug reactions and drug-disease interactions [55]. It can help optimise adherence, not only through the reduction in the number of medications but also from forming a therapeutic alliance with the clinician. It can reduce the risk of adverse events including renal impairment, cognitive impairment and falls and also reduce morbidity and mortality [56]. This may be of particular value in frail older people at greater risk of adverse events. Overall, as well as reducing healthcare costs, it can importantly lead to an improved quality of life.

Deprescribing is not a simple task. It requires time, good communication and buy-in from the older person and the clinician [57], plus any family carer involved. Many deprescribing frameworks exist; some examples are provided in Box 1. Frameworks may be explicit, where they are based on criteria and focus on the medication, or may be implicit, where they are more judgement based and focus on the person and their wishes. A person-centred approach to deprescribing is likely to be most successful and have better outcomes [56]. Whilst the frameworks can be useful tools, the older person's own values and beliefs about medicines need to be an integral part of the process. Incorporating both the person's and the clinician's priorities for medical treatment contributes to shared decision-making and hopefully a successful outcome [56]. As mentioned, deprescribing is not a simple task; Box 2 highlights a case where medication review and deprescribing should be considered.

Box 1: Examples of Deprescribing Frameworks and Tools

- STOPP/START [58].
- PrescQIPP Polypharmacy and Deprescribing Webkit [59].
- NO TEARS [60].
- NHS Scotland 7 steps [61].
- Bruyère Research Institute guidelines and algorithms available via https://deprescribing.org
  - Proton pump inhibitors [62].
  - Anti-hyperglycaemic agents [63].
  - Antipsychotics [64].
  - Cholinesterase inhibitors and memantine [65].
  - Benzodiazepine receptor agonists [66].

Box 2: Deprescribing Case Example

AP is 76-year-old who has been complaining of feeling dizzy and unstable on her feet. This morning, she was found on the floor by her daughter between the bedroom and the bathroom. She is brought to the hospital via ambulance for assessment.

Excerpt from observations on admission:

Blood pressure – 108/71 mmHg.
Heart rate – 62 BPM.
Temperature – 37.6 °C.
Random blood glucose – 6.3 mmol/L.

Past medical history (PMH):

Hypertension, urinary incontinence, back pain, insomnia following the recent death of her partner.

Medication history:

Amlodipine 5 mg tablets: 1 tablet once daily.
Bisoprolol 2.5 mg tablets: 1 tablet once daily.
Omeprazole 20 mg capsules: 1 capsule once daily.

Solifenacin 5 mg tablets: 1 tablet once daily.
Co-codamol 30/500 tablets: 2 tablets up to four times a day when required.
Tramadol 50 mg capsules: 1–2 tablets up to four times a day when required.
Senna 7.5 mg tablets: 2 tablets at night.
Lactulose solution: 15 mL twice a day.
Atorvastatin 20mg tablets: 1 tablet once daily.
Zopiclone 7.5 mg tablets: 1 tablet one at night when required.
No known drug allergies.

Some suggested discussion points:

If AP consents to a medication review, any decisions should involve her input. Her views and wants should be the centre of any decision jointly made. If appropriate and consent was given, her daughter could also be involved (informal carer).

More information would be needed. This may need to be obtained from the GP and any other healthcare professionals involved in prescribing, and more broadly medication management, for AP.

Review of antihypertensives should be undertaken as AP's systolic blood pressure was 108 mmHg on admission. A low blood pressure could contribute to further falls. Consider if appropriate to continue these medications, monitoring blood pressure. If indicated, could blood pressure be controlled on amlodipine alone. This could be titrated to maximum dose if needed once bisoprolol is stopped safely. Need to check PMH to ensure the bisoprolol is not for another indication not documented as beta-blockers not routinely used for blood pressure control under current guidelines.

Two opioid pain relief medications are prescribed, both as 'as required'. Confirm how often AP uses these medications. They can both cause sedation and confusion. Consider stopping at least one, tramadol if possible and agreed, as it can have more central nervous system side effects. Make plan to reduce slowly.

Confirm why AP is prescribed omeprazole. There appears to be no indication for this in her history and it is also associated with osteoporosis long-term, as well as other adverse effects.

Confirm how long AP has been taking zopiclone. This could be cause of a 'hangover' effect in the morning. This should only be used short term. A dose of 3.75 mg would have been more appropriate given AP's age. Enquire if she has been offered psychological support for her partner's death and whether this is something she would like to consider.

Review solifenacin as it could precipitate confusion. Consider best option for urinary incontinence.

2.7 Dementia

Dementia is a disease primarily associated with old age; the older that we get, the more likely we are to suffer from dementia. According to the Alzheimer's Society in 2019, there were 850,000 people with dementia in the UK representing 1 in every 14 of people aged 65 plus [67]. Based on the current prevalence, this will increase to more than 1.5 million people by 2040 [67].

Thus, it is vital to avoid, whenever possible, medications that may worsen the symptoms of dementia. Particular concern has been expressed at the use of medications with anticholinergic activity [68–70]. The risks appear to be greatest with new users of potent anticholinergics, particularly anticholinergics that act therapeutically via the cholinergic system, such as oxybutynin, and in people with mild dementia or mild cognitive impairment [50, 68, 69].

One of the key symptoms of dementia is cognitive impairment. This can be manifest as memory problems, confusion or difficulty with higher level so-called 'executive task' [71]. Managing a complex medication regimen is an executive task. Thus, dementia can impact the ability of the older person to manage their medication regimen [72]. This may increase the risk of medication error because one of the barriers to a medication error is potentially removed. Box 3 illustrates an example.

Box 3: Cognitive Impairment and Medication Error

LJ is 78 years old and suffers from atrial fibrillation, a raised cholesterol, high blood pressure and depression. He lives alone. He has recently been diagnosed with MCI (mild cognitive impairment) – a precursor for dementia – and is waiting to see the old-age psychiatric team.

He receives the following treatments:

Digoxin 62.5 microgram tablets: 1 tablet once daily.
Apixaban 5 mg tablets: 1 tablet twice a day.
Atorvastatin 20 mg tablets: 1 tablet once daily.
Citalopram 20 mg tablets: 1 tablet once daily.

Prescriptions are dispensed every month by his local community pharmacy. One busy month, by mistake, the local pharmacy dispensed 250 micrograms tablets of digoxin. Due the confusion associated with the cognitive impairment in dementia, LJ fails to identify the medication error and takes the higher dose. He starts to suffer from nausea, dizziness and visual disturbances. His daughter rings his doctor for advice, and LJ is admitted to the hospital where blood tests reveal he has digoxin toxicity, for which he receives supportive treatment.

With increasing cognitive impairment, medication management usually moves from the responsibility of the older person to a carer (either formal or informal) (see section 'Social Issues'). However, many older people live alone, or if they have a family member they live with, this family member may also have cognitive impairment.

2.8 Social Issues

A complex team of practitioners and family and friends support older people with medication management. Both informal (family and friends) and formal (paid) carers have a key role, particularly if the older person suffers from dementia [73, 74]. Paid carers in this context include those working in residential and nursing homes, and carers going into an older person's home to support activities of daily living including medication management.

Informal carers frequently struggle with medication management. Unlike formal carers, they receive little or no formal training and yet conversely may be asked to undertake more complex tasks [24, 25]. The MEMORABLE study, funded by the National Institute for Health Research (NIHR, who are effectively the research arm of the NHS), mapped out medication management as a five-stage process and identified the importance of burden as seen in Fig. 1 [24, 25, 74].

One key role for community pharmacy can be supporting family carers – this could be as simple as organising home delivery of medication or supplying medication in non-children proof containers [75].

The capacity of the older person and any family carer plus the total medication management workload are key drivers in burden experienced [24, 25]. Both fluctuate. High burden, when informal carers and older people struggle to cope, is likely to be experienced when capacity is reduced, for example, the older person or informal carer suffers from dementia, or when workload is high, for example, the older person is taking a complex regimen. This burden is often hidden from clinicians, and the consequences include non-adherence and medication-related adverse events [7, 24, 25, 72].

Social care (e.g. formal carers) can play a key role in medication optimisation. Formal carers work both in care homes (residential and nursing) and in the community (when they visit older home people's homes to provide support). A key role for pharmacy can be supporting formal carers, and this includes providing compliance aids, education and training and providing medication-related information.

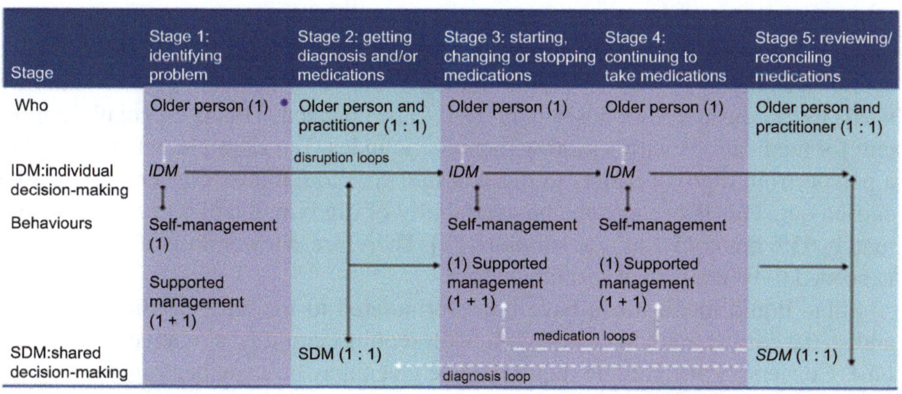

Fig. 1 Five-stage process of medication management in older people. (Reproduced with permission from Maidment et al. [24])

3 Importance of Formulation in Medication Optimisation

3.1 Formulation Considerations for People with Swallowing Difficulties

Swallowing difficulties, or oropharyngeal dysphagia as it known medically, become more common as people age [76]. Oral frailty has been shown to have a link between general frailty and decline in general health. There currently does not seem to be an agreement on defining oral frailty, but it includes consideration of poor oral health and poor oral functioning, e.g. ability to chew and oral motor function [77, 78]. As oral frailty increases, there is an increased need for support with eating and oral health partly related impairments in chewing and swallowing difficulties [77]. Dysphagia is an issue within itself as it is associated with malnutrition and dehydration and increases the risk of pneumonia [79]. Ultimately it leads to increased morbidity and mortality.

Most medications are given orally as solid oral dosage forms, and dysphagia may therefore present a problem for the safe administration of medication [80]. People with swallowing difficulties should be reviewed by a speech and language therapist and if appropriate be initiated on a modified diet, which may include things such as pureed food and thickened fluids. At the same time, it is vital that thought is given to any medication a person with dysphagia takes, and ideally, a pharmacist should be involved to review the person's medications.

Prescribing a liquid medication or dispersible tablet, for example, may seem like an easy solution for a person with dysphagia, but there are several considerations that need to be heeded before making this switch. Firstly, this may not still be safe as the person may have been assessed to need thickened fluids; therefore, there could be a risk of aspiration if their medication is in a formulation that is not thick enough. Some medications can be mixed with thickeners, although it must be noted unless this is recommended by the manufacturer, this will be outside the product licence. This should always be checked as some thickeners used interact with medication and could reduce their efficacy [81]. Dispersible and effervescent tablets may also have a high sodium content, which can be unsuitable in older people, especially those with certain comorbidities such as hypertension or heart failure.

When changing to a different formulation of the same medication, bioavailability needs to be considered as doses may need to be changed. This is especially important for medications with a narrow therapeutic index. For example, when changing a person from digoxin tablets to digoxin oral solution, a dose conversion must be carried out. This is because the bioavailability of digoxin in tablet form is approximately 63% but is 75% as oral solution [82]. Therefore, a lower dose of oral solution is needed to be equivalent to a tablet dose.

Some liquid medications have been formulated to facilitate paediatric dosing, and so the correct amount of liquid an adult requires could be a relatively large volume. Conversely, the volume required to be given may be a smaller volume than we might typically think of for liquid medications (e.g. less than a 5 mL spoonful).

Using digoxin again as an example, the oral solution is 50 micrograms per millilitre (mL) [82], so an adult maintenance dose is most likely to be around 1–4 mL, with older people usually being at the lower end of the range. It is important that the person or the person giving them the medication is counselled appropriately on how to measure the medication and the amount that should be given [83]. Care should also be given if liquid medications are available in different strengths as people may be used to measuring their dose, and this could lead to an under or overdose if they are given a different strength and not counselled.

Another consideration is that a medication may not be available in a liquid or dispersible preparation, and then thought needs to be given to an alternative method of administration, for example, transdermal. If this is not appropriate or it is not available in a different, suitable formulation, then the medication may need to be switched to a different one. An alternative that is used commonly in practice would be to ascertain if the solid oral dosage form can be manipulated for administration, for example, crushing a tablet, dispersing a tablet in water, or opening a capsule. This is not appropriate for all solid oral dosage forms, and pharmacists are well placed to advise on this. Using this method for administration (unless stated by the manufacturer) will mean the formulation is unlicensed [84], and this should be discussed with not only the prescriber but also the person taking the medication. There may also be an option for a special-order liquid medication; however, these are also unlicensed and can be expensive.

Acceptance of solid forms may also be important for adherence in older people without a formal diagnosis of dysphagia or a particular problem with swallowing. For example, older people may break large tablets in half to make them easier to swallow, or they may use food to mask taste, and so manufacturers should consider dimensions and palatability when designing the final product [85, 86].

3.2 Other Factors

Appearance of Solid Oral Dosage Forms Older people may use the appearance of medication, for example, colour or shape, to help with identification of medications [85, 86]. People with visual impairments may mix up medications that look similar and be at increased risk of errors with their medication [87].

Dexterity and Strength Although smaller tablets may be preferable to ease of swallowing, they can be difficult to remove from the blisters they are packaged in [85], and this could impact on adherence.

Dexterity also needs to be considered for non-oral formulations such as inhalers. Increasing age has been associated with a decline in correct inhaler technique [88]. This is multi-factorial, but dexterity and physical strength are implicated. For example, some older people cannot exert the minimum strength required to actuate a metered dose inhaler [89].

3.3 *Personalised Medication for Older People*

Pharmacogenomics offers an exciting opportunity for people to receive personalised medicine that will be the best choice for them by using genetic information to predict benefit and also likelihood of side effects from a specific medication. There may be more genomic variation in old age in part due to decline in function, in part due to some of the other factors discussed in this chapter, for example, polypharmacy and multimorbidity [90]. Further research into these areas could ultimately aid prescribing decisions and lead to better outcomes for older people [90].

Various methods of 3D printing can be utilised to offer personalised medication [91]. This could be based on pharmacogenomics enabling precision dosing of medication for a person. Bespoke medication could also be formulated that incorporated more than one medication. This would be useful for older people on multiple medications to reduce pill burden and hopefully to improve adherence if this is an issue. However, there are many considerations and challenges that need to be addressed. These include licensing of the 3D printed pharmaceuticals, quality assurance and the facilities for on-demand manufacturing of medications in pharmacies [91].

4 Conclusion

Many factors, as discussed in this chapter, contribute to medication optimisation in older people. Whilst we have discussed what we believe are some of the key factors, there may be more that need to be considered depending on the person. Research must continue to search for evidence on how to ensure the appropriate use of medication in older people, considering not only age but factors such as frailty and multimorbidity. The importance of person-centred care cannot be over-emphasised in helping to achieve the best outcomes for older people.

References

1. Age UK. Later life in the United Kingdom 2019 – Factsheet; 2019. p. 27.
2. Fortin M, Bravo G, Hudon C, Vanasse A, Lapointe L. Prevalence of multimorbidity among adults seen in family practice. Ann Fam Med. 2005;2:223–8.
3. Barnett K, Mercer SW, Norbury M, Watt G, Wyke S, Guthrie B. Epidemiology of multimorbidity and implications for health care, research, and medical education: a cross-sectional study. Lancet. 2012;380(9836):37–43. https://doi.org/10.1016/S0140-6736(12)60240-2.
4. Masnoon N, Shakib S, Kalisch-Ellett L, Caughey GE. What is polypharmacy? A systematic review of definitions. BMC Geriatr. 2017;17(1):1–10. https://doi.org/10.1186/s12877-017-0621-2.
5. National Institute for Health and Care Excellence (NICE). Medicines optimisation: the safe and effective use of medicines to enable the best possible outcomes. NICE guideline NG5. Available at: https://www.nice.org.uk/guidance/ng5 (2015). Accessed 19 July 2021.

6. Gao L, Maidment I, Matthews FE, Robinson L, Brayne C, Medical Research Council Cognitive Function and Ageing Study. Medication usage change in older people (65+) in England over 20 years: findings from CFAS I and CFAS II. Age Ageing. 2018;47(2):220–5. https://doi.org/10.1093/ageing/afx158.
7. Maidment I, Huckerby C, Shukla D. Medication management in older people – a hidden burden. Prescriber. 2020;31(11–12):30–3. https://doi.org/10.1002/psb.1881.
8. Audit Commission. A spoonful of sugar: medicines management in NHS hospitals. London; 2001.
9. National Institute for Health and Clinical Excellence (NICE). Medicines Adherence: Involving patiens in decisions about prescribed medicines and supporting adherence. Clinical guideline CG76. Available at: https://www.nice.org.uk/guidance/cg76 (2009). Accessed 19 July 2021.
10. Balint E. The possibilities of patient-centered medicine. J R Coll Gen Pract. 1969;17(82)
11. Ekman I, Swedberg K, Taft C, Lindseth A, Norberg A, Brink E, Carlsson J, Dahlin-Ivanoff S, Johansson IL, Kjellgren K, Lidén E. Person-centered care – ready for prime time. Eur J Cardiovasc Nurs. 2011;10(4):248–51. https://doi.org/10.1016/j.ejcnurse.2011.06.008.
12. Santana MJ, Manalili K, Jolley RJ, Zelinsky S, Quan H, Lu M. How to practice person-centred care: a conceptual framework. Health Expect. 2018;21(2):429–40. https://doi.org/10.1111/hex.12640.
13. Rockwood K, Mitnitski A. Frailty in relation to the accumulation of deficits. J Gerontol Ser A Biol Sci Med Sci. 2007:722–7. https://doi.org/10.1093/gerona/62.7.722.
14. Rockwood K, Song X, MacKnight C, Bergman H, Hogan DB, McDowell I, Mitnitski A. A global clinical measure of fitness and frailty in elderly people. CMAJ. 2005;173(5) https://doi.org/10.1503/cmaj.050051.
15. Clegg A, Young J, Iliffe S, Rikkert MO, Rockwood K. Frailty in elderly people. Lancet. 2013;381(9868):752–62. https://doi.org/10.1016/S0140-6736(12)62167-9.
16. Dent E, Kowal P, Hoogendijk EO. Frailty measurement in research and clinical practice: a review. Eur J Intern Med. 2016; https://doi.org/10.1016/j.ejim.2016.03.007.
17. NHS England. NHS England identifying frailty. [online] NHS England. Available at: https://www.england.nhs.uk/ourwork/clinical-policy/older-people/frailty/frailty-risk-identification/ (2021). Accessed 23 Oct 2021.
18. Hilmer SN, Gnjidic D. Prescribing for frail older people. Aust Prescr. 2017;40(5):174–8. https://doi.org/10.18773/austprescr.2017.055.
19. Peklar J, O'Halloran AM, Maidment ID, Henman MC, Kenny RA, Kos M. Sedative load and frailty among community-dwelling population aged ≥65 years. J Am Med Dir Assoc. 2015;16(4):282–9. https://doi.org/10.1016/j.jamda.2014.10.010.
20. Thanapluetiwong S, Chewcharat A, Takkavatakarn K, Praditpornsilpa K, Eiam-Ong S, Susantitaphong P. Vitamin D supplement on prevention of fall and fracture: a meta-analysis of randomized controlled trials. Medicine. 2020;99(34):e21506. https://doi.org/10.1097/MD.0000000000021506.
21. British Geriatrics Society. Fit for Frailty- Consensus best practice guidance for the care of older people living in community and outpatient settings' [online] Available at: https://www.bgs.org.uk/sites/default/files/content/resources/files/2018-05-23/fff_full.pdf (2014). Accessed 4 Sept 2021.
22. NHS England. Network contract directed enhanced service: contract specification 2020/21 – PCN requirements and entitlements. [online] Available at: https://www.england.nhs.uk/publication/des-contract-specification-2020-21-pcn-entitlements-and-requirements/ (2020). Accessed: 12 Sept 2021.
23. British Geriatrics Society. End of life care in frailty: medicines management [online] Available at: https://www.bgs.org.uk/resources/end-of-life-care-in-frailty-medicines-management (2020). Accessed: 4 Sept 2021.
24. Maidment I, Lawson S, Wong G, Booth A, Watson A, Zaman H, Mullan J, McKeown J, Bailey S. Towards an understanding of the burdens of medication management affecting older people: the MEMORABLE realist synthesis. BMC Geriatr. 2020;20(1):183. https://doi.org/10.1186/s12877-020-01568-x.

25. Maidment ID, Lawson S, Wong G, Booth A, Watson A, McKeown J, Zaman H, Mullan J, Bailey S. Medication management in older people: realist approaches based on literature and evaluation (MEMORABLE): a realist synthesis. NIHR J Library. 2020;8(26)
26. Nakham, A., Myint, P.K., Bond, C.M., Newlands, R., Loke, Y.K. and Cruickshank, M. (2020) 'Interventions to reduce anticholinergic burden in adults aged 65 and older: a systematic review', J Am Med Dir Assoc 172–180.e5. https://doi.org/10.1016/j.jamda.2019.06.001.
27. Richardson K, Bennett K, Maidment ID, Fox C, Smithard D, Kenny RA. Use of medications with anticholinergic activity and self-reported injurious falls in older community-dwelling adults. J Am Geriatr Soc. 2015;63(8):1561–9. https://doi.org/10.1111/jgs.13543.
28. Richardson K, Loke YK, Fox C, Maidment I, Howard R, Steel N, Arthur A, Boyd PJ, Aldus C, Ballard C, Savva GM. Adverse effects of Z-drugs for sleep disturbance in people living with dementia: a population-based cohort study. BMC Med. 2020;18(1):1–15. https://doi.org/10.1186/s12916-020-01821-5.
29. Coupland C, Hill T, Morriss R, Moore M, Arthur A, Hippisley-Cox J. Antidepressant use and risk of adverse outcomes in people aged 20-64 years: cohort study using a primary care database. BMC Med. 2018;16(1):1–24. https://doi.org/10.1186/s12916-018-1022-x.
30. Ng BJ, Le Couteur DG, Hilmer SN. Deprescribing benzodiazepines in older patients: impact of interventions targeting physicians, pharmacists, and patients. Drugs Aging. 2018;35(6):493–521. https://doi.org/10.1007/s40266-018-0544-4.
31. Juraschek SP, Simpson LM, Davis BR, Beach JL, Ishak A, Mukamal KJ. Effects of antihypertensive class on falls, syncope, and orthostatic hypotension in older adults: the ALLHAT trial. Hypertension. 2019;74(4):1033–40. https://doi.org/10.1161/HYPERTENSIONAHA.119.13445.
32. Maidment ID, Fox C, Boustani M. A review of studies describing the use of acetyl cholinesterase inhibitors in Parkinson's disease dementia. Acta Psychiatr Scand. 2005;111(6):403–9. https://doi.org/10.1111/j.1600-0447.2005.00516.x.
33. Richardson K, Savva GM, Boyd PJ, Aldus C, Maidment I, Pakpahan E, Loke YK, Arthur A, Steel N, Ballard C, Howard R. Non-benzodiazepine hypnotic use for sleep disturbance in people aged over 55 years living with dementia: a series of cohort studies. Health Technol Assess. 2021;25(1) https://doi.org/10.3310/hta25010.
34. Hilmer SN, Gnjidic D. The effects of polypharmacy in older adults. Clin Pharmacol Ther. 2009:86–8. https://doi.org/10.1038/clpt.2008.224.
35. Cruz-Jentoft AJ, Gutiérrez B. Upper age limits in studies submitted to a research ethics committee. Aging Clin Exp Res. 2010;22(2) https://doi.org/10.1007/BF03324793.
36. Hubbard RE, O'Mahony MS, Woodhouse KW. Medication prescribing in frail older people. Eur J Clin Pharmacol. 2013:319–26. https://doi.org/10.1007/s00228-012-1387-2.
37. Cherubini A, Oristrell J, Pla X, Ruggiero C, Ferretti R, Diestre G, Clarfield AM, Crome P, Hertogh C, Lesauskaite V, Prada GI. The persistent exclusion of older patients from ongoing clinical trials regarding heart failure. Arch Intern Med. 2011;171(6):550–6. https://doi.org/10.1001/archinternmed.2011.31.
38. Bartlam B, Lally F, Crome P, Topinkova E, Clarfield AM, Cherubini A, Lesauskaite V, Hertogh C, Szczerbińska K, Prada GI, Farre A. The PREDICT study. Patients and carers views of older peoples participation in clinical trials. Age Ageing. 2011;40(Suppl 1):i73.
39. Crome P, Lally F, Cherubini A, Oristrell J, Beswick AD, Clarfield AM, Hertogh C, Lesauskaite V, Prada GI, Szczerbińska K, Topinkova E. Exclusion of older people from clinical trials: professional views from nine European countries participating in the PREDICT study. Drugs Aging. 2011;28(8):667–77. https://doi.org/10.2165/11591990-000000000-00000.
40. NHS Digital. National diabetes audit, 2020–21 quarterly report [online] Available at: https://digital.nhs.uk/data-and-information/publications/statistical/national-diabetes-audit/e4-data (2021). Accessed: 30 Oct 2021.
41. Bellary S, Kyrou I, Brown JE, Bailey CJ. Type 2 diabetes mellitus in older adults: clinical considerations and management. Nat Rev Endocrinol. 2021; https://doi.org/10.1038/s41574-021-00512-2.

42. Strain WD, Down S, Brown P, Puttanna A, Sinclair A. Diabetes and frailty: an expert consensus statement on the management of older adults with type 2 diabetes. Diabetes Ther. 2021:1–21. https://doi.org/10.1007/s13300-021-01035-9.
43. Duerden M, Payne R. Polypharmacy and medicines optimisation. The King's Fund, pp. 1–68. Available at: http://www.kingsfund.org.uk/publications (2013). Accessed 14 Aug 2021.
44. Wastesson JW, Morin L, Tan EC, Johnell K. An update on the clinical consequences of polypharmacy in older adults: a narrative review. Expert Opin Drug Saf. 2018;17(12):1185–96. https://doi.org/10.1080/14740338.2018.1546841.
45. Foley L, Larkin J, Lombard-Vance R, Murphy AW, Hynes L, Galvin E, Molloy GJ. Prevalence and predictors of medication non-adherence among people living with multimorbidity: a systematic review and meta-analysis. BMJ Open. 2021;11(9):e044987. https://doi.org/10.1136/bmjopen-2020-044987.
46. Paisansirikul A, Ketprayoon A, Ittiwattanakul W, Petchlorlian A. Prevalence and associated factors of drug-related problems among older people: a cross-sectional study at King Chulalongkorn Memorial Hospital in Bangkok. Drugs-Real World Outcomes. 2021;8(1):73–84. https://doi.org/10.1007/s40801-020-00219-2.
47. Department of Health and Social Care. Good for you, good for us, good for everybody: a plan to reduce overprescribing to make patient care better and safer, support the NHS, and reduce carbon emissions [online] Available at: https://www.gov.uk/government/publications/national-overprescribing-review-report (2021). Accessed: 23 Oct 2021.
48. Grossi CM, Richardson K, Fox C, Maidment I, Steel N, Loke YK, Arthur A, Myint PK, Campbell N, Boustani M, Robinson L. Anticholinergic and benzodiazepine medication use and risk of incident dementia: a UK cohort study. BMC Geriatr. 2019;19(1) https://doi.org/10.1186/s12877-019-1280-2.
49. Pieper NT, Grossi CM, Chan WY, Loke YK, Savva GM, Haroulis C, Steel N, Fox C, Maidment ID, Arthur AJ, Myint PK. Anticholinergic drugs and incident dementia, mild cognitive impairment and cognitive decline: a meta-analysis. Age Ageing. 2020;49(6) https://doi.org/10.1093/ageing/afaa090.
50. Moriarty F, Savva GM, Grossi CM, Bennett K, Fox C, Maidment I, Loke YK, Steel N, Kenny RA, Richardson K. Cognitive decline associated with anticholinergics, benzodiazepines and Z-drugs: findings from the Irish longitudinal study on ageing (TILDA). Br J Clin Pharmacol. 2020; https://doi.org/10.1111/bcp.14687.
51. Boustani M, Campbell N, Munger S, Maidment I, Fox C. Impact of anticholinergics on the aging brain: a review and practical application. Aging Health. 2008;4(3):311–20. https://doi.org/10.2217/1745509X.4.3.311.
52. Machado-Alba JE, Gaviria-Mendoza A, Machado-Duque ME, Chica L. Deprescribing: a new goal focused on the patient. Expert Opin Drug Saf. 2017;16(2):111–2.
53. Thompson W, Farrell B. Deprescribing: what is it and what does the evidence tell us? Can J Hosp Pharm. 2013;66(3):201.
54. Thompson W, Farrell B, Welch V, Tugwell P, Way C, Richardson L, Bjerre LM. Continuation or deprescribing of proton pump inhibitors: a consult patient decision aid. Can Pharm J. 2019;152(1) https://doi.org/10.1177/1715163518816719.
55. Scott IA, Hilmer SN, Reeve E, Potter K, Le Couteur D, Rigby D, Gnjidic D, Del Mar CB, Roughead EE, Page A, Jansen J. Reducing inappropriate polypharmacy: the process of deprescribing. JAMA Intern Med. 2015;175(5):827–34. https://doi.org/10.1001/jamainternmed.2015.0324.
56. Smith H, Miller K, Barnett N, Oboh L, Jones E, Darcy C, McKee H, Agnew J, Crawford P. Person-centred care including deprescribing for older people. Pharmacy. 2019;7(3):101. https://doi.org/10.3390/pharmacy7030101.
57. Bokhof B, Junius-Walker U. Reducing polypharmacy from the perspectives of general practitioners and older patients: a synthesis of qualitative studies. Drugs Aging. 2016; https://doi.org/10.1007/s40266-016-0354-5.

58. O'Mahony D, O'Sullivan D, Byrne S, O'Connor MN, Ryan C, Gallagher P. STOPP/START criteria for potentially inappropriate prescribing in older people: version 2. Age Ageing. 2014;44(2):213–8. https://doi.org/10.1093/ageing/afu145.
59. PrescQIPP. Ensuring appropriate polypharmacy: a practical guide to deprescribing. Bulletin 136 (2017, April):1–8. Available at: https://www.prescqipp.info/component/jdownloads/send/356-polypharmacy-practical-guide-to-deprescribing/3410-bulletin-176-p-d-practical-guide
60. Lewis T. 10-minute consultation: using the NO TEARS tool for medication review. Br Med J. 2004:434. https://doi.org/10.1136/sbmj.0410362.
61. NHS Scotland. Scottish government polypharmacy model of care group. Polypharmacy gidance, realistic prescribing. 3rd ed. NHS Scotland; 2018.
62. Farrell B, Pottie K, Thompson W, Boghossian T, Pizzola L, Rashid FJ, Rojas-Fernandez C, Walsh K, Welch V, Moayyedi P. Deprescribing proton pump inhibitors: evidence-based clinical practice guideline. Can Fam Physician. 2017;63(5):354–64.
63. Farrell B, Black C, Thompson W, McCarthy L, Rojas-Fernandez C, Lochnan H, Shamji S, Upshur R, Bouchard M, Welch V. Deprescribing antihyperglycemic agents in older persons: evidence-based clinical practice guideline. Can Fam Physician. 2017;63(11):832–43.
64. Bjerre LM, Farrell B, Hogel M, Graham L, Lemay G, McCarthy L, Raman-Wilms L, Rojas-Fernandez C, Sinha S, Thompson W, Welch V. Deprescribing antipsychotics for behavioural and psychological symptoms of dementia and insomnia: evidence-based clinical practice guideline. Can Fam Physician. 2018;64(1):17–27. PMID: 29358245
65. Reeve E, Farrell B, Thompson W, Herrmann N, Sketris I, Magin P, Chenoweth L, Gorman M, Quirke L, Bethune G, Forbes F. Evidence-based clinical practice guideline for deprescribing cholinesterase inhibitors and memantine guidelines. In: Recommendations. Sydney: The University of Sydney; 2018.
66. Pottie K, Thompson W, Davies S, Grenier J, Sadowski CA, Welch V, Holbrook A, Boyd C, Swenson R, Ma A, Farrell B. Deprescribing benzodiazepine receptor agonists: evidence-based clinical practice guideline. Can Fam Physician. 2018;64(5):339–51.
67. Wittenberg R Wittenberg R, Hu B, Barraza-Araiza L, Rehill A. Projections of older people with dementia and costs of dementia care in the United Kingdom, 2019–2040. CPEC working paper 5, (November). London: London School of Economics; 2019.
68. Fox C, Richardson K, Maidment ID, Savva GM, Matthews FE, Smithard D, Coulton S, Katona C, Boustani MA, Brayne C. Anticholinergic medication use and cognitive impairment in the older population: the Medical Research Council cognitive function and ageing study. J Am Geriatr Soc. 2011;59(8):1477–83. https://doi.org/10.1111/j.1532-5415.2011.03491.x.
69. Fox C, Livingston G, Maidment ID, Coulton S, Smithard DG, Boustani M, Katona C. The impact of anticholinergic burden in Alzheimer's dementia-the laser-AD study. Age Ageing. 2011;40(6):730–5. https://doi.org/10.1093/ageing/afr102.
70. Richardson K, Fox C, Maidment I, Steel N, Loke YK, Arthur A, Myint PK, Grossi CM, Mattishent K, Bennett K, Campbell NL. Anticholinergic drugs and risk of dementia: case-control study. BMJ. 2018;361:k1315. https://doi.org/10.1136/BMJ.K1315.
71. Maidment, I. Medication -related adverse events in older people with dementia causes and possible solutions. Doctoral dissertation, Aston University; 2013.
72. Maidment ID, Haw C, Stubbs J, Fox C, Katona C, Franklin BD. Medication errors in older people with mental health problems: a review. Int J Geriatr Psychiatry. 2008;23(6):564–73. https://doi.org/10.1002/gps.1943.
73. Aston L, Hilton A, Moutela T, Shaw R, Maidment I. Exploring the evidence base for how people with dementia and their informal carers manage their medication in the community: a mixed studies review. BMC Geriatr. 2017;17(1):242. https://doi.org/10.1186/s12877-017-0638-6.
74. Maidment I, et al. Developing a framework for a novel multi-disciplinary, multi-agency intervention(s), to improve medication management in community-dwelling older people on complex medication regimens (MEMORABLE) – a realist synthesis. Syst Rev. 2017;6(1) https://doi.org/10.1186/s13643-017-0528-1.

75. Maidment ID, Aston L, Moutela T, Fox CG, Hilton A. A qualitative study exploring medication management in people with dementia living in the community and the potential role of the community pharmacist. Health Expect. 2017;20(5) https://doi.org/10.1111/hex.12534.
76. Baijens LW, Clavé P, Cras P, Ekberg O, Forster A, Kolb GF, Leners JC, Masiero S, Mateos-Nozal J, Ortega O, Smithard DG. European society for swallowing disorders - European union geriatric medicine society white paper: oropharyngeal dysphagia as a geriatric syndrome. Clin Interv Aging. 2016; https://doi.org/10.2147/CIA.S107750.
77. Hiltunen K, Saarela RK, Kautiainen H, Roitto HM, Pitkälä KH, Mäntylä P. Relationship between Fried's frailty phenotype and oral frailty in long-term care residents. Age Ageing. 2021;afab177 https://doi.org/10.1093/ageing/afab177.
78. Tanaka T, Takahashi K, Hirano H, Kikutani T, Watanabe Y, Ohara Y, Furuya H, Tetsuo T, Akishita M, Iijima K. Oral frailty as a risk factor for physical frailty and mortality in community-dwelling elderly. J Gerontol Ser A. 2018;73(12):1661–7. https://doi.org/10.1093/gerona/glx225.
79. Ortega O, Cabre M, Clave P. Oropharyngeal dysphagia: aetiology and effects of ageing. J Gastroenterol Hepatol Res. 2014;3(5):1049–54.
80. Drumond N, Stegemann S. Better medicines for older patients: considerations between patient characteristics and solid oral dosage form designs to improve swallowing experience. Pharmaceutics. 2021;13(1):32. https://doi.org/10.3390/pharmaceutics13010032.
81. Brennan K. Preparing and giving medicines for people with swallowing difficulties [online] Available at: https://www.sps.nhs.uk/articles/preparing-and-giving-medicines-for-people-with-swallowing-difficulties/ (2021). Accessed 23 Oct 2021.
82. Aspen. Lanoxin PG Elixir SmPC [online] Available at: https://www.medicines.org.uk/emc/product/5463/smpc (2020). Accessed 24 Oct 2021.
83. Ryu GS, Lee YJ. Analysis of liquid medication dose errors made by patients and caregivers using alternative measuring devices. J Manag Care Pharm. 2012;18(6):439–45. https://doi.org/10.18553/jmcp.2012.18.6.439.
84. Barnett N, Parmar P. Tailoring medication formulations for patients with dysphagia. Pharm J. 2016;297(7892):106–8. https://doi.org/10.1211/PJ.2016.20201498.
85. Shariff Z, Kirby D, Missaghi S, Rajabi-Siahboomi A, Maidment I. Patient-centric medicine design: key characteristics of oral solid dosage forms that improve adherence and acceptance in older people. Pharmaceutics. 2020;12(10) https://doi.org/10.3390/pharmaceutics12100905.
86. Shariff ZB, Dahmash DT, Kirby DJ, Missaghi S, Rajabi-Siahboomi A, Maidment ID. Does the formulation of Oral solid dosage forms affect acceptance and adherence in older patients? A mixed methods systematic review. J Am Med Dir Assoc. 2020; https://doi.org/10.1016/j.jamda.2020.01.108.
87. Zhi-Han L, Hui-Yin Y, Makmor-Bakry M. Medication. Handling challenges among visually impaired population. Arch Pharm Pract. 2017;8(1):8–14.
88. Barbara S, Kritikos V, Anticevich SB. Inhaler technique: does age matter? A systematic review. Eur Respir Rev. 2017; https://doi.org/10.1183/16000617.0055-2017.
89. Gray SL. Characteristics predicting incorrect metered-dose inhaler technique in older subjects. Arch Intern Med. 1996;156(9):984. https://doi.org/10.1001/archinte.1996.00440090084008.
90. Brockmöller J, Stingl JC. Multimorbidity, polypharmacy and pharmacogenomics in old age. Pharmacogenomics. 2017;18(6) https://doi.org/10.2217/pgs-2017-0026.
91. Curti C, Kirby DJ, Russell CA. Current formulation approaches in design and development of solid oral dosage forms through three-dimensional printing. Prog Addit Manuf. 2020;5(2):111–23. https://doi.org/10.1007/s40964-020-00127-5.

Medication Safety in Older People

Nkiruka Umaru and Shirley Sau Yin Ip

Contents

Abstract This chapter considers implications for medication safety due to pharmacokinetic and pharmacodynamic changes affecting the handling of medicines in older people and further considerations when switching medicine formulations with differing bioequivalences. This chapter outlines concerns around supply and administration of medicines, with attention paid to cognitive capacity, accessibility in relation to oral medications and skin, eye, and other routes of drug delivery. Monitoring of medication therapy and outcomes by patients, carers and healthcare professionals is discussed.

Support strategies to facilitate the safer use of medicines throughout the prescribing, supply, administration and monitoring process are provided. The wider, holistic aspects that need to be considered on initiation and cessation (deprescribing) of medicines, including the appropriateness of using single-condition guidelines in

N. Umaru (✉) · S. S. Y. Ip
Department of Clinical, Pharmaceutical and Biological Sciences, School of Life and Medical Sciences, University of Hertfordshire, Hatfield, UK
e-mail: n.e.umaru@herts.ac.uk

M. Orlu, F. Liu (eds.), *Pharmaceutical Formulations for Older Patients*, AAPS Advances in the Pharmaceutical Sciences Series 51,
https://doi.org/10.1007/978-3-031-35811-1_2

people suffering from multimorbidity, are provided. Important aspects of medicine use around shared decision-making and patient empowerment are briefly discussed. Finally, a case study reflecting on the medicine use process with a focus on formulation is provided.

Keywords Medication safety · Medicines use · Shared decision-making · Deprescribing · Carers

1 Background

1.1 Medication Safety Considerations in Older People

Between 2015 and 2050, the proportion of the world's over 60-year-olds is estimated to increase from 12% to 22%, with the number of people aged 60 years and older outnumbering children younger than 5 years by 2020 [1]. There were estimated 15,120 centenarians (people aged 100 years and over) in the UK in 2020, an increase of almost a fifth (18%) from 2019 [2]. Overall, the pace of population ageing is rising, and all countries face major challenges to ensure that their health and social systems are ready to cater for this demographic shift, and this, in turn, applies to the formulation of drugs for use in older people. The older population is not an insignificant one but, rather, a section of the population that needs to be increasingly catered for in all aspects of drug formulation.

Safety considerations of medicine use in older people broadly fall into two categories: general consideration in relation to the prescribing, usage, storage, administration and monitoring of the medicines that are in use and more specific considerations around the ageing physiology of older people, which may affect how they handle medicines.

Clinical trials of drugs are typically carried out in younger adults under 65 years of age and often in people with no or few comorbidities. Thus, the adverse effects profile reported in clinical trials may not accurately reflect the profile of that drug in real-world usage; the pharmacological effect of these drugs in older people may be less well understood, and incidence of adverse effects and drug-drug interactions in older people with multimorbidity may be more pronounced. Within the UK, there has been a recent call from the Chief Pharmaceutical Officer for increased monitoring of health outcomes across *all* patient groups once a medicine is in widespread use (post-marketing surveillance) to gather a wealth of evidence to be made available in addition to that from clinical trials [3].

1.2 General Support Strategies for Safer Use of Medicines

All patients should be offered support to manage their medicines as best as they can, and this means creating and enhancing opportunities for medicine optimisation. The National Institute for Health and Care Excellence (NICE) in the UK has defined medicine optimisation as 'a person-centred approach to safe and effective medicine use, to ensure people obtain the best possible outcomes from their medicines' [4]. All healthcare professionals responsible for patient care provision are expected to demonstrate the principles of shared decision-making and medicine optimisation [5, 6]. This starts with creating an environment where patients' choices about the extent of their involvement in decisions about their care are explored and their choices respected without judgement [7]. Older people should be supported in shared decision-making to understand personalised factors around the use of medicines including their beliefs, values, social contexts, experiences related to illness and preferences to treatment options [8, 9]. Healthcare professionals should aim to ensure medicine use is as safe as possible, which includes an exploration of drug formulations with older patients and their carers during the medicine use process [10].

Consultation with patients and their carers about the patient's medicine use can occur at the point of prescribing or recommendation (over-the-counter (OTC) medicines), supply, administration and monitoring or review [4]. Discussions around the rationale for prescribed or recommended medicines should take place, to provide the patient with confidence and knowledge about new or ongoing medicinal therapy. Agreement on the desired outcome of the intervention provides both prescriber and patient, a shared outcome based on the expertise of the healthcare professional and the expertise of the patient or their carer. During the prescribing process, drug formulation, route of administration and convenience should be considered to support adherence to medicine use. Prescribing and recommendation of medicines should only be undertaken by qualified and competent healthcare professionals. The decision to take or use the medicines usually ultimately rests with the patient. Drug formulation, route of administration and convenience may be important factors a patient considers when assessing the benefits and risks associated with new or ongoing treatment with medicines [11].

Checking and confirming the patient and their carer's understanding of their medicines and how to use is vital during the supply process. Consultation with patients and their carers provides the opportunity to ask questions or clarify uncertain prior information or knowledge. Older patients may wish to explore other drug formulation options available. Issues with drug formulation may only become apparent upon sighting the medicine for the first time or following initial use of the medication. Pharmaceutical companies are bound by regulations that demand that their product meet the requirements of the relevant marketing authorisation or product specification. The same drug can be manufactured by several manufacturers but differ in characteristics such as size. Older patients can be offered alternate formulations such as the smallest available size of a given medication manufactured by several pharmaceutical companies, for a more comfortable swallowing experience. Equally, a capsule or liquid formulation may be preferable to a tablet for the same reason. Other formulation choices to support adherence include sustained-release preparations and controlled-release tablets. In all cases, attention should be paid to the bioequivalence in each formulation as discussed further on in this chapter.

Supply may involve contact with a healthcare facility such as a pharmacy, dispensing general practitioner practice or hospital premises, and on other occasions, patients also have medicines delivered at home. In all circumstances, communication with relevant members of the healthcare team should be in place to provide opportunities for discussions about the safe use of medicines. Older patients may have representatives and carers collecting medicines on their behalf due to mobility or cognitive ability challenges. Relevant information should be provided to such representatives to support safer medicine use [11]. Packaging design is often dependent on the properties of the medication to ensure durability and stability of the medicine. Issues around accessibility to medicines can be discussed during the supply process, for example, difficulty accessing outer and immediate packaging of medicines such as blister packs and child-resistant bottles, which can be challenging for older patients with reduced finger-grip strength [12, 13]. Supply of medicines in medicine administration aids such as dosette or pill boxes can support the safer use of medicines; however, sometimes, due to formulation characteristics, dosette boxes are not suitable for every type of medicine [14, 15]. All healthcare professionals have a responsibility to ensure appropriate processes are in place to support the safe supply of medicines, including healthcare professionals making the supply and those receiving the supply. Exploring options to support older patients with their medicine use is vital at this stage of the medicine use process.

Medicine administration can be undertaken by patients, carers, healthcare professionals or other suitable trained support personnel, dependent on setting and circumstances [16]. For example, some hospitals and healthcare facilities have policies in place, which allow patients to self-administer some or all their medicines following a risk assessment [17]. This supports self-care and provides initial assessment opportunity and education on how to use specified medicines or devices. Equally, some patients living in their own homes are supported with administration of medicines by healthcare professionals such as district nurses. The given route for the administration will often depend on the medicinal properties, pharmacokinetics, convenience and patient and system factors. It is helpful during discussions to gather information on self-management plans that patient and their carers have developed to administer medicines. The appropriate healthcare professional can tailor advise to the needs, which have been identified.

All healthcare professionals have a responsibility to ensure they have the appropriate knowledge and skill relevant to the administration of the medicines where it falls within their scope of practice. In most healthcare settings in the UK, pharmacists are accessible to healthcare professionals, to provide advice on the appropriate administration of medicines.

The focus of monitoring can be on patient outcomes, therapeutic drug levels or a combination of both. Monitoring can be undertaken by patients, carers or members of their healthcare team. The responsibility for monitoring by older patients may be less realistic compared with the same expectations in the general population, without due support from carers or family members. Monitoring requirements may form the basis for selection of drug formulation. In type 2 diabetes, patients can be managed using both oral antidiabetic drugs and insulin. After due consideration to evidence-based practice and patient choice, the use of insulin may be disregarded if this presents increased problems with manipulating the devise used to administer the subcutaneous injection and decreased cognitive ability to recognise and act on hypoglycaemic levels of blood glucose. A shared decision about the treatment options for the patient is key here. For therapeutic drug level monitoring, changes due to the ageing process should be considered when drawing conclusions on results obtained.

2 Medication Safety Concerns in Older People

The use of medicines in older people requires careful management to provide optimum care [18]. Careful consideration should be given to several areas as detailed further on.

2.1 Impact of Pharmacokinetic and Pharmacodynamics Changes

As people age, their body systems change in several ways [19]. The ageing process results in anatomical and physiological changes. These may result in pharmacodynamic changes to the way in which drugs affect the ageing body, and target organ sensitivity may be affected. Age-related changes cause altered (usually increased) sensitivity to several drug classes: sensitivity to sedation and psychomotor impairment with benzodiazepines, level and duration of pain relief with opioids, drowsiness and lateral sway with alcohol, sensitivity to anticholinergic agents, cardiac sensitivity to digoxin and heart rate response to beta-blockers.

Within the ageing body, there is progressive loss of functional capacities of most organs, including reduced response to receptor stimulation and homeostatic mechanisms, loss of water content and increased fat and changes in neuroendocrine and cardiovascular systems.

Within the cardiovascular system, the ageing body typically has increased arterial and myocardial stiffness, with a less stable vascular nerve control. Drugs formulated to lower blood pressure may overtreat the problem, causing hypotension, or may exacerbate the blunted orthostatic response, causing orthostatic hypotension upon standing.

Within the gastrointestinal tract, gastric acid production is decreased, which may significantly impair absorption of weakly basic drugs, which show low solubility at high pH, leading to reduced or variable bioavailability. Increased gastric emptying time may affect drug absorption by affecting the time that the drug is present at the absorption site. There may also be reduced absorption surface in the small intestine, affecting the extent to which a drug is orally bioavailable.

Aside from the decreased absorption of drugs owing to gastric changes, thoughts should be given to changes in skin physiology in older people affecting topical absorption. The epidermis becomes thinner and more susceptible to mild mechanical injury forces of moisture, friction and trauma – this needs to be considered when formulating the adhesives on patches or dressings.

In addition, there is a 20% reduction in the thickness of the dermis, giving the skin its paper-thin appearance, potentially increasing bioavailability of topically delivered drugs and increasing the incidence of topical adverse reactions [20].

With the increase in body fat and decrease in body water in the ageing body and decrease in lean body mass, the distribution of drugs may change with age.

The liver also declines with age, with the gradual loss of liver mass, reduced hepatic blood flow and reduced liver enzyme activity. This results in the reduced metabolic capability of the liver, thus potentially prolonging drug half-life. Reduced hepatic blood flow causes a reduction in the amount of drug delivered to the liver, whilst the reduced first-pass metabolic capability of the liver will allow increased concentrations of drugs to enter the circulation, causing increased, and possibly toxic, effects.

It is widely appreciated that changes to the renal system affect drug handling. As people age, kidney function naturally declines. There is a gradual reduction in renal mass and renal blood flow adversely affecting the glomerular flow rate. This results in accumulation of drugs that can cause toxic effects to the kidneys and other body organs.

2.2 Switching and Bioequivalence

In the older population, there may be a need to switch between different formulations of drugs for a variety of reasons, and certain considerations should be considered when doing so.

Firstly, differences in bioavailability between different brands or between branded and generic formulations should be considered. Also, administration instructions between different oral formulations may also occur, i.e. whether a drug needs to be taken before, with or after food.

Attention needs to be paid to drugs where there is difference in clinical effect between each different manufacturer's versions of the drug entity, and these should be prescribed by brand name. And in some cases, the licensed indications/contraindications for one formulation of a drug may be different from another formulation. Prescribers should ensure that the required formulation is licensed for the relevant indication.

Some older people may have difficulty in swallowing large solid oral formulations and may wish to switch from solid oral form (tablet/capsule) to a liquid.* it should be noted that not all oral liquid formulations are bioequivalent to their solid oral counterparts. Some examples include the following [21]:

Citalopram 40 mg/mL oral drops have around 25% higher oral bioavailability than tablets; hence, a 10 mg tablet is equivalent to 8 mg (or 4 drops) of oral drops.

Digoxin: The bioavailability of orally administered digoxin is approximately 63% in tablet form and 75% as elixir. This equates to one 62.5microgram tablet being approximately equivalent to 50 mg (1 mL) elixir. However, the manufacturer is aware of one study showing no clinically relevant difference in bioavailability of digoxin tablets and elixir and stating that most clinicians suggest that these dosage forms can usually be used interchangeably. But it should be noted that digoxin has a low therapeutic index, so cautious dose determination is essential, and individual patient factors must be considered during formulation changes. Patients' plasma concentrations should be monitored appropriately.

Lithium: All lithium preparations vary widely in bioavailability. Prescriptions should specify brand and formulation; changing the preparation requires the same precautions as initiation of treatment. Most lithium tablets are modified release; therefore, when lithium is given as a liquid, the total daily dose of lithium will need to be given in divided doses.

Phenytoin: Preparations containing *phenytoin sodium* (capsules and tablets) are not bioequivalent to those containing phenytoin base (such as Epanutin Infatabs and Epanutin suspension); 100 mg phenytoin sodium is approximately equivalent to 92 mg phenytoin base. When switching between these products, the difference in phenytoin content may be clinically significant. Therefore, plasma-phenytoin concentration monitoring is recommended.

Sodium fusidate tablets and *fusidic acid* oral suspension: Whilst sodium fusidate tablets are absorbed throughout the whole length of the gastrointestinal tract, fusidic acid oral suspension is absorbed only in the stomach and upper gut. Therefore, doses recommended for the suspension are proportionately higher than those for sodium fusidate tablets. Each 5 mL fusidic acid 250 mg/5 mL suspension (250 mg) is therapeutically equivalent to 175 mg sodium fusidate.

Additionally, be mindful that the relative bioavailability of drugs when switching between different formulations may differ from that which is published owing to ageing physiology and that therefore the clinical effects of switching formulations or administration routes may be unpredictable. Increased monitoring of the patient and relevant patient parameters should be carried out.

*Please note that switching to liquid medications owing to patients experiencing swallowing difficulties can have its own dangers. If a patient's swallowing function has been assessed by a speech and language therapist as requiring thickened fluids to prevent aspiration into the respiratory tract, additional safety considerations should be given to liquid medicines. If a certain fluid thickness (indication by the IDDSI level [22]) is required, then any liquid medicines need to be at this same IDDSI level to prevent aspiration of the liquid medicine, which could contribute to aspiration pneumonia. At present, liquid formulations do not indicate IDDSI levels. Considering that older people, along with paediatric patients, may be one of the larger cohort of patients who may readily use liquid medicines, there should be some consideration given to routine categorisation of liquid medicine IDDSI levels, which would aid greatly in terms of the safer use of liquid medicines in the older population.

2.3 Provision and Monitoring Concerns

When formulating medicinal products, due consideration should be given to potential provision and supply concerns of medicinal products that may make medicines less accessible to older people, in terms of physical accessibility to the medicinal product and accessibility of regular monitoring of drug therapy.

2.4 Accessibility

Certain medicines may be less accessible to older people owing to the normal physiological changes that occur in the older population – this may include changes in swallowing capability (for instance, post-stroke, in late-stage Parkinson's disease, or in late-stage dementia), reducing the ability to take drugs owing to the large sizes of some tablet. Impaired cognition may impact on the understanding of complex medication regimens (such as weekly dosing, variable or titration doses) or exacting storage requirements. There may be reduced physical ability to manipulate packaging, e.g. patient with arthritis may not be able to open blisters or have difficulty in opening child-resistant lids. A higher proportion of older people may also have poor eyesight, which may pose problems for this in reading instructions and patient information leaflets, which are provided with medicinal products.

In addition to this, the prevalence of long-term conditions increases with age, which means that as comorbidities increase, the number of medicines older people take increases. Multiple medical conditions very often equate to use of multiple drugs (polypharmacy), resulting in complex drug regimens. Polypharmacy may impact on the ability of patients to take their medicines, resulting in missed doses, improper adherence to administration instructions and drug/drug interactions.

2.5 Monitoring by Patient, Carer and Healthcare Professionals

As presented within this chapter, changes in formulation in drugs may require additional monitoring to ensure that the older patient does not suffer undue adverse effects from increased or decreased levels of drugs within the body.

Additionally, complicated monitoring regimens may put undue stress on older patients in terms of mobilising to appointments outside of their residence. More specialised drug monitoring requiring hospital attendance may put even further strain on older people as they may need to access appropriate public transport to enable them to present themselves at these appointments.

Self-monitoring of drug treatment (for instance, capillary blood glucose monitoring of diabetes) by the patients themselves may also be problematic in patients in poor physical health or with decreased cognitive ability.

Finally, it should be noted that the monitoring parameters that are applied to a younger population may not be appropriate in those patients who are older. For instance, in the UK, hypertension is defined as a blood pressure measuring 140/90 mmHg or higher, but for people over 80 years old, the range is 150/90 mmHg or higher. This may be confusing for clinicians not accustomed to treating older patients, and unnecessarily aggressive treatment targets may be inappropriately applied to this population, which may result in serious adverse clinical outcomes, such as the overtreatment of hypertension causing hypotension and falls.

3 Medication Safety Support Strategies

A plethora of medicine use information exists; however, awareness and understanding of this information can vary greatly due to various circumstances including level of health literacy [23]. Regardless, involving patients and their carers as partners in decision-making about their treatment is vital to understand their experiences and safeguard against medicine-related problems. A multidisciplinary team approach to support older people who usually have multiple long-term conditions and take multiple medicines is important.

The choice of available drug formulations where appropriate should be explored with patients and their carers. Factors to consider for older people will include the evidence base for use in this population. Most clinical trials routinely exclude members from the older population due to restricted sampling inclusion criteria. The risk vs benefits of the medicine should be considered. This applies to drug formulation where a particular choice may present more risk compared to an alternate formulation. For example, older people prescribed non-steroidal anti-inflammatory medicines should also be prescribed a proton pump inhibitor to prevent the risk of oesophageal ulceration. Considering the implications of prophylactic prescription medicines, the offset to this may be to prescribe enteric coated tablets or custom-made liquid formulation, but costs are more considerable [24]. Conversely measuring out the correct amount of liquid may be challenging due to diminished fine motor skills. An older patient may prefer to use a long-acting patch for pain relief compared to taking a tablet due to patient factors such as dysphagia or reduced unintentional concordance.

The drug formulation should be made explicit on prescriptions to give clear directions for supply. This may not be applicable to medicines that only exist in one formulation. Delays in contacting the prescriber to make alterations to the prescription can cause unwanted delay in treatment.

Where non-conventional alternative formulations are prescribed due to inaccessibility such as swallowing difficulty, the risks associated with such decision should be made clear to patients and their carers and other members of the healthcare team where necessary. Not all tablets can be crushed and dissolved in water for easier administration, as doing so will invariably change the licensing status of the medicine. Prescribers, suppliers and healthcare professionals involved in administration should be aware of the unlicensed status of the new formulation.

Medication reviews and medicine use reviews offer the opportunity to assess how patients are responding to their treatment and managing their medicines. Formulation issues may be raised at these consultations, providing the opportunity to review need for the drug and alternatives if still required. Some inhalers may be easier to manipulate compared to others. It may be more convenient to prescribe prefilled pens for insulin administration rather than cartridges, which require manipulation to prepare. Where an alternative devise is available, the option should be discussed and offered.

The field of frail and older people specialism has rapidly expanded in recent years. Members of the healthcare team have increased access to such specialists to obtain advice on prescribing and administration. Pharmacists are key in the provision of medicines information throughout the medicine use process and should be used a primary resource to support prescribing and administration in older people.

As mentioned earlier in this chapter, older people are more likely to need the support of carers or healthcare professionals to manage their medicines. Consideration should be given to the drug formulation of choice in relation to handling, storage, administration and monitoring. Some formulations require specific directions for storage, for example, in specified vessels which may be challenging to manipulate. Due to reduced cognitive ability and to support adherence, some older patients use dosette boxes prepared in a pharmacy setting, by themselves or their informal carer. Dosette boxes prepared in a pharmacy would have considered the properties of each medication and its appropriateness before inclusion in a dosette box. Issues of degradation may lead the responsible pharmacist to liaise with the prescriber to prescribe a more stable formulation considering that in some cases, dosette boxes are made up weeks in advance and medicines may be out of their primary container for a prolonged period [14, 15].

Some medicine formulations require fridge storage even after opening or first use. Cream formulation compared to ointment formulation of a particular medicines may require fridge storage. Equally, this requirement may be based on the size in the primary container, where larger sizes require fridge storage. Older patients and their carers need to be given adequate information about the safe storage of medicines.

Pill cutters may aid manipulation to get the appropriate dose; however, where a smaller dose entity is available, this should be provided. Manipulation of the pill cutter also must be considered, particularly with tablets that are very small.

Whilst most medicine administration require access to the primary container, some require further preparation to following this process. For example, a powder may need to be reconstituted with water into liquid form or dissolution of a tablet in water before administration. Consideration should be given to the practicality of undertaking these processes, and any mitigation to avoid this step would support safer medicine use. Where an older person can undertake these processes, then they should be encouraged to do so. In general, patients are often advised to return any unused medicines to a pharmacy for safe disposal. Older patients have been known to hoard medicines and therefore at risk of using expired medication. There is usually no indication that a medication has expired without a date check, and although not something most patients would do, this becomes even more challenging for those with impaired vision.

Older people and their carers are also more likely to use dosette boxes to support safe medicine administration. As mentioned earlier, pharmacies can provide this service at no cost and are able to advise on how to store and use medicines which should not be placed in a dosette box.

Adequate monitoring can support the achievement of therapeutic goals. Some formulations may require close monitoring due to a narrow therapeutic window. For example, some intravenous antibiotics are closely monitored by the healthcare team to ensure drug levels are at the required range before subsequent administration. On most occasions, this route of administration is justified but require regular review to convert to oral route, which does not require close monitoring of drug levels. Patient factors are usually considered including renal function and pharmacodynamic characteristics.

Older people and their carers can undertake some monitoring of medicines outcomes such as blood pressure, blood glucose levels and peak expiratory flow. Aids to support monitoring can be beneficial for older patients. A patient with type 1 diabetes is often limited to the use of insulin to manage their blood glucose levels. A continuous glucose monitoring system that needs replacement every 2 weeks may be more practical than daily finger pricking to monitor capillary blood glucose.

Every opportunity to engage with older people to support medicine optimisation should be undertaken including discussions about drug formulation to address challenges raised.

4 Risk-Benefit Considerations in Initiating and Deprescribing Medication

4.1 Use of Guidelines

As previously stated, older people are not well represented in clinical trials, and guidelines are drawn from trial data, which results in whole rafts of guidelines that may not adequately reflect the needs of the older population. Not only is the data for the trials not representative of the older population, the parameters in which the outcomes of the pharmacological treatment are measured by, but may not reflect the differences that are exhibited in the ageing body.

A more pressing problem may be, though, that as people age, there is an increased incidence of multimorbidity. NICE defines multimorbidity as the presence of two or more long-term health conditions, which can include defined physical or mental health conditions, such as diabetes or schizophrenia; ongoing conditions, such as learning disability; symptom complexes, such as frailty or chronic pain; sensory impairment, such as sight or hearing loss; and alcohol or substance misuse [25].

In older people, the presence of multimorbidity is almost universal. In a retrospective cohort study in England [26] ($n = 403{,}985$), the prevalence of multimorbidity was 27.2%, with prevalence increasing significantly with age. In a Scottish study [27] ($n = 1{,}751{,}841$), the prevalence of multimorbidity was estimated to be 23.2%, with prevalence of multimorbidity increasing substantially with age – half of the people had at least one morbidity by age 50 years, and by age 65, most were multimorbid.

In England, some of the most prevalent morbidities include conditions such as hypertension, diabetes, chronic pain, coronary heart disease and asthma, amongst others [28]. Each of these morbidities has its own guideline:

Hypertension in adults: diagnosis and management (NG136).
Type 1 diabetes in adults: diagnosis and management (NG17).
Type 2 diabetes in adults: management (NG28).
Chronic pain (primary and secondary) in over 16 s: assessment of all chronic pain and management of chronic primary pain (NG193).
Cardiovascular disease: risk assessment and reduction, including lipid modification (CG181).
Asthma: diagnosis, monitoring and chronic asthma management (NG80).

With an increasing number of morbidities, the likelihood of people being commenced and maintained on multiple medications increases. In England, more than one in ten people aged over 65 years, take at least eight different prescribed medications each week. This increases to nearly one in four people aged over 85 years [28].

This use of single guidelines to treat people with multimorbidity has its problems, as with increasing numbers of morbidities, the likelihood of treatment recommendations for one morbidity opposing the advice from the guidance for another morbidity increases. Therefore, a different approach needs to be adopted for people living with multimorbidity.

In people with competing multiple morbidities, the therapeutic goals of any treatment must be driven by the things that matter the most to them. A 95-year-old woman may be more preoccupied with maintaining good pain control so that she is able to mobilise to her local shop daily to shop for groceries to maintain good mental health and independence, but not so concerned about her hypertension, which gives her no symptoms and secondary prevention of cardiovascular complications 10 years, hence may be less important. Once this focus has been elucidated in individualised patient care plans, and the risks and benefits of the course of action have been made clear, the needs of the patient can be prioritised ahead of the wants of the single-condition guidelines.

Since, in many cases for older people, the risk-benefit ratio of drug treatment shifts towards risk, the therapeutic outcomes for drug treatment need to be well defined and may tip towards immediate symptomatic relief as opposed to secondary prevention of delayed adverse clinical outcomes, as these delayed outcomes may never be realised in this population. The adverse effect profile of any drug may be more tolerable if being used for immediate symptomatic relief as opposed to the attempted attainment of a treatment goal where the patient can see no real-world benefit (e.g. reduction in blood pressure or HbA1C). If the adverse effect profile of a drug is not tolerable for a patient, then of course this may lead to non-adherence to treatment.

Once the risk-benefit ratio of long-term medications is elucidated in line with the patient's wishes, poorly tolerated or drugs for prevention of delayed adverse clinical outcomes can be deprescribed, in partnership with the patient. Medications should be deprescribed singly and systematically, with an increased monitoring period after each drug is deprescribed. For medications with potential withdrawal effects, doses should be slowly tapered down over several weeks with increased monitoring of patient effects. Good communication between prescriber and patient is key in establishing the expected outcomes of the deprescribing of long-term medications.

4.2 Shared Decision-Making and Patient Empowerment

NICE defines shared decision-making as "a collaborative process that involves a person and their healthcare professional working together to reach a joint decision about care….this joint process empowers people to make decisions about care that is right for them at that time" [29]. This includes decisions about drug formulation choices where appropriate.

Decision-making for older people becomes more complex due to a cumulation of several factors including multiple long-term conditions and social care needs, polypharmacy and reduced coping mechanisms and resources brought on by demanding situations. In addition, partnership working between several health and care professionals, as well as family and carers, needs to be in place to improve chances of achieving optimal care for the older patient [30]. Understanding an older patient's preferences in relation to drug formulation or the decision to progress with a treatment where only one formulation of the drug exists may be a linear process or one-off consultation.

An integrated approach to health and care provision can partly inform members of the healthcare team to progress shared decision-making processes. For example, knowledge of the older patient's social care circumstances, medical and medication history and current access to resources and support networks, alongside direct patient and carer informed contributions to discussions about medicine choices and use, can enable a shared choice of a particular drug formulation. Periodic medication and medicine use review provide further opportunities to review therapeutic outcomes and changes to treatment options where needed.

5 Case Study

Case study with a focus on medication safety relevant to formulation.

Mrs. J. Smith is an 89-year-old patient with multiple long-term conditions:

Osteoarthritis affecting multiple joints resulting in chronic pain
Osteoporosis
Hypertension
Type 2 Diabetes
Mild cognitive impairment
Atrial fibrillation

Relevant patient parameters include the following:

Weight 45 kg
Height 155 cm
BMI (body mass index) 18.7 kg/m^2
Potassium 3.2
Urea: 6.8
Creatinine: 80 when admitted (baseline creatinine runs at 60)
Calculated creatinine clearance: 30 mL/min (baseline CrCl = 40 mL/min)
Capillary blood glucose (random): 4.6
HbA1c: 46 (6.4%)
BP lying: 160/90
BP standing 1 min: 100/70
BP standing 3 min: 120/75
Heart Rate 45 bpm
Digoxin level 1.9 μg/L (0.5–2.0 μg/L) on admission

She takes the following list of regular medication:

Buprenorphine 10 mcg/hour weekly patch – 1 patch weekly on Sundays
Paracetamol 1 g QDS
Metformin 500 mg BD
Adcal D3 effervescent 1 tablet BD
Alendronic acid 70 mg tablet weekly – 1 tablet weekly on Thursdays, taken together with her morning medicines
Macrogol sachets 1 BD
Amlodipine 5 mg OM
Atorvastatin 40 mg ON
Digoxin 125 micrograms OM recent change to 2.5 mL of 50mcg/mL liquid at same dose
Warfarin for atrial fibrillation – most recent dose 7 mg every evening Lifelong treatment. INR range 2–3. Monitored at GP surgery but often unable to attend. GP reports labile INRs with multiple adjustments to therapy required. INR is often subtherapeutic <2. Recent increase to dose as recent INR was 1.3

Mrs. J. Smith is an 89-year-old lady who lives alone in a two-storey house. She is fiercely independent and does not accept help from nearby concerned family and friends. She is normally independent with all her activities of daily living including washing, dressing, cooking and shopping and pays for a cleaner fortnightly.

Mrs. J. Smith has an unwitnessed fall in her own home and sustains a minor superficial skin laceration on her face. For the past few days, Mrs. J. Smith has been feeling non-specifically unwell and has not been eating and drinking well. She had been feeling increasingly drowsy and dizzy on standing. She had also not been taking all her medicines, as she felt that there were too many to manage whilst she has been feeling unwell and has only taken her metformin, amlodipine and warfarin, as she has been told in the past never to omit these as they are very important. Unfortunately, as she has been a little confused, she had taken her previous warfarin dose of 5 mg OD.

She has been admitted to hospital with dysuria and flank pain. On presentation to hospital, Mrs. J. Smith is diagnosed with a urinary tract infection and acute kidney injury. As she has fallen and hit her head, a CT head is performed, with results being NAD (no abnormality detected).

Her medicines are reviewed in light of her observations and laboratory results as detailed above.

Mrs. J. Smith has been increasingly drowsy and confused – possibly from the adverse effects of her buprenorphine patch, exacerbated by her recent decline in renal function. The decision is made to reduce her patch strength to 5mcg/hr. patch. On withdrawal of the 10mcg/hr. patch, it is expected that buprenorphine concentrations decline, decreasing approximately 50% in 12 h (range 10–24 h) [31]. Within 2 days, Mrs. J. Smith is subsequently less drowsy and confused. Mrs. J. Smith's pain is monitored and does not seem to worsen with the reduction in buprenorphine dose.

Paracetamol has been given at too high a dose for Mrs. J. Smith who is 45 kg. Chronic overdosing of paracetamol can result in hepatotoxicity with serious clinical sequelae [32]. Dosing of paracetamol should be capped at 15 mg/kg every 4–6 h, maximum 60 mg/kg per day. For Mrs. J. Smith, the maximum daily dose is calculated at 2700 mg per day, which works out as 675 mg every 4–6 h.

If given as tablets, paracetamol can only be given at 500 mg (1 tablet) 4–6 hourly, as the tablets will not allow for a 675 mg dosing. If given as the 250 mg/5 mL oral suspension, 13.5 mL (675 mg) can be given, but this may be an unwieldy amount for a patient who is already slightly confused. It would be practicably given at 10 mL (500 mg) 4–6 hourly.

Metformin: It is advised to avoid metformin if eGFR is less than 30 mL/minute/1.73 m^2 [33]. As Mrs. J. Smith is not currently eating well and her renal function has declined, the decision is made to withhold metformin. With poor renal function, there is the danger developing lactic acidosis.

Alendronic acid: Mrs. J. Smith has been taking her dose at the same time as her other morning medicines (including her Adcal-D3). To permit adequate absorption of alendronic acid, the dose must be taken at least 30 min before the first food, beverage or medicinal product of the day with plain water only. Other beverages (including mineral water), food and some medicinal products are likely to reduce the absorption of alendronic acid. If taken at the same time, it is likely that food and beverages (including mineral water), calcium supplements, antacids and some oral medicinal products will interfere with absorption of alendronic acid. Therefore, patients must wait at least 30 min after taking alendronic acid before taking any other oral medicinal product. Relative to an intravenous reference dose, the oral mean bioavailability of alendronic acid in women was 0.64% for doses ranging from 5 to 70 mg when administered after an overnight fast and 2 h before a standardised breakfast. Bioavailability was decreased similarly to an estimated 0.46% and 0.39% when alendronic acid was administered 1 h or half an hour before a standardised breakfast. In osteoporosis studies, alendronic acid was effective when administered at least 30 min before the first food or beverage of the day. Bioavailability was negligible whether alendronic acid was administered with, or up to 2 h after, a standardised breakfast. Concomitant administration of alendronic acid with coffee or orange juice reduced bioavailability by approximately 60% [34].

To facilitate delivery to the stomach and thus reduce the potential for local and oesophageal irritation/adverse experiences, alendronic acid should only be swallowed upon arising for the day with a full glass of water (not less than 200 mL). Patients should only swallow alendronic acid whole. Patients should not crush or chew the tablet or allow the tablet to dissolve in their mouths because of a potential for oropharyngeal ulceration. Patients should not lie down for at least 30 min after taking alendronic acid and until after the first food of the day. Alendronic acid should not be taken at bedtime or before arising for the day [34].

As Mrs. J. Smith's renal function has declined, and the recommendation from the manufacturers state that alendronic acid should be avoided in renal impairment where creatinine clearance is less than 35 mL/min, therefore, alendronic acid is withheld [34]. If renal function recovers to her previous baseline level, a decision should be made at a later date as to whether to resume treatment. Mrs. J. Smith should be counselled on the correct administration requirements of alendronic acid to ensure an adequate dose is delivered and of the possible severe oesophageal adverse consequences if instructions are not followed fully.

Amlodipine: As Mrs. J. Smith is suffering from postural hypotension, exacerbated by inadequate hydration, and the blunted orthostatic response in ageing physiology, amlodipine is stopped.

Digoxin: Mrs. J. Smith was recently changed from tablets to 50 mcg/mL oral elixir. Her digoxin level on admission is towards the top end of normal range. Her heart rate is currently 45 bpm. She also has a low potassium level, which can worsen digoxin toxicity because digoxin normally binds to the ATPase pump on the same site as potassium. When potassium levels are low, digoxin can more easily bind to the ATPase pump [35], exerting an increased effect. In addition, some of her recent symptoms such as bradycardia, confusion and anorexia are highly suggestive of digoxin toxicity. The combination of impaired renal function, low potassium and the recent change in formulation from tablet to oral elixir resulting in increased bioavailability has most likely caused digoxin toxicity. The decision is made to stop digoxin.

Warfarin: The patient is having difficulty with the monitoring requirements of warfarin and the variable dose range, which sometimes confuses her. She is changed to a direct oral anticoagulant (DOAC) edoxaban at a reduced fixed dose of 30 mg OD, owing to her poor renal function and low weight [36].

6 Conclusion

Older and frail people are one of several vulnerable groups likely to have reduced capacity to bear untoward effects of medicine-related problems due to changes in body systems with the ageing process. Patient safety is of utmost importance in the delivery of healthcare services. All healthcare providers and professionals must ensure due processes are undertaken to promote medication safety by supporting patients, carers and members of the public with medicine optimisation and safe use underlined by the principles of shared decision-making.

References

1. World Health Organization. Ageing and health. 2021. Available at: https://www.who.int/newsroom/fact-sheets/detail/ageing-and-health. Accessed 1 June 2022.
2. Office for National Statistics. Estimates of the very old, including centenarians, UK: 2002 to 2020. 2021. Available at https://www.ons.gov.uk/peoplepopulationandcommunity/birthsdeathsandmarriages/ageing/bulletins/estimatesoftheveryoldincludingcentenarians/2002to2020#:~:text=There%20were%20an%20estimated%2015%2C120,in%20100%2Dyear%2Dolds. Accessed 1 June 2022.
3. Department of Health and Social Care. Good for you, good for us, good for everybody: a plan to reduce overprescribing to make patient care better and safer, support the NHS, and reduce carbon emissions. 2021. Available at https://assets.publishing.service.gov.uk/government/uploads/system/uploads/attachment_data/file/1019475/good-for-you-good-for-us-good-for-everybody.pdf. Accessed 1 June 2022.
4. National Institute for Health and Care Excellence. Medicines optimisation: the safe and effective use of medicines to enable the best possible outcomes. NICE guideline NG5. 2015. Available at https://www.nice.org.uk/guidance/ng5. Accessed 2 June 2022.
5. Department of Health. Equity and excellence: liberating the NHS. 2010. Available at http://www.dh.gov.uk/prod_consum_dh/groups/dh_digitalassets/@dh/@en/@ps/documents/digitalasset/dh_117794.pdf. Accessed 2 June 2022.
6. Department of Health. Liberating the NHS: no decision about me, without me – government response to the consultation. 2012. Available at https://assets.publishing.service.gov.uk/government/uploads/system/uploads/attachment_data/file/216980/Liberating-the-NHS-No-decision-about-me-without-me-Government-response.pdf. Accessed 2 June 2022.
7. Levinson W, Kao W, Kuby A, Thisted RA. Not all patients want to participate in decision making. J Gen Intern Med. 2005;20:531–5.
8. Rathbone AP, Todd A, Jamie K, Bonam M, Banks L, Husband AK. A systematic review and thematic synthesis of patients' experience of medicines adherence. Res Soc Adm Pharm. 2017;13(3):403–39.
9. Bagge M, Tordoff J, Norris P, Heydon S. Older people's attitudes towards their regular medicines. J Prim Health Care. 2013;5(3):234–42.
10. Shariff ZB, Dahmash DT, Kirby DJ, Missaghi S, Rajabi-Siahboomi A, Maidment ID. Does the formulation of oral solid dosage forms affect acceptance and adherence in older patients? A mixed methods systematic review. J Am Med Dir Assoc. 2020;21(8):1015–23.
11. Francis S-A, Smith F, Gray N, Denham M. Partnerships between older people and their carers in the management of medication. Int J Older People Nursing. 2006;1(4):201–7.

12. Notenboom K, Beers E, van Riet-Nales DA, Egberts TC, Leufkens HG, Jansen PA, Bouvy ML. Practical problems with medication use that older people experience: a qualitative study. J Am Geriatr Soc. 2014;62(12):2339–44.
13. Drenth-van Maanen AC, Wilting I, Jansen PAF. Prescribing medicines to older people – how to consider the impact of ageing on human organ and body functions. Br J Clin Pharmacol. 2020;86:921–30.
14. Haywood A, Glass BD. Evidence of stability of medicines repackaged in compliance aids: a review. Curr Drug Saf. 2016;11:69–77.
15. Husain I, Minshull J. Summary of guidance and evidence for use of multi-compartment compliance aids (MCCAs). Specialist Pharmacy Service. 2019. Available at https://www.sps.nhs.uk/wp-content/uploads/2019/05/SPS_MCCA_briefing_May2019_final.pdf. Accessed 7 June 2022.
16. Goldstein R, Rivers P. The medication role of informal carers. Health Soc Care Community. 1996;4(3):150–8.
17. Vanwesemael T, Boussery K, Dilles T. Self-administration of medication in hospital: A literature review. Nurs Sci Q. 2020;33(3):249–57.
18. Ude-Okeleke RC, Aslanpour Z, Dhillon S, Umaru N. Medicines related problems (MRPs) originating in primary care settings in older adults – a systematic review. J Pharm Pract. 2021;36:357. https://doi.org/10.1177/08971900211023638.
19. Mangoni AA, Jackson SHD. Age-related changes in pharmacokinetics and pharmacodynamics: basic principles and practical applications. Br J Clin Pharmacol. 2004;57(1):614.
20. Nigam Y, Knight J. (2017) Anatomy and physiology of ageing 11: the skin. Nursing Times [online]; 113: 12, 51–55. Available at https://www.nursingtimes.net/roles/older-people-nurses-roles/anatomy-and-physiology-of-ageing-11-the-skin-27-11-2017/. Accessed 2 June 2022.
21. Lewis G. UKMI QA Switching between liquid and tablet/capsule formulations – which medicines require extra care? 2021. Available at https://www.sps.nhs.uk/articles/switching-between-liquid-and-tablet-capsule-formulations-which-medicines-require-extra-care/. Accessed 4 May 2022.
22. International Dysphagia Diet Standardisation Initiative. 2019). Available at https://iddsi.org/framework/. Accessed 4 May 2022.
23. Panagioti M, Skevington SM, Hann M, Howells K, Blakemore A, Reeves D, Bower P. Effect of health literacy on the quality of life of older patients with long-term conditions: a large cohort study in UK general practice. Qual Life Res. 2018;27:1257–68.
24. Maes ML, Fixen DR, Linnebur SA. Adverse effects of proton-pump inhibitor use in older adults: a review of the evidence. Ther Adv Drug Saf. 2017;8(9):273–97.
25. National Institute for Health and Care Excellence. Multimorbidity: clinical assessment and management [NG56]. 2016. Available at https://www.nice.org.uk/guidance/ng56. Accessed 2 May 2022.
26. Cassell A, Edwards D, Harshfield A, Rhodes K, Brimicombe J, Payne R, griffin S. The epidemiology of multimorbidity in primary care: a retrospective cohort study. Br J Gen Pract. 2018;68(699):245–51.
27. Barnett K, Mercer SW, Norbury M, et al. Epidemiology of multimorbidity and implications for health care, research, and medical education: a cross-sectional study. Lancet. 2012;2012(380):37–43.
28. Age UK. More harm than good – why more isn't always better with older people's medicines. 2019. Available at https://www.ageuk.org.uk/globalassets/age-uk/documents/reports-and-publications/reports-and-briefings/health%2D%2Dwellbeing/medication/190819_more_harm_than_good.pdf. Accessed 3 May 2022.
29. National Institute for Health and Care Excellence. Shared decision making. NICE guideline NG197. 2021. Available at https://www.nice.org.uk/guidance/ng5. Accessed 1 July 2022.
30. Bunn F, Goodman C, Russell B, Wilson P, Manthorpe J, Rait G, Hodkinson I, Durand M-A. Supporting shared decision-making for older people with multiple health and social care needs: a realist synthesis. Health Serv Deliv Res. 2018;18(1):165.

31. Electronic Medicines Compendium. BuTrans 5 microgram/hour transdermal patch summary of product characteristics. 2021. Available at https://www.medicines.org.uk/emc/product/124/smpc#gref. Accessed 11 May 2022.
32. Healthcare Safety Investigation Branch. Unintentional paracetamol overdose in adult inpatients with low bodyweight. 2022. Available at https://www.hsib.org.uk/investigations-and-reports/unintentional-overdose-of-paracetamol-in-adults-with-low-bodyweight/. Accessed 11 May 2022.
33. Electronic Medicines Compendium. Metformin 500 mg tablets summary of product characteristics. 2019. Available at https://www.medicines.org.uk/emc/product/13160/smpc. Accessed 11 May 2022.
34. Electronic Medicines Compendium. Alendronic Acid 70 mg tablets summary of product characteristics. 2022. Available at https://www.medicines.org.uk/emc/product/5206. Accessed 11 May 2022.
35. Electronic Medicines Compendium. Digoxin Tablets BP 62.5 micrograms summary of product characteristics. 2020. Available at https://www.medicines.org.uk/emc/product/5771. Accessed 11 May 2022.
36. Electronic Medicines Compendium. Lixiana 30mg film-coated tablets summary of product characteristics. 2020. Available at https://www.medicines.org.uk/emc/product/6906/smpc. Accessed 11 May 2022.

[illegible]

Medicine Acceptability: A Key Aspect in the Older Population

Thibault Vallet and Fabrice Ruiz

Contents

Abstract Older people have particular needs that require special considerations concerning healthcare due to the effects of ageing, such as the deterioration of physical and cognitive abilities, multimorbidity, polypharmacy, or frailty. In recent years, acceptability has emerged as a key factor for patient adherence in vulnerable populations and, consequently, for treatment effectiveness. Acceptability has been defined as the ability and willingness of a patient to self-administer and also of any of his/her lay or professional caregivers, to administer a medicinal product as intended. Although this multi-faceted concept is likely to be driven by many factors such as product or user characteristics, knowledge on acceptability drivers remains fragmented. Herein we present a data-driven approach based on a large set of real-life observer-reported outcomes. Using examples, this chapter illustrates how this multivariate approach can contribute to provide the scientific community with relevant knowledge on medicine acceptability drivers in the older population. Evidence base on acceptability drivers should facilitate the identification of the medicine

T. Vallet · F. Ruiz (✉)
ClinSearch, Malakoff, France
e-mail: Thibault.vallet@clinsearch.net; fabrice.ruiz@clinsearch.net

M. Orlu, F. Liu (eds.), *Pharmaceutical Formulations for Older Patients*, AAPS Advances in the Pharmaceutical Sciences Series 51,
https://doi.org/10.1007/978-3-031-35811-1_3

features that best fit user characteristics in order to ensure the choice of appropriate formulations for adequate acceptability.

Keywords Acceptability · Evaluation · Multivariate analysis · Product design · Prescription

1 Why Medicine Acceptability Is Essential in the Elderly?

The proportion of the world's population over 65 years is expected to double between 2019 and 2050, from 1 in 11 people in the world to 1 in 6 [1]. Although neither linear nor consistent, changes due to ageing lead to a gradual deterioration of physical and cognitive abilities and an increased risk of disease [2]. Furthermore, older people are more likely to experience several health problems simultaneously, especially chronic and degenerative disorders. Therefore, drug prescriptions play an important role for these older people who are the main users of medicines [3–5].

Multiple components of ageing may negatively impact the safety and efficacy of some drugs. At the biological level, decreased gastrointestinal motility and alterations of kidney and liver function might affect absorption, distribution, metabolism, and elimination of the active pharmaceutical ingredient [6]. Another example is dysphagia, which is one of the most common age-related alterations. In some institutions for the elderly, up to 60% of patients are affected by difficulty in swallowing [7–9]. This condition has a negative impact on oral medication intake [10–12]. Indeed, this disorder may lead to pain, choking, or serious life-threatening aspiration pneumonia during the oropharyngeal swallowing phase [13, 14]. End-user alteration of oral solid dosage forms (SODF) prior to administration, such as opening a capsule or crushing/splitting a tablet, is a widespread practice [15]. A French study found that in reported cases of capsule opening or tablet crushing, 42% were of medications for which this was contraindicated [16]. Such medicinal product manipulations could seriously interfere with dosage accuracy, medicine stability, pharmacokinetics, and pharmacodynamics, thus compromising treatment efficiency or even inducing toxic effects for the patients as well as for the caregivers [17–27]. Some neurodegenerative diseases, such as Parkinson's and Alzheimer's diseases, are more common in older people. Leading to behavioural issues, these disorders may negatively affect the patient's ability and willingness to take their medication as intended. Understanding instructions, handling containers, and product identification are all essential for completing the prescribed course of any treatment, tasks made more difficult by the loss of physical and mental capacities so frequent among the elderly [28, 29]. Due to a high rate of multimorbidity, the negative impact of inappropriate dosage forms is exacerbated by polypharmacy in the older population, exposing patients to adverse drug reactions [30–32].

Older people, the main users of medicines, appear to be a vulnerable population that faces increasing risks regarding medicine use. Patient adherence, which is the extent to which a patient follows their physician's instructions, is essential to achieve all of the expected therapeutic benefits. Inappropriate dosage forms may negatively impact patient adherence, because very simply, "drugs don't work in patients who don't take them" [33]. For a long time in the field of pharmaceutics, the crucial endpoints of efficacy and safety concealed other users' needs such as product acceptability. In the context of medicinal product administration, the earliest occurrence of the word "acceptability" that we have found in the literature was published in 1965 [34]. Since, the number of manuscripts addressing this concept has steadily increased, especially in paediatrics. In recent years, the European Medicine Agency (EMA) [35], the Food and Drug Administration (FDA) [36, 37], the World Health Organization (WHO) [38], and the International Council for Harmonization (ICH) [39] evoked the importance of developing medicinal products well suited to the characteristics of these young patients. Although essential in this population, the importance of acceptability is not limited to paediatrics, as highlighted as early as 2009 by ICH Q8 guideline: "in all cases, the product should be designed to meet patients' needs and the intended product performance" [40]. In 2020, the EMA endorsed the significant role of acceptability on patient adherence in the elderly through a reflection paper on the pharmaceutical development of medicines for use in this population [41]. In this document, acceptability has been defined as "the ability and willingness of a patient to self-administer, and also of any of his/her lay or professional caregivers, to administer a medicinal product as intended". This multifaceted concept is driven by patients' characteristics, e.g. age, health status, behaviour, disabilities, and culture, as well as by medicines' features, e.g. route of administration, type of dosage form, swallowability, appearance, required dose, required dosing frequency, duration of treatment, ease of use of the preparation, and selected administration. Although the EMA stated that "adequate patient acceptability is an essential aspect of the pharmaceutical development of a medicinal product and its post-authorization life cycle", an internationally harmonized method of evaluation is long overdue, and knowledge remains fragmented.

2 How Medicine Acceptability Assessment Can Be Standardized?

Herein we present a methodology that was first developed and validated in paediatrics [42, 43] and then transposed to the older population [44–49], to acquire a better knowledge of medicine acceptability drivers in these vulnerable populations. Figure 1 presents the main steps for designing the acceptability reference framework allowing knowledge enhancement.

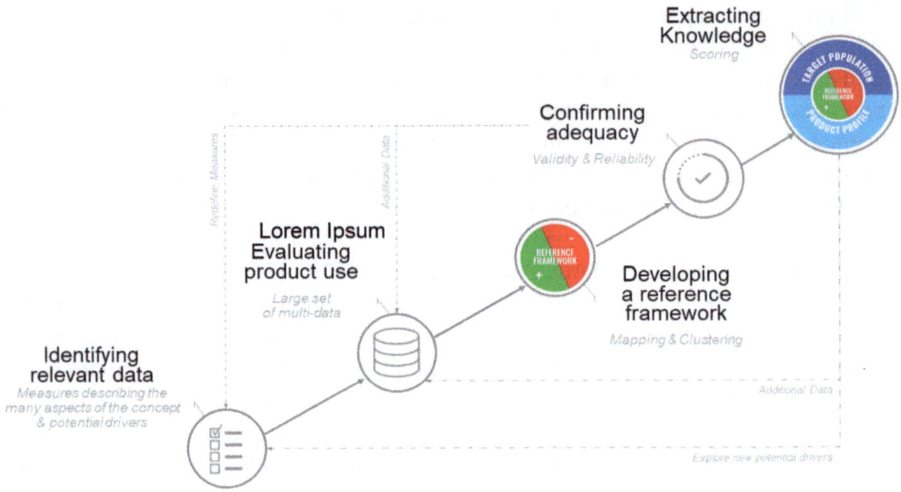

Fig. 1 Multivariate approach to better understand the multi-faceted concept of medicine acceptability in vulnerable populations [50]

2.1 Identifying Relevant Data

Direct assessment of the patient perspective is needed to thoroughly understand patients' experiences of treatment. Many patients in the paediatric and the elderly populations are unable to provide reliable and valid self-evaluations due to their development status or due to the deterioration of physical and cognitive abilities. Thus, identifying relevant data to be collected is the first issue to be addressed with vulnerable populations. Observer-reported outcomes (ObsRO) appear quite relevant to standardize data collection in such a case. The FDA encourages privileging observer reports that include only those events or behaviours that can be observed rather than proxy measures [51]. In contrast to proxy measures, observational measures exclude inferences about the patient's subjective experience. For example, observers are unable to provide reliable and valid reports on a patient's pain intensity but can report patient behaviour caused by pain. Different observed measures that were previously used to assess medicines' acceptability, especially in paediatrics, the most studied population, include the result of the intake, the patient reaction, the time needed to prepare and administer the medicine, and the recourse of any methods to ease/achieve administration such as dividing the intake of a dose that cannot be taken as a whole, altering the intended use (modify dosage form such as capsule opened or tablet crushed; use another route/mode of administration), using food/drink to mask a bad taste, and using a device not provided or using restraint.

Beyond validity and reliability of measures, combining different measures appears to be crucial to assess the many aspects of acceptability. If the required dose of a medication were to be fully taken without any negative reaction, that might appear to be well accepted. Is this truly the case if the caregiver needed to crush the tablet, mix the drug with food to mask its aversive taste, and, consequently, required a long time to prepare and administer the mixture? Furthermore, this complex concept cannot be quantified on its own [52], and consequently, assessment would need to be relative. Therefore, simultaneous consideration of those different measures of a single intake, as well as measures recorded from many patients and medicines from the wide spectrum of products on the global market, would appear to be necessary to fully grasp medicine acceptability and investigate its drivers.

2.2 *Evaluating Product Use*

Based on this, a large set of real-life observer-reported outcomes collected for many medicine intakes in various older patients (65 years and over) was formed. The data were collected in a multicentre, prospective, cross-sectional, and observational study conducted in France and the UK from October 2016. Twelve care homes and elderly care wards of nine hospitals were involved in the study. The Greater Paris University Hospitals (AP-HP), University College London (UCL) hospitals, the UCL school of pharmacy, and the university of Hertfordshire were involved in the study. A trained member of the site study team observing the first medicine use following study inclusion filled in a standardized web-questionnaire (Fig. 2). The questionnaire consists of objective measures of acceptability including events/behaviours that can be observed during the use of the medicine (observer-reported outcomes), as well as contextual information on the user, the medicine being assessed, and the administration setting.

Fig. 2 Data collection questionnaire

In total, 2008 evaluations were collected in the study. The evaluations were predominantly collected in France (86%) and in hospitals (76%). The mean age of the patients was 86.1 years (SD, 7.5; range, 65–108), and 67% were women.

The evaluations were related to 569 distinct medicinal products. There were 207 chemical substances classified in 45 therapeutic subgroups according to the fifth level and the second level of the Anatomical Therapeutic Chemical (ATC) classification system, respectively. Analgesics (11%) and psychoanaleptics (9%) were the most commonly prescribed therapeutic subgroups of medicines followed by psycholeptics (8%) and antiepileptics (8%). All the 14 anatomical main groups, corresponding to the first level of this classification system, were represented. There were 51 pharmaceutical forms included, highlighting the diversity of the medicines assessed. Most of them were oral or buccal medicines (90%), but other routes of administration were assessed such as ocular, pulmonary, parenteral, topical, or rectal. Tablet (55%), which could be coated or divisible, and capsule (16%) were the most frequently prescribed and administered forms. The use of solid for oral liquid preparations (8%), oral liquid ready to use (7%), and orally disintegrating forms (4%) was also observed.

Each evaluation of one medicinal product, taken by one patient, corresponded to a particular combination of an observed measure (e.g. fully taken) for each of the eight aforementioned observational variables (e.g. result of intake). Evaluations recorded 173 distinct combinations out of the 864 mathematically possible. It seems that the variability in the source population allowed identification of existing users' main behaviours corresponding to the relevant combinations of measures. As the use of new combinations of measures decreases as data collection progressed and consequently, key user behaviours should have been captured from the source population.

2.3 *Developing a Reference Framework*

Unsupervised methods were used to visualize the key information from this large dataset. First, a multiple correspondence analysis summarized in a low-dimensional Euclidean space the main underlying structures of the dataset. The three dimensions forming the map revealed the associations and dissociations of observed measures that contributed the most to explaining variability observed in the data in an intelligible form. The evaluations – distinct combinations of an observed measure for each of the eight observational variables – and the 19 observed measures themselves were positioned together on the map. Interpretation of the map is based on distances between elements: proximities on the map express similarities. Observed measuvres that were closest on the map were often selected together in the evaluations. Similarly, evaluations completed in a comparable manner converged on the map. Subsequently, hierarchical clustering on principal components and *k*-means consolidation partitioned the set of evaluations into clusters according to their Euclidean distances on the map. The clusters defined coherent and meaningful acceptability profiles, which were described by the observed measures over-represented into their

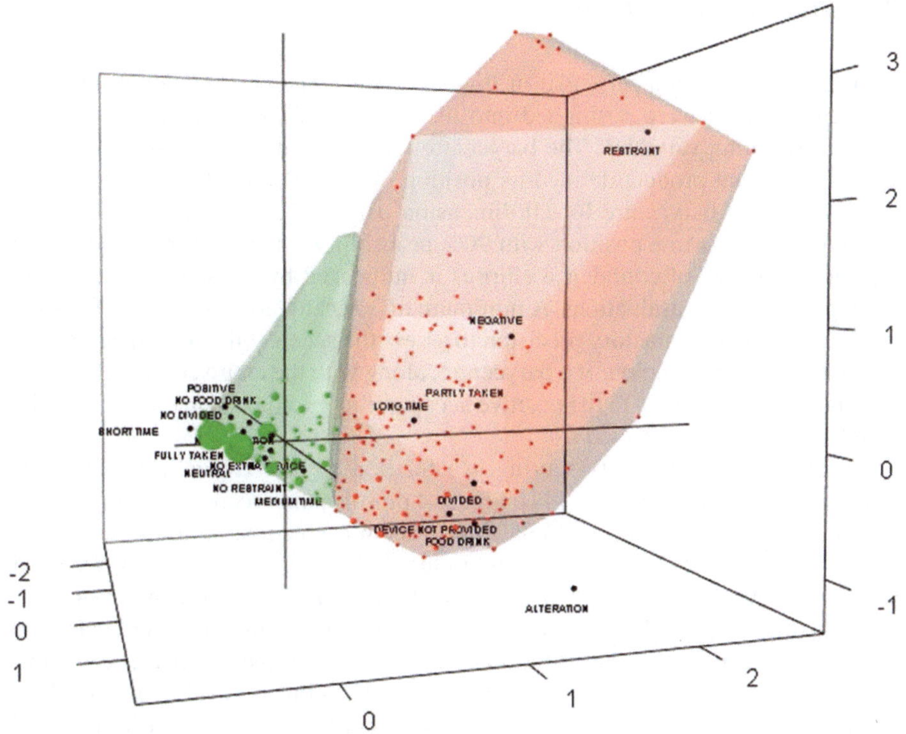

Fig. 3 Acceptability reference framework (3D)

subset of evaluations. Thus, this multivariate analysis gave rise to the acceptability reference framework: a three-dimensional map (3D-map) juxtaposing two distinct acceptability profiles – "positively accepted" and "negatively accepted" – materialized by the green and the red areas on the 3D map, respectively (Fig. 3).

2.4 *Confirming Adequacy*

Validity and reliability of the multivariate approach have been demonstrated [50]. As the model fit with the set of evaluations that had been collected at random, the influence of variations in source data on modelling processes had to be explored before using this tool to score acceptability. Procedure investigating statistical reliability is succinctly described hereafter. Resampling statistics were used to validate the statistical reliability of the acceptability reference framework [44]. Permutation testing validated robustness and significance of the relationships among categories highlighted by the acceptability map, and bootstrapping demonstrated a high stability of the map and a good recovery of the two clusters regardless of variations in the set of evaluations.

2.5 Scoring Acceptability

All the evaluations of a particular medicinal product taken by individual patients in a specific context were positioned into the reference framework according to the eight constituting variables. The barycentre (centroid) of the evaluations of a particular medicinal product defined its position on the 3D map. Confidence ellipses surrounding the barycentre for all dimension pairs [1-2, 1-3, and 2-3] defined an area containing its true position with 90% probability if the experiment were to be repeated. The size of confidence ellipses is influenced by the sample size and the homogeneity of the evaluations. A minimum of 30 evaluations is required for a reliable score with a satisfactory precision to assess the acceptability of a specific product in a targeted population. If a barycentre, along with the entire confidence ellipsis surrounding it, belonged to the positive acceptability profile, the medicine was classified as "positively accepted".

To illustrate acceptability scoring, we focused on a medicinal product anonymously labelled medicine "Y" [44]. Figure 4 presents the acceptability score of this psycholeptic drug assessed in 36 patients. The 36 evaluations were located in both acceptability profiles: 61% into the first cluster defining the "positively accepted" profile and 39% into the second cluster materialized by the red zone on the map defining the "negatively accepted" profile. The barycentre of these evaluations (square mark) was assigned to the first cluster defining the positive acceptability profile. However, a significant part of the confidence ellipses (oval shape) resides in the second cluster; thus, this product cannot be classified as "positively accepted". Indeed, the barycentre, along with the entire confidence ellipsis surrounding it, must fall within the positive acceptability profile materialized by the green zone on the map, to be classified as acceptable.

Regarding medicine "Y", although the ideal combination of measures was the most frequently observed, accounting for 31% of the evaluations, almost two-thirds of the 19 combinations collected fell into the second cluster. This heterogeneity of evaluations led us to explore acceptability differences in subpopulations of patients [44]. Figure 5 presents the acceptability scores depending on whether the patient swallowed with or without difficulty. Due to the insufficient number of patients in each subset of evaluations ($n < 30$), the results remain inconclusive. It was nonetheless observed that the barycentre of responses for patients with swallowing disorders fell in the "negatively accepted" area. This preliminary finding illustrated how the reference framework allows explorations of acceptability drivers in the elderly.

The evaluations were successively partitioned according to the many explanatory variables (e.g. medicinal product, pharmaceutical form, patient health status) for obtaining relevant acceptability scores and thus exploring acceptability drivers using the reference framework.

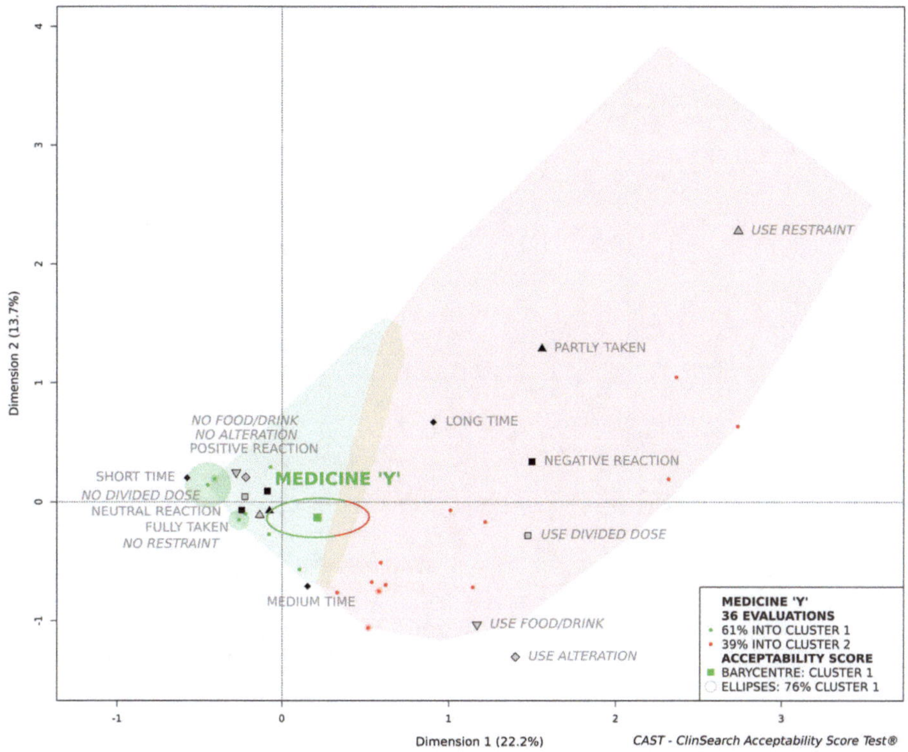

Fig. 4 Acceptability of a medicinal product anonymously labelled medicine "Y" [44]

3 Exploring Medicine Acceptability Drivers in the Elderly

3.1 Medicine Acceptability in the Elderly with Swallowing Disorders

As described in the first section of the chapter, altered swallowing function is a common disability in the older population, compromising the use of SODF, which are widely manufactured and frequently prescribed in older populations regardless of the specific characteristics of patients.

In older inpatients without dysphagia, SODF were found to be well accepted even for patients over 95 years of age that were still able to consume two capsules in a single sitting [45]. Although tablets could be classified as "positively accepted" in older patients without swallowing disorders, as expected this was not the case in older patients with dysphagia (Fig. 6) [48].

By reviewing studies on the effect of the SODF size on acceptability, the FDA observed that increasing the size of tablets and capsules is associated with an increase in the number of complaints related to swallowing difficulties [53].

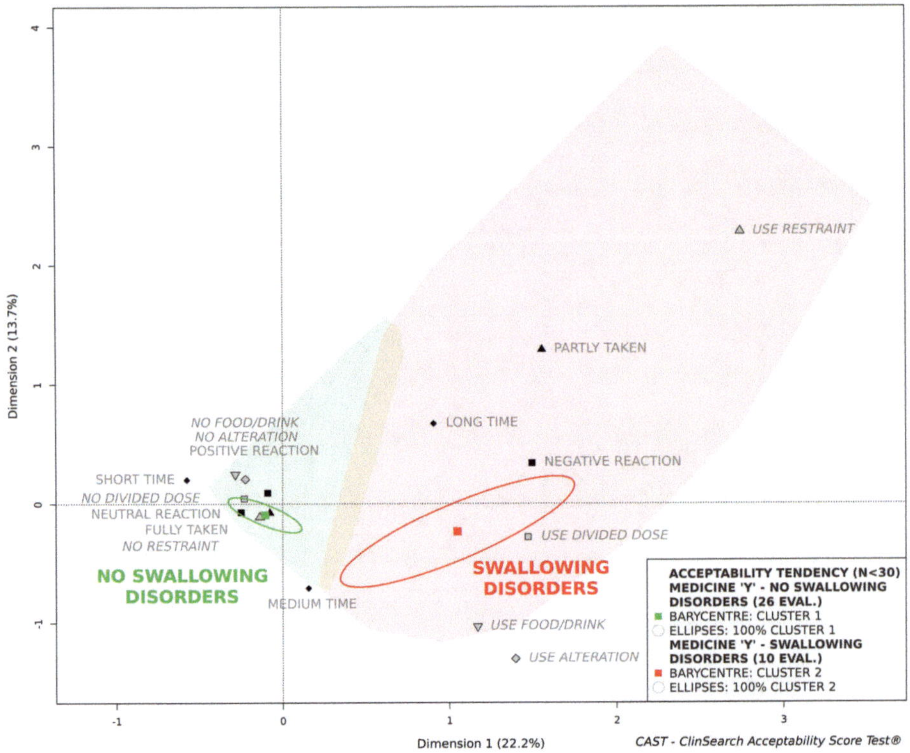

Fig. 5 Influence of swallowing disorders on acceptability of a medicinal product anonymously labelled medicine "Y" [44]

According to their guidelines on "size, shape and other physical attributes of generic tablets and capsules", medications greater than approximately 8 mm in diameter seem to be more likely to be related to swallowability issues [53]. Focusing on patients with swallowing disorders, we demonstrated that reducing tablet size had a significant impact on acceptability in this subpopulation: tablets smaller than 6.5 mm – the largest dimension of the tablet referring to the length of oval or oblong tablets or the diameter of round tablets – appeared to be accepted by patients with swallowing disorders [48]. This threshold of 6.5-mm-sized tablets was the largest for which the barycentre and its entire confidence ellipse remain completely within the "positively accepted" cluster. According to Fields et al. [54], among patients who blame the size of SODF, three-quarters experienced difficulties in taking tablets that were too large, while a quarter complained that tablets were too small. In institutions, where the study was carried out, the preparation and administration of medications are primarily conducted by nurses and trained healthcare staff, which is not necessarily the case in domestic settings. Consequently, some handling issues may arise with smaller tablets, especially for the aging population living in their own home, resulting in small tablets being poorly accepted. Adequate devices and

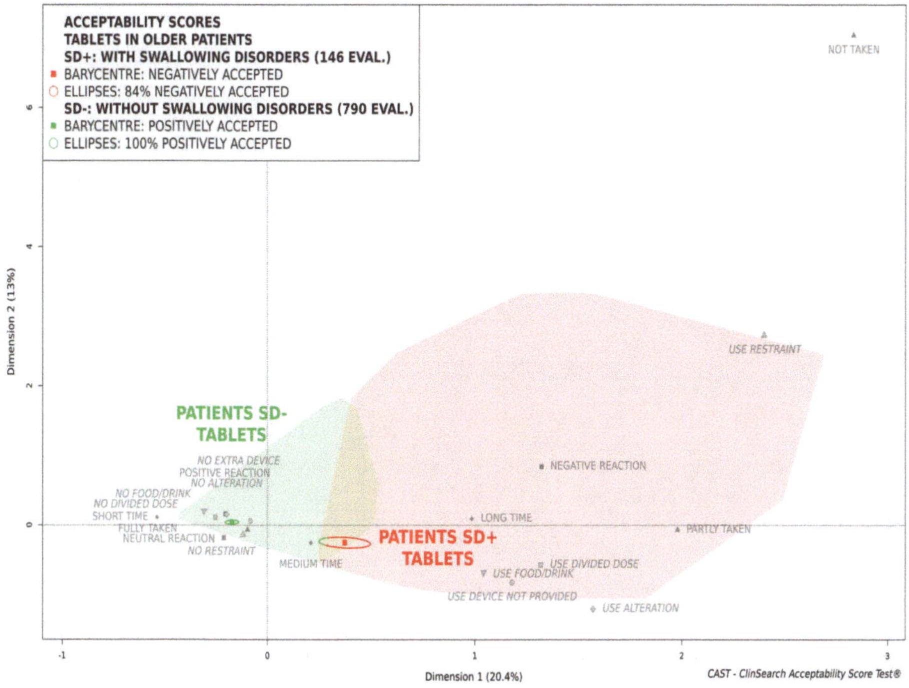

Fig. 6 Acceptability of tablets in older patients with swallowing disorders (SD+) and without swallowing disorders (SD−) [48]

containers will be essential to alleviate these issues [53]. Medicine acceptability remains a multi-faceted concept, and it is therefore crucial to consider all its different aspects, including context, along with user and product characteristics.

Focusing on data from the UK, we highlighted a difference in the acceptability of SODF administered in a hospital compared with a care home setting [49]. Surprisingly, these medicines appeared to be better accepted in hospital (Fig. 7).

As expected, SODF were poorly accepted by patients with swallowing disorders in both settings, but by focusing on patients without swallowing alteration, we have shown that SODF were better accepted in hospitals than in care homes (Fig. 8). This difference was due to variations in care practices between administration settings. Indeed, in the care home, a consideration of patient preferences, regardless of their ability to swallow, increased tablet crushing, mixing the drug with food or drink, or using an extra administration device. Of the SODF medicines that were modified prior to administration in patients without swallowing disorders in the care home, 83% percent were not explicitly permitted within the summary of product characteristics [49]. This example demonstrates the roles that different contexts play in medicine administration, corroborating the multi-dimensional nature of acceptability.

In real-life conditions, oral liquid medicinal products are a widely used alternative to SODF in institutional care in order to ease medicine administration for

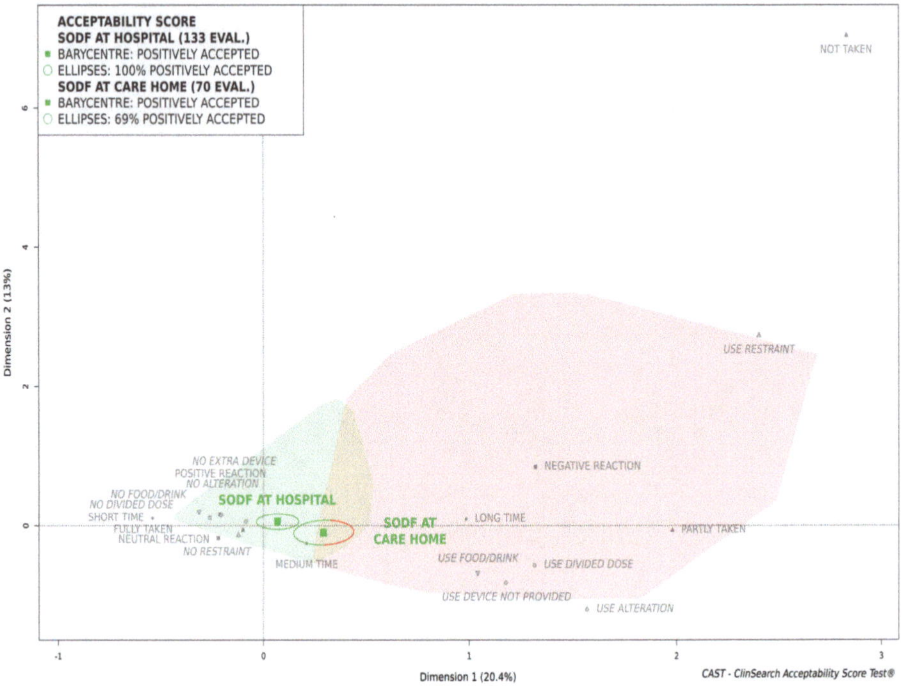

Fig. 7 Acceptability of solid oral dosage form (SODF) in older patients in the UK depending on settings: hospital and care home [49]

patients with swallowing alteration. However, as shown by Fig. 9, these products could not be classified as accepted in older patients with swallowing disorders. This was not very surprising, considering that the "drink" test – ability to swallow 80 mL of water with no interruptions or difficulties – is a method of diagnosing dysphagia.

Acceptability of different dosage forms of paracetamol in the older inpatient population was investigated [45]. Capsule, orally disintegrating tablet (ODT), and powder for oral solution were classified as "positively accepted" (Fig. 10). However, the barycentre of the evaluations of powder for oral solution fell into the furthest limit of the green area on the 3D map. Although sample sizes were quite similar, the confidence ellipses surrounding the barycentre for powder for oral solution were larger than the confidence ellipses surrounding the barycentre for ODT, reflecting the heterogeneity in the evaluations of powder for oral solution. This dosage form appeared to have been preferentially prescribed in older inpatients with swallowing disorders. Unsurprisingly, these patients, unable to drink a half-glass of water with no interruptions or difficulties, were less amenable to the administration of paracetamol powder dissolved in water.

Although additional evaluations will be needed, ODT represents an interesting option in terms of acceptability for those patients [45]. Beyond ODT, other dosage forms such as dispersible tablets, chewable tablets, and mini-tablets are alternative

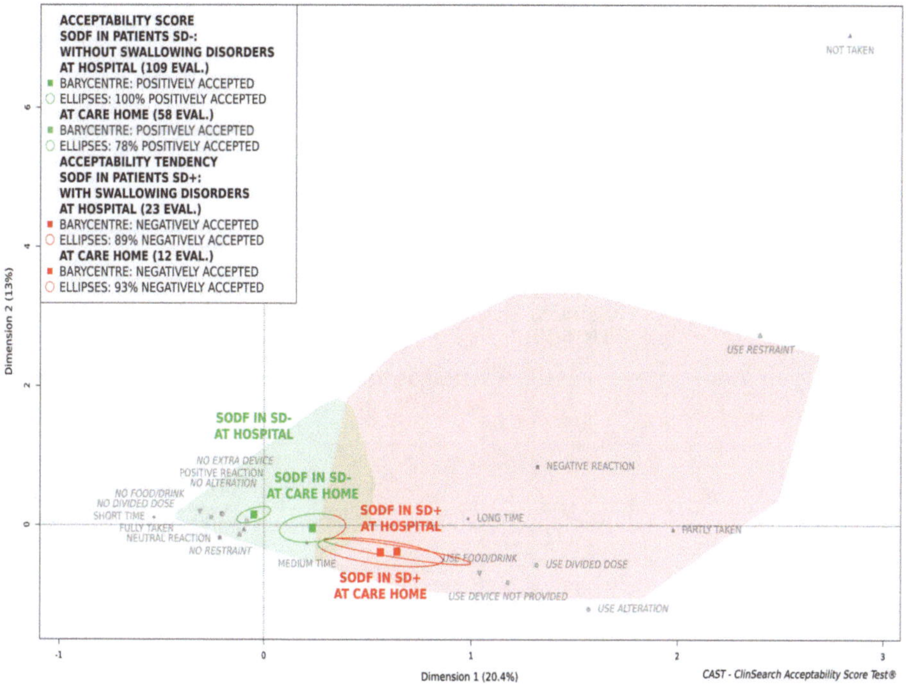

Fig. 8 Acceptability of solid oral dosage form (SODF) in older patients in the UK with (SD+) and without (SD−) swallowing disorders depending on settings: hospital and care home [49]

dosage forms to conventional tablets and solutions that might be explored to improve acceptability in the older patients [14, 55]. In addition, modified-release multi-particulates, which could be incorporated into easy-to-swallow dosage forms such as dry suspensions and oral jellies, could be of interest [56, 57].

3.2 *Palatability Issues in the Elderly*

Although palatability is known to be an important acceptability driver in paediatrics, it is usually not considered to be a major issue for older patients due to the increasing prevalence of dysgeusia with age and poly-medication [58–60]. The acceptability reference framework has demonstrated, however, that palatability issues remain essential for the acceptability of oral liquid medicinal products in older people hospitalized or receiving institutional care, especially in women [46, 47].

Figures 11 and 12 investigated palatability issues by patient sex. Oral liquids formulated with flavour seemed to be accepted, regardless of the sex of patients, as the barycentre, along with the entire confidence ellipses, with the exception of a

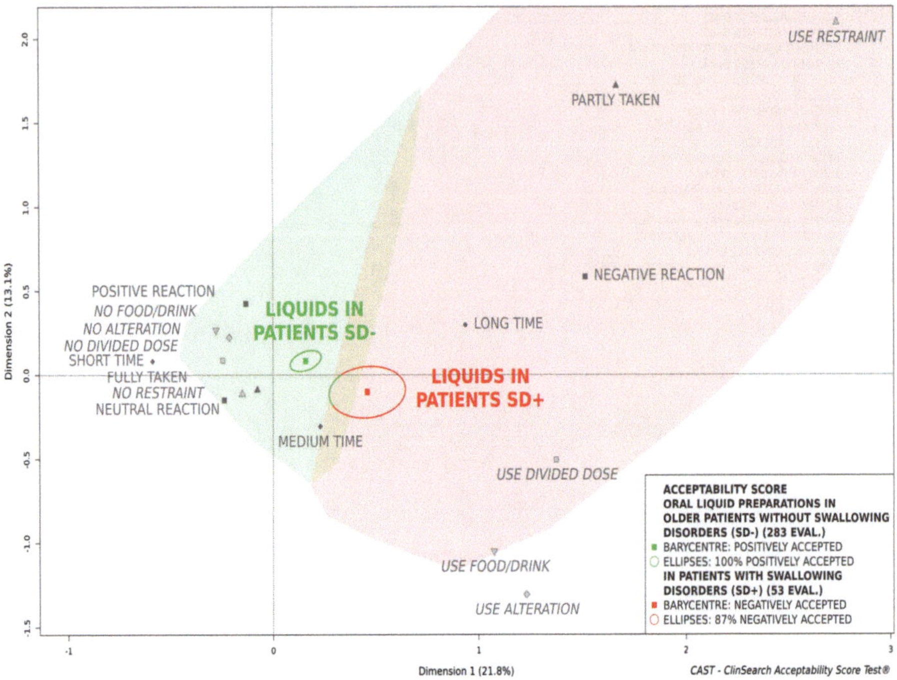

Fig. 9 Acceptability of oral liquid pharmaceutical products in older patients with swallowing disorders (SD+) and without swallowing disorders (SD−) [47]

very limited part of the ellipse in the third dimension for women, belonged to the "positively accepted" profile (Fig. 11), while those formulated without flavour could be classified as accepted in men, but not in women (Fig. 12).

Such differences regarding palatability were confirmed in a study focusing on an Alzheimer's treatment: Ebixa®, memantine hydrochloride [46]. Memantine hydrochloride was originally formulated only as a coated tablet (10 and 20 mg). The need for a more appropriate formulation for older patients with swallowing disorders resulted in the development of an oral solution (10 mg/mL). As Alzheimer's disease exacerbates taste and smell impairments in the older population [61], we might assume that a lack of interest for product palatability resulted in a formulation without any flavouring agent.

Using the acceptability reference framework, both forms, tablet and oral solution, were classified as accepted in men. In women, the coated tablet, providing a physical barrier between the active pharmaceutical ingredient (API) and the patient's taste buds, was classified as accepted, while the oral solution formulated without any flavouring agent appeared to be "negatively accepted" (Fig. 13).

This has also been explored using the rodent Brief Access Taste Aversion (BATA) model, which objectively detects the aversive taste of API [62–65]. This model utilizes a lickometer, which records the number of licks that rodents (placed in an intermittent water-restriction schedule) take of different test solutions in a

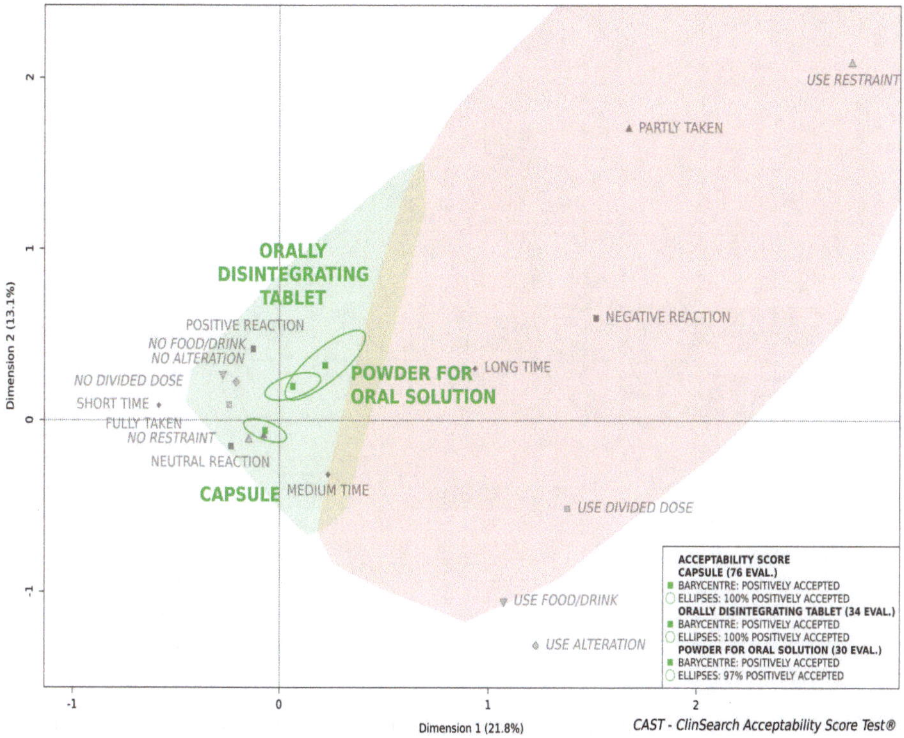

Fig. 10 Acceptability of three dosage forms of 500 mg paracetamol in older inpatients: capsule, orally disintegrating tablet, and powder for oral solution [45]

standardized protocol. Using the BATA model, the original oral solution of memantine was categorized as highly aversive/untolerated as indicated by a significant reduction in lick number in comparison with water [46]. The number of licks was also significantly lower in female than in male rats (Fig. 14).

Such converging results from human (CAST) and animal (BATA) models illustrated acceptability issues with the original oral solution of memantine (Ebixa®) driven by palatability. Furthermore, it appeared that palatability profiles can significantly vary between the sexes, as a higher sensitivity of the females to this unpalatable oral formulation was observed. These findings indicated that palatability issues remain an important driver of acceptability in the older population, especially in women. Therefore, formulation scientists and the pharmaceutical industry should consider taste masking as an important quality attribute of products developed for the elderly.

When the EMA drafted a reflection paper for the older population in 2017, palatability was not addressed. Findings such as this impelled the inclusion of palatability as one of the product characteristics influencing patient acceptability in the reflection paper adopted by the Committee for Medicinal Products for Human Use (CHMP) in 2020 [41].

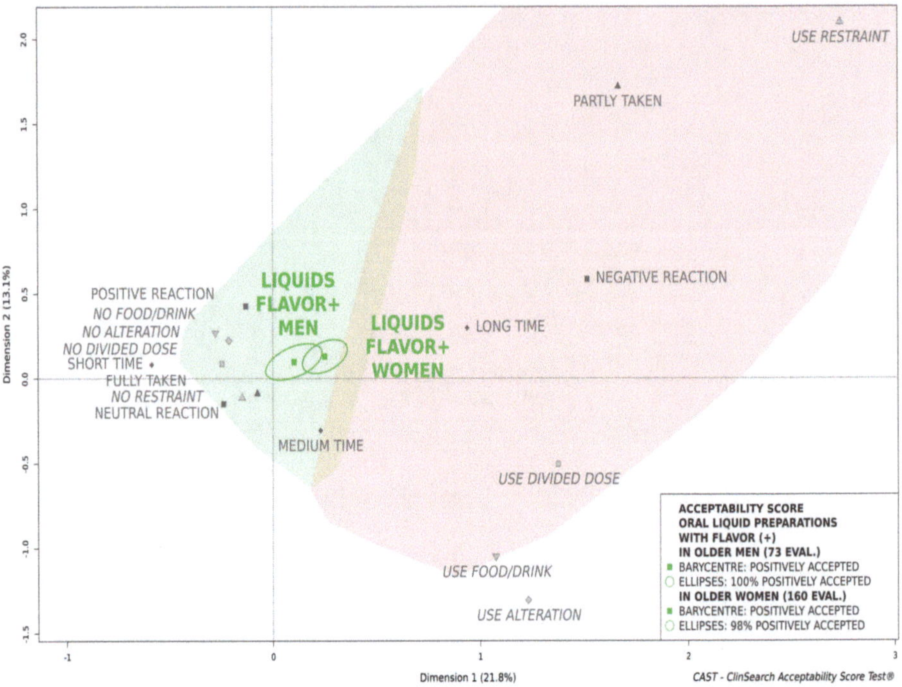

Fig. 11 Acceptability of oral liquid pharmaceutical products with flavour (FLAVOR+) in older men and women [47]

4 Improving Knowledge on Acceptability Drivers

The reference framework integrating the many aspects of acceptability allows comparison of standardized evaluation of this multi-faceted concept. This tool was thus of real benefit for improving our understanding of acceptability drivers in the older population. Using the reference framework, we were able to objectively demonstrate acceptability variations due to changes among any user or product characteristics, as presented in this chapter. Based on a significant number of evaluations, relevant acceptability scores were obtained, and significant differences based on patient swallowing ability or sex, medicine size or flavour, as well as the context or setting of administration were highlighted. However, many other characteristics may influence acceptability. Although essential in the older population, swallowability and palatability are only two product characteristics that may influence acceptability in older people [41]. The type of patient environment is likely to affect medicine use by the patient and/or the caregiver. While medicine preparation and administration are mainly conducted by trained healthcare staff in institutions, the patient may face specific barriers at home. Furthermore, those barriers may differ according to the patient's culture. In this study, the data were collected in hospitals and care homes in France and the UK, and consequently, further evaluations in outpatients from different countries are needed. Due to the wide range of medicines available on the global market,

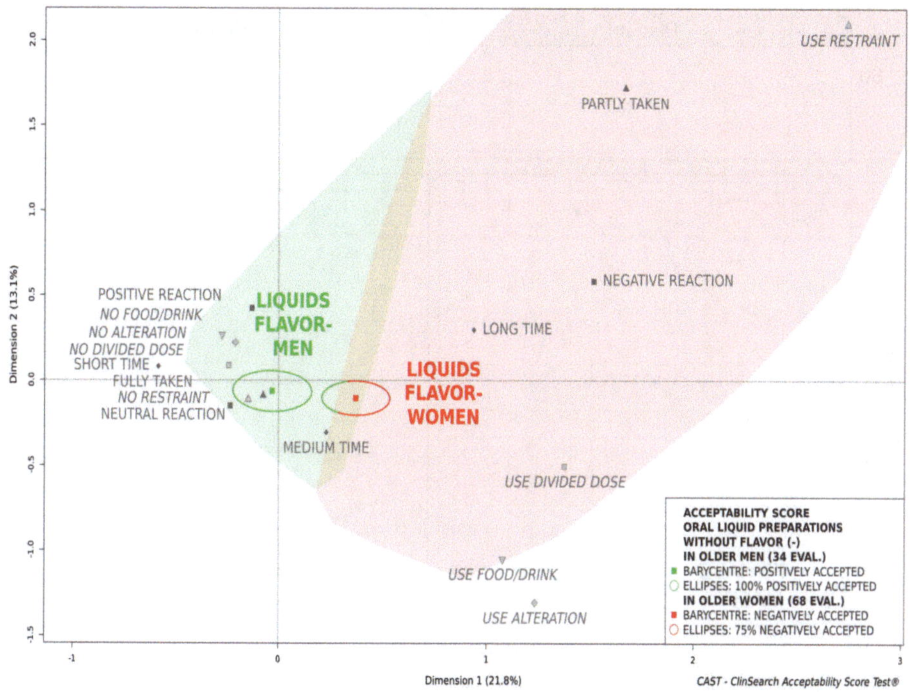

Fig. 12 Acceptability of oral liquid pharmaceutical products without flavour (FLAVOR−) in older men and women [47]

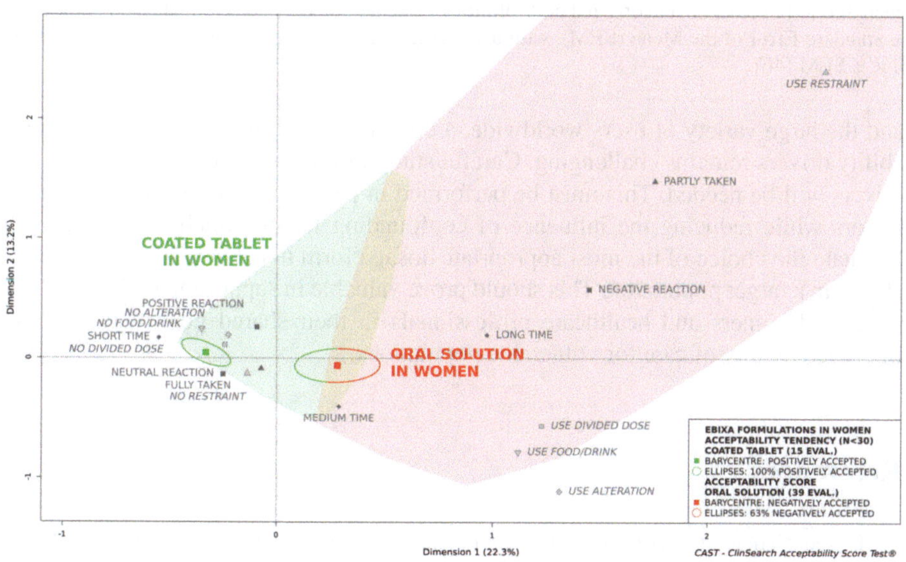

Fig. 13 Acceptability of the original formulations of memantine (Ebixa®) in older women [46]

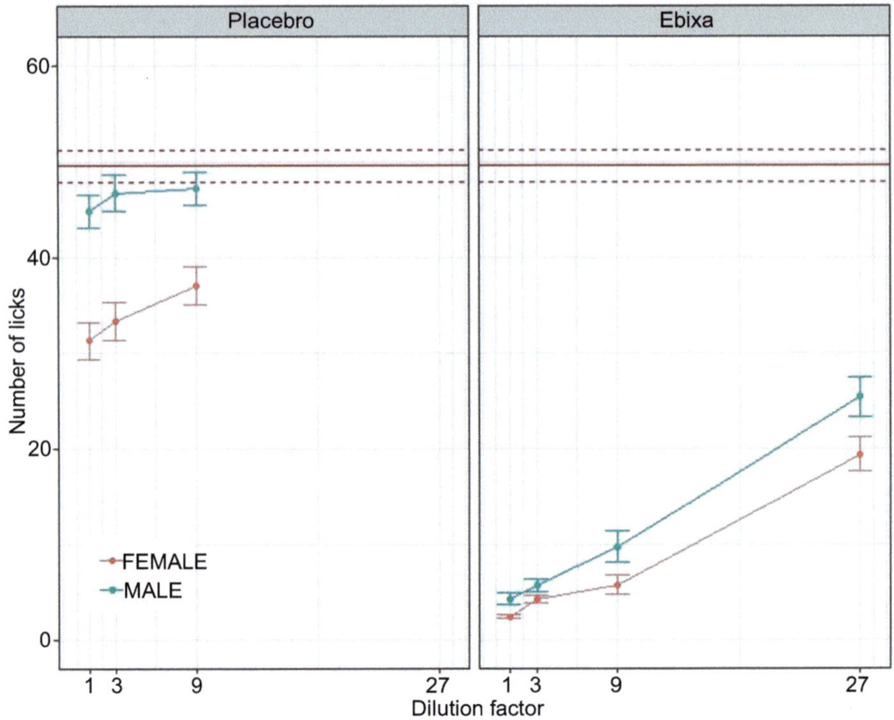

Fig. 14 The aversiveness of test solutions for male and female rats. The average lick number for each dilution of placebo and original oral solution of memantine (Ebixa®) for male and female rats ± Standard Error of the Mean (SEM), with the average lick number of water shown as a solid red line ± SEM [46]

and the large variety of users worldwide, a comprehensive understanding of acceptability drivers remains challenging. Careful study of the combined effects of distinct drivers will be needed. This must be performed in parallel to the exploration of new factors while reducing the influence of confounding factors. Such knowledge will facilitate the choice of the most appropriate dosage form to achieve adequate acceptability in a target population. This should prove valuable in supporting both medicinal product designers and healthcare professionals in their shared goal to provide an improved quality of care for vulnerable populations.

References

1. United Nations. Department of Economic and Social Affairs. Population Division. World Population Prospects 2019: Highlights (ST/ESA/SER.A/423). New York: United Nations; 2019.
2. World Health Organization. World report on ageing and health. World Health Organization; 2015.

3. Rajaei-Dehkordi Z, McPherson G. The effects of multiple medication in the elderly. Nurs Times. 1996;93(27):56–8.
4. Kaufman DW, Kelly JP, Rosenberg L, Anderson TE, Mitchell AA. Recent patterns of medication use in the ambulatory adult population of the United States: the Slone survey. JAMA. 2002;287(3):337–44.
5. Dunbar-Jacob J, Bohachick P, Mortimer MK, Sereika SM, Foley SM. Medication adherence in persons with cardiovascular disease. J Cardiovasc Nurs. 2003;18(3):209.
6. Drenth-van Maanen AC, Wilting I, Jansen PA. Prescribing medicines to older people – how to consider the impact of ageing on human organ and body functions. Br J Clin Pharmacol. 2020;86(10):1921–30.
7. Puisieux F, D'Andrea C, Baconnier P, Bui-Dinh D, Castaings-Pelet S, Crestani B, et al. Swallowing disorders, pneumonia and respiratory tract infectious disease in the elderly. Rev Mal Respir. 2011;28(8):e76–93.
8. Clavé P, Arreola V, Romea M, Medina L, Palomera E, Serra-Prat M. Accuracy of the volume-viscosity swallow test for clinical screening of oropharyngeal dysphagia and aspiration. Clin Nutr. 2008;27(6):806–15.
9. Wirth R, Dziewas R, Beck AM, Clave P, Hamdy S, Heppner HJ, et al. Oropharyngeal dysphagia in older persons–from pathophysiology to adequate intervention: a review and summary of an international expert meeting. Clin Interv Aging. 2016;11:189.
10. Lindgren S, Janzon L. Prevalence of swallowing complaints and clinical findings among 50-79-year-old men and women in an urban population. Dysphagia. 1991;6(4):187–92.
11. Nilsson H, Ekberg O, Olsson R, Hindfelt B. Quantitative aspects of swallowing in an elderly nondysphagic population. Dysphagia. 1996;11(3):180–4.
12. Stegemann S, Gosch M, Breitkreutz J. Swallowing dysfunction and dysphagia is an unrecognized challenge for oral drug therapy. Int J Pharm. 2012;430(1):197–206.
13. Schiele JT, Penner H, Schneider H, Quinzler R, Reich G, Wezler N, et al. Swallowing tablets and capsules increases the risk of penetration and aspiration in patients with stroke-induced dysphagia. Dysphagia. 2015;30(5):571–82.
14. Liu F, Ranmal S, Batchelor HK, Orlu-Gul M, Ernest TB, Thomas IW, et al. Patient-centred pharmaceutical design to improve acceptability of medicines: similarities and differences in paediatric and geriatric populations. Drugs. 2014;74(16):1871–89.
15. Kelly J, Wright D. Administering medication to adult patients with dysphagia. Nurs Stand. 2009;23(29):62–8.
16. Caussin M, Mourier W, Philippe S, Capet C, Adam M, Reynero N, et al. L'écrasement des médicaments en gériatrie: une pratique «artisanale» avec de fréquentes erreurs qui nécessitait des recommandations. Rev Med Interne. 2012;33(10):546–51.
17. Mc Gillicuddy A, Crean AM, Sahm LJ. Older adults with difficulty swallowing oral medicines: a systematic review of the literature. Eur J Clin Pharmacol. 2016;72(2):141–51.
18. Marquis J, Schneider M-P, Payot V, Cordonier A-C, Bugnon O, Hersberger KE, et al. Swallowing difficulties with oral drugs among polypharmacy patients attending community pharmacies. Int J Clin Pharm. 2013;35(6):1130–6.
19. Sestili M, Logrippo S, Cespi M, Bonacucina G, Ferrara L, Busco S, et al. Potentially inappropriate prescribing of oral solid medications in elderly dysphagic patients. Pharmaceutics. 2018;10(4):280.
20. Kelly J, Wright D. Administering medication to adult patients with dysphagia. Nurs Stand. 2009;23(29):62.
21. Logrippo S, Ricci G, Sestili M, Cespi M, Ferrara L, Palmieri GF, et al. Oral drug therapy in elderly with dysphagia: between a rock and a hard place! Clin Interv Aging. 2017;12:241.
22. Fodil M, Fillette A, Trivalle C. Considérations portant sur l'écrasement des comprimés en gériatrie. NPG. 2013;13(73):35–40.
23. Johnson DA, Roach AC, Carlsson AS, Karlsson AA, Behr DE. Stability of esomeprazole capsule contents after in vitro suspension in common soft foods and beverages. Pharm J Hum Pharmacol Drug Therapy. 2003;23(6):731–4.

24. Carrier M-N, Garinot O, Vitzling C. Stability and compatibility of tegaserod from crushed tablets mixed in beverages and foods. Am J Health Syst Pharm. 2004;61(11):1135–42.
25. Lippert C, Gbenado S, Qiu C, Lavin B, Kovacs SJ. The bioequivalence of telithromycin administered orally as crushed tablets versus tablets swallowed whole. J Clin Pharmacol. 2005;45(9):1025–31.
26. Simojoki K, Lillsunde P, Lintzeris N, Alho H. Bioavailability of buprenorphine from crushed and whole buprenorphine (subutex) tablets. Eur Addict Res. 2010;16(2):85–90.
27. Verrue C, Mehuys E, Boussery K, Remon JP, Petrovic M. Tablet-splitting: a common yet not so innocent practice. J Adv Nurs. 2011;67(1):26–32.
28. Ruscin JM, Semla TP. Assessment of medication management skills in older outpatients. Ann Pharmacother. 1996;30(10):1083–8.
29. Notenboom K, Beers E, Riet-Nales DA, Egberts TC, Leufkens HG, Jansen PA, et al. Practical problems with medication use that older people experience: a qualitative study. J Am Geriatr Soc. 2014;62(12):2339–44.
30. Maher RL, Hanlon J, Hajjar ER. Clinical consequences of polypharmacy in elderly. Expert Opin Drug Saf. 2014;13(1):57–65.
31. Nguyen JK, Fouts MM, Kotabe SE, Lo E. Polypharmacy as a risk factor for adverse drug reactions in geriatric nursing home residents. Am J Geriatr Pharmacother. 2006;4(1):36–41.
32. Bourgeois FT, Shannon MW, Valim C, Mandl KD. Adverse drug events in the outpatient setting: an 11-year national analysis. Pharmacoepidemiol Drug Saf. 2010;19(9):901–10.
33. Koop CE. Keynote address. In improving medication compliance: Proceedings of a symposium. Reston, Va. (1894 Preston White Dr., Reston 22090): National Pharmaceutical Council; 1985. 1–4 p.
34. Jones DJ, Kronemer N. Parents' response to a drug study: a clinical study of the acceptability and efficacy of pediatric forms of erythromycin. Clin Pediatr. 1965;4(3):137–40.
35. European Medicine Agency. Guideline on pharmaceutical development of medicines for paediatric use. EMA/CHMP/QWP/805880/2012 Rev. 2. London, UK: EMA; 2013.
36. Food & Drug Administration. Food and Drug Administration Safety and Innovation Act (FDASIA); 2012.
37. U.S. Department of Health and Human Services. Food and Drug Administration. Pediatric Study Plans: Content of and Process for Submitting Initial Pediatric Study Plans and Amended Initial Pediatric Study Plans Guidance for Industry; 2020.
38. World Health Organization. Toolkit for research and development of paediatric antiretroviral drugs and formulations. Module 5: Acceptability; 2018.
39. International Council for Harmonisation. ICH E11(R1) guideline on clinical investigation of medicinal products in the pediatric population; 2016.
40. International Council for Harmonisation. Pharmaceutical development Q8(R2); 2009.
41. European Medicine Agency. Reflection paper on the pharmaceutical development of medicines for use in the older population. EMA/CHMP/QWP/292439/2017. Amsterdam, The Netherlands: EMA; 2020.
42. Ruiz F, Vallet T, Pense-Lheritier AM, Aoussat A. Standardized method to assess medicines' acceptability: focus on paediatric population. J Pharm Pharmacol. 2017;69(4):406–16.
43. Vallet T, Ruiz F, Lavarde M, Pense-Lheritier AM, Aoussat A. Standardized evaluation of medicine acceptability in paediatric population: reliability of a model. J Pharm Pharmacol. 2018;70(1):42–50.
44. Vallet T, Belissa E, Laribe-Caget S, Chevallier A, Chedhomme FX, Leglise P, et al. A decision support tool facilitating medicine design for optimal acceptability in the older population. Pharm Res. 2018;35(7):136.
45. Ruiz F, Vallet T, Dufay Wojcicki A, Belissa É, Fontan J-E, de Pontual L, et al. Dosage form suitability in vulnerable populations: a focus on paracetamol acceptability from infants to centenarians. PLoS One. 2019;14(8):e0221261.
46. Ruiz F, Keeley A, Léglise P, Tuleu C, Lachuer C, Rwabihama J-P, et al. Sex differences in medicine acceptability: a new factor to be considered in medicine formulation. Pharmaceutics. 2019;11(8):368.

47. Belissa E, Vallet T, Laribe-Caget S, Chevallier A, Chedhomme FX, Abdallah F, et al. Acceptability of oral liquid pharmaceutical products in older adults: palatability and swallowability issues. BMC Geriatr. 2019;19(1):1–9.
48. Vallet T, Michelon H, Orlu M, Jani Y, Leglise P, Laribe-Caget S, et al. Acceptability in the older population: the importance of an appropriate tablet size. Pharmaceutics. 2020;12(8):746.
49. Jani YH, Liu F, Orlu M, Desai N, du Chayla F, Ruiz F, et al. Medicine acceptability for older people in hospital and care home: the influence of setting. Int J Pharm Pract. 2021;30:67.
50. Vallet T. Conception d'un outil d'évaluation de l'acceptabilité des médicaments [Génie industriel]: Ecole Nationale Supérieure d'Arts et Métiers (ENSAM); 2017.
51. U.S. Department of Health and Human Services. Food and Drug Administration. Guidance for industry: patient-reported outcome measures: use in medical product development to support labeling claims. Fed Regist. 2009;74(235):65132–3.
52. Drumond N, van Riet-Nales DA, Karapinar-Carkit F, Stegemann S. Patients' appropriateness, acceptability, usability and preferences for pharmaceutical preparations: results from a literature review on clinical evidence. Int J Pharm. 2017;521:294.
53. U.S. Department of Health and Human Services. Food and Drug Administration. Size, shape, and other physical attributes of generic tablets and capsules guidance for industry. Fed Regist. 2015:35366–7.
54. Fields J, Go JT, Schulze KS. Pill properties that cause dysphagia and treatment failure. Curr Ther Res. 2015;77:79–82.
55. Liu F, Ghaffur A, Bains J, Hamdy S. Acceptability of oral solid medicines in older adults with and without dysphagia: a nested pilot validation questionnaire based observational study. Int J Pharm. 2016;9(16):30197–1.
56. Al-Hashimi N, Begg N, Alany RG, Hassanin H, Elshaer A. Oral modified release multiple-unit particulate systems: compressed pellets, microparticles and nanoparticles. Pharmaceutics. 2018;10(4):176.
57. Patel S, Scott N, Patel K, Mohylyuk V, McAuley WJ, Liu F. Easy to swallow "instant" jelly formulations for sustained release gliclazide delivery. J Pharm Sci. 2020;109(8):2474–84.
58. Mojet J, Christ-Hazelhof E, Heidema J. Taste perception with age: generic or specific losses in threshold sensitivity to the five basic tastes? Chem Senses. 2001;26(7):845–60.
59. Imoscopi A, Inelmen EM, Sergi G, Miotto F, Manzato E. Taste loss in the elderly: epidemiology, causes and consequences. Aging Clin Exp Res. 2012;24(6):570–9.
60. Syed Q, Hendler KT, Koncilja K. The impact of aging and medical status on dysgeusia. Am J Med. 2016;129(7):753.
61. Schiffman SS. Taste and smell losses in normal aging and disease. JAMA. 1997;278(16):1357–62.
62. Soto J. Assessing the feasibility of using an animal model for in vivo taste assessment of pharmaceutical compounds and formulations: UCL (University College London); 2016.
63. Devantier HR, Long DJ, Brennan FX, Carlucci SA, Hendrix C, Bryant RW, et al. Quantitative assessment of TRPM5-dependent oral aversiveness of pharmaceuticals using a mouse brief-access taste aversion assay. Behav Pharmacol. 2008;19(7):673–82.
64. Rudnitskaya A, Kirsanov D, Blinova Y, Legin E, Seleznev B, Clapham D, et al. Assessment of bitter taste of pharmaceuticals with multisensor system employing 3 way PLS regression. Anal Chim Acta. 2013;770:45–52.
65. Bhat MG, Jordt RM, Khan MA, Foley CE, Gilbertson TA. Validation of a rat behavioral avoidance model from a drug delivery perspective. Int J Pharm. 2005;303(1):31–6.

The Design of Patient-centric Dosage Forms for Older Adults

Susanne Page, Sabrina Bras Da Costa, Cordula Stillhart, Carsten Timpe, and Leonie Wagner

Contents

Abstract The older population is very diverse with regard to the health condition, as well as the perception and physical and cognitive ability of each individual subject. Furthermore, the prevalence of multi- morbidity and polypharmacy is increasing in the aging population, and while some patients are able to manage their medication independently, others require a significant level of caregiver assistance. From a pharmaceutical industry perspective, the development of a patient-centric drug product is a key objective to ensure safe, effective, and efficacious treatments, thus reducing healthcare costs and increasing the patient's quality of life. This mission requires a cross-functional effort within the companies, a constructive dialogue

S. Page (✉) · S. B. Da Costa · C. Stillhart · C. Timpe · L. Wagner
Pharma Technical Development, F. Hoffmann-La Roche Ltd., Basel, Switzerland
e-mail: susanne.page@roche.com; sabrina.bras_da_costa@roche.com; cordula.stillhart@roche.com; leonie.wagner@roche.com

M. Orlu, F. Liu (eds.), *Pharmaceutical Formulations for Older Patients*, AAPS Advances in the Pharmaceutical Sciences Series 51,
https://doi.org/10.1007/978-3-031-35811-1_4

with health authorities, as well as the involvement of patient populations, caregivers, physicians, hospitals, nursing homes, and health insurance companies. It starts from a sound understanding of patient needs, requirements, and disease conditions, which then feed into the design of age-appropriate dosage forms, the use of patient-specific biopharmaceutics prediction tools, as well as the application of human factors engineering to carefully consider and test the safe and effective use of the drug product by its intended user populations under real life conditions.

Keywords Age appropriateness · Medication adherence · Human factors · Physiological changes · Dosage form · Patient centricity · Pharmaceutical industry

1 Introduction

Over many years, the pharmaceutical industry has mainly focused on the development of novel blockbuster drugs as being safe, efficacious, and complying with global quality standards. Beside regulatory requirements, the cost-effective manufacturability of high-volume products was a central element of industrial drug development. This blockbuster-driven concept for developing new medicines against diseases that could not be treated before had started with the discovery of innovative medicines like insulin for the treatment of type I diabetes in the early twentieth century, followed by penicillin and derivatives as lifesaving treatment against bacterial infections in the 1930s and the market introduction of drugs like the anxiolytic diazepam in the 1960s.

In 2010, Dolgin [1] published an article that described the move of industry away from the old to a new "niche blockbuster" concept where companies develop drugs for rare diseases with smaller patient populations. Specifically, the US Orphan Drug Act from 1983 led to the development of more than 350 new drugs for around 200 rare diseases mainly by small biotech startups.

With this new trend, the industry started focusing on the individual patient and specific disease and drug product aspects much more than ever before, as symbolized in Fig. 1 [2].

Demographics of modern societies clearly show a significant growth of the older population: Over the next decades until 2050, the number of people aged 90 years and older will increase from 8.1 to 37.7 million in the developed countries and from 12.9 to 123.5 million in the developing countries representing an increase of 463% and 954%, respectively [2]. The increasing numbers of older people stress the importance of optimal pharmacotherapies considering the high prevalence of diseases and potential physical or cognitive impairments in this age group, e.g.:

- Intake of multiple drugs
  - 30–50% of all prescription drugs taken by older adults (aged 65 years and older) [3].

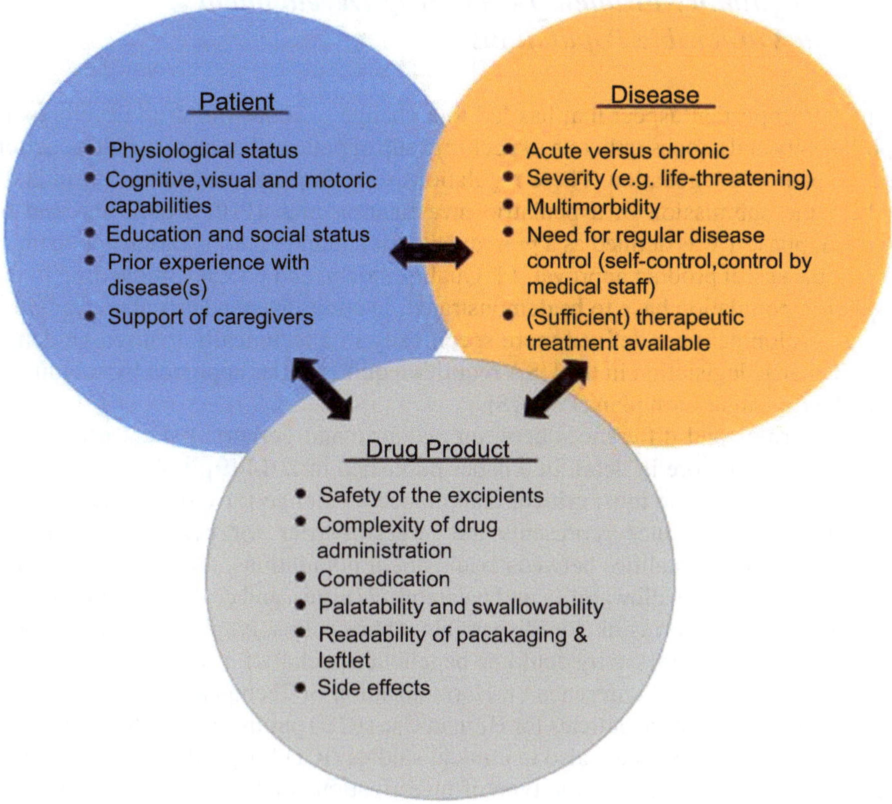

Fig. 1 Relationship between patient, disease, and drug product

  - On average, geriatric patients take 2–6 prescription drugs and 1–3 non-prescription drugs on a regular basis [4].
- Significant compliance issues, such as
  - 29% non-compliance overall.
  - 51% take additional medications (not in the report from the physicians).
  - 20% of the patients modify the dose strength [5].
- Addiction to alcohol, nicotine, and medicines
- Reduced kidney and liver functions as well as dehydration
- Reduced audiovisual and ergonomic capabilities

1.1 Regulatory Framework for Drug Development in Vulnerable Populations

Another important aspect that has led to a change of industrial drug development philosophy and a focus rather on a specific group of patients has been the introduction of the pediatric legislation. These regulations were implemented in the EU in 2007 where the submission of a pediatric investigation plan (PIP) around the end of clinical phase 1 was defined as a prerequisite for a market authorization applicant to get a successful product approval [6]. Quality, safety, and efficacy of the drug in the pediatric population have to be demonstrated. In addition, standards about pediatric drug development are defined more specifically in a technically focused guideline [7]. Pediatric legislation in the USA requires a quite similar approach by submitting a pediatric submission plan (PSP) [8].

Similarities and differences between pediatric and geriatric patient needs have been discussed more in detail in a paper published in 2014 [9]: Polypharmacy has been identified as the most critical compliance challenge for older patients, while taste-related acceptance represents the largest barrier for pediatric medicines. Among the commonalities between both patient populations, there are aspects like excipients safety, swallowability and palatability (taste), and co-administration with soft food. Dosage forms like orodispersible tablets (films, lyophilisates) that would disintegrate in the oral cavity could be beneficial specially for older patients.

The International Conference on Harmonisation of Technical Requirements for Registration of Pharmaceuticals for Human Use (ICH) published in 1994 a guideline to involve geriatric patients also in clinical studies (ICH E7 [10]). It focuses from a legislative point of view on the type of investigations that should be performed in older people and what information needs to be reported in the pre-authorization dossier of a new medicinal product [11]. Despite being not mandatory and the need for further updates, pharmaceutical companies have to justify in case they are not following these recommendations.

In addition to the ICH E7 guideline, the EMA Committee for Medicinal Products for Human Use (CHMP) has set up a geriatric expert group giving scientific advice regarding the population of older patients. A strategic framework has been defined by the expert group that presents the vision of the EMA for the next years, ensuring high quality of medicines for older patients over the product life cycle supported by adequate research and evaluation including the availability of sufficient information [12]. This paper also foresees the implementation of regulatory guidelines and identification of validated methods and tools assessing effect and safety of medicines in "frail" patients as a post-authorization approach.

1.2 *Defining a Target Product Profile for a Specific Indication and Patient Population*

To better define the drug product intended for treatment of a specific disease, the FDA had published in 2007 a draft guidance "Target Product Profile – A Strategic Development Process Tool" [13]. The purpose of a TPP (target product profile) is to provide a format for discussion between sponsor and FDA that can be used throughout the drug development process to pursue new indications or other substantial changes in the label. The TPP can include information from each discipline. Typically, it briefly summarizes planned and completed clinical studies supporting a labeling concept. Key contents include, for example, indications and usage, dosage strengths, dosage forms and route of administration, contraindications, warnings and precautions, adverse reactions, the use in specific populations, supply modes including storage and handling, and patient counseling information.

1.3 *Defining the Drug Product Quality Profiles*

Patient centricity expects a much stronger focus on the patients' needs than ever before with a thorough understanding of the physiology and the disease specific needs of the patient. This also requires the integration of the appropriate drug product profile with respect to efficacy, safety, quality, and effectiveness in drug administration aspects.

In 2005, the Quality by Design (QbD) concept was introduced into pharmaceutical drug development as a tripartite, harmonized ICH Q8 guideline [14]. It provides guidance for Europe, Japan, and the USA regarding contents of drug product development and application of formulation and manufacturing sciences. At a later point, the quality risk assessments were introduced in drug development by ICH Q9 [15]. The strategy behind these guidelines is to identify the impact of critical material and process factors on critical quality attributes, which should become part of the regulatory documentation in submissions. The basis for the drug product formulation development during clinical testing is the "Quality Target Product Profile" (QTPP) [14], as it defines the route of administration, the selection of the dosage form, the release principle, the dose strengths, and requirements of the container closure system and analytical criteria like stability, sterility, and drug release.

The annex to ICH Q8 Pharmaceutical Development directly focuses on the patients' needs: "...in all cases, the product should be designed to meet patients' needs and the intended product performance." The QTPP should serve as a QbD-based tool and help pharmaceutical companies to design and develop their products in a patient-centric way. An example for such a profile is presented in a paper from Timpe et al. [16]. It illustrates the link between product quality acceptance criteria (including critical quality attributes like drug release and stability) and the impact that these might have on the patient [16].

Highlighting the already recognized importance of the focus on geriatric patients in 2017, EMA published a reflection paper [17] on the pharmaceutical development of medicines specifically for use in the older population. Disease, patient, and drug product aspects are appropriately considered, and this paper provides for the first time guidance to pharmaceutical industry experts on how to tailor medicines much better for the use in the vulnerable population of older adults. Important aspects and learnings from pediatric drug development have been included, e.g., different administration routes (oral, parenteral, etc.), administration of medicines via enteral feeding tubes, modification of dosage forms to facilitate drug intake or lower the dose, dosing frequency, excipients aspects, container closure/packaging, and device topics beside product information and medication management.

The development of patient-centric oral drug products in the pharmaceutical industry will be discussed more specifically in the following subchapters. Key aspects include the consideration of differences in gastrointestinal (GI) tract physiology and oral absorption between young and older adults, the selection of suitable drug product designs, as well as the evaluation of acceptability and usability in the targeted population.

2 Intended Product Performance: Development of Safe and Efficacious Medicines

The development of safe and efficacious medicines is one of the key deliverables in the pharmaceutical industry. Clinical safety and efficacy assessments are conducted throughout the development process, from early safety pharmacology and exposure-response predictions to dedicated clinical studies and post-approval pharmacovigilance. The human pharmacokinetic profile of a new drug is determined in early clinical studies and typically serves as a surrogate for drug safety and efficacy. However, while these early studies are normally conducted in a population of healthy young adults, there is increasing evidence that the processes involved in drug absorption, distribution, metabolism, and elimination may be significantly different in the targeted patient population. Such differences can root in various reasons including age-, disease-, or co-medication-specific changes in organ size and function and may result in substantial alterations in pharmacokinetic behavior when compared to the profiles measured in healthy young adults. Therefore, when administering the drug to the targeted patient population, there is an inherent risk of over- or underdosing an individual subject, thus leading to potential safety or efficacy issues.

The human GI tract and its impact on oral drug absorption has been extensively studied in young and healthy human individuals [18, 19]. This knowledge allows a more advanced understanding of the in vivo performance of oral drug products and was considered in the setup of in vitro protocols, which are meanwhile well-established (e.g., fasted and fed-state simulated gastric and intestinal fluids) and in silico tools used to simulate drug release and absorption (e.g., GastroPlus®, Simcyp®). Nowadays, these tools are routinely applied in formulation performance testing and represent generally accepted methods for predicting oral drug exposure in the context of drug development (e.g., formulation selection, dose finding, clinical study design). However, in older patients as well as in other patient sub-populations, an alteration of the anatomy and/or physiology of the GI tract can be observed, which can be related to demographic factors (e.g., age, ethnicity, sex), the disease, or the intake of co-medications [20]. Very often, alterations result from a combination of multiple factors. Hence, the rate and extent of drug absorption may be different, potentially resulting in under- or overexposure. The use of standard biopharmaceutics in vitro and in silico tools may therefore not be appropriate to predict the relevant conditions in the targeted population.

In addition to pharmacokinetic differences, also pharmacodynamic changes can occur and often include increased sensitivity to drugs such as anticoagulants and cardiovascular and psychotropic drugs in older patients [21, 22]. A good understanding of age-, disease-, and co-medication-related changes is therefore desirable to ensure safe and effective use of medications.

2.1 Physiological Changes in the GI Tract Due to Age and Disease

The current knowledge on age-dependent alterations in the GI tract and their impact on oral drug absorption has been summarized in several review publications [23–27]. There still exist significant knowledge gaps, and the ethical constraints associated with the conduct of research in patient populations are among the main reasons. In addition, older adults are very heterogeneous in terms of disease conditions, co-medications, and comorbidities. Thus, the characterization of a standard population is very challenging.

Despite many unknowns, some properties of the human gastrointestinal tract are known to change with age and have to be considered in the development of oral drug products. The gastric pH in the stomach of older subjects is acidic under fasted conditions and similar to values measured in healthy young adults [28]. In fact, the capacity to secrete gastric acid is maintained with age [29]; however, the restoration of low gastric acidity following meal intake is slower in older compared to young individuals [30]. Also, there is a higher prevalence of reduced or even absence of gastric acid secretion in older compared to young subjects, which is typically a result of chronic atrophic gastritis. Since hypo- or achlorhydria often is asymptomatic, many patients remain undiagnosed, and thus, it is difficult to determine the actual prevalence. It is estimated that approximately 10–20% of older adults suffer from achlorhydria compared to <1% in young adults [31]. The prevalence of high gastric pH is even higher when including cases of iatrogenic hypochlorhydria, for example, due to intake of acid-reducing agents such as proton pump inhibitors, which is very common in older patients [32, 33].

There is very scarce data on the pH and composition of intestinal fluids in older subjects. Einarsson et al. reported a bile acid synthesis rate of 1.75 mmol/day in young adults and 0.91 mmol/day in subjects >60 years [34], with an increase in cholesterol secretion from 53 mmol/h to 73 mmol/h in older adults. Other studies reported total bile secretions and composition of intestinal fluids with high intersubject variability, but no statistically significant difference between age groups [35, 36].

Differences have also been reported with respect to the gastrointestinal motility. The gastric emptying time appears to increase in older subjects, with an average of 6 minutes every 10 years increase in age [37]. Also, a reduction in postprandial peristalsis and gastric contractile force was observed with aging [38]. While the overall motility pattern and transit time in the small intestine were reported to be independent of age [39], a more recent study using a wireless motility capsule reported a decrease in small intestinal transit time of around 12 minutes for every 10 years increase in age [37]. The age-dependency of colonic transit time appears to be controversial, with some studies reporting a prolongation of colonic transit time, while others reported no change with aging [39, 40].

The effects of aging on the intestinal barrier in humans are very limited and often conflicting [41]. Wilms et al. studied the effects of aging on the human intestinal barrier function by determining intestinal permeability and the expression of barrier-related genes. This study suggested that the functional capacity of the intestinal barrier is maintained in older adults [42]. Thus, while passive absorption may remain unchanged with age, there is still little knowledge about the effect of ageing on the expression and function of gastrointestinal transporter proteins.

Several studies have examined morphological changes in the intestinal surface. While investigations on animal models have suggested a decline in intestinal surface area with age [43–45], this effect appears to be less clear in humans [46–48]. Also, the clinical relevance of a decrease in intestinal surface area remains unclear. Other physiological properties of the gastrointestinal tract in older adults, such as buffer capacity, osmolality, and volume of gastrointestinal fluids, are still largely unexplored.

While ageing per se seems to have limited impact on the gastrointestinal functions, there is a high prevalence of gastrointestinal disorders in older patients, which directly affect the function of the digestive tract. These disorders include gastroesophageal reflux, anorexia, small-intestinal bacterial overgrowth (SIBO), and constipation.

The prevalence of gastroesophageal reflux disorder is approximately 20% and presents more severe etiopathology and more frequent complications in the older population compared to younger adults [49]. Gastroesophageal reflux is typically treated with acid-reducing agents, such as H2-receptor antagonists and proton pump inhibitors, which reduce stomach acid and alleviate heartburn symptoms. The higher prevalence of hypo- and achlorhydria in older patients can have a direct influence on the bioavailability of weakly basic drugs, which, due to their pH-dependent solubility profile, tend to have a higher solubility at low pH (acidic stomach conditions). An increase in stomach pH, such as under hypo- or achlorhydric conditions, results in poor solubilization in gastric fluids followed by a reduced absorption from the intestinal lumen. This effect has been observed, for example, with ketoconazole [50], a weakly basic antimicrobial drug, which resulted in reduced exposure in hypochlorhydria patients compared to subjects with normal gastric function. As opposed to weakly basic compounds, acid-labile drugs may exhibit higher exposure in older patients with hypo- or achlorhydria. The use of advanced formulation strategies such as acidifying agents for weakly basic compounds and enteric coated dosage forms for acid-labile drugs could be beneficial to reduce interpatient variability due to changes in absorption. Hypo- or achlorhydria has also been correlated to the compromised absorption of nutritional components such as iron [51].

Anorexia, which is defined as poor appetite and hunger combined with decreased physical activity, is often caused by depression, other comorbidities, malignancies, and comedications [52]. Anorexia typically can have a severe impact on drug disposition processes, for example, due to changes in serum albumin concentrations, but is also known to affect the physiology of the gastrointestinal tract by delaying the gastric emptying time [53]. Also, while the reduced intake of food has a direct impact on the nutritional state, this condition can also become critical for orally administered poorly soluble drugs, which require the co-administration with food to ensure adequate bioavailability. The predisposition to SIBO is likely favored by the prolongation of intestinal transit time and reduced gastric acidity in older adults [54]. SIBO symptoms typically include nausea, vomiting, and diarrhea. Oral drug substances, which are metabolized by the intestinal microbiome, may thus be affected by the changes in type, activity, and number of microorganisms in patients affected by SIBO.

In addition to the increased prevalence of gastrointestinal disorders, the older population is also characterized by a higher prevalence of comorbidities [55], which are not primarily of gastrointestinal origin but can be accompanied by several gastrointestinal symptoms.

Wollmer and Klein recently reviewed the gastrointestinal disorders associated with Parkinson's disease [56]. Although being a progressive neurodegenerative disease presenting primarily motoric symptoms such as rigidity, bradykinesia, tremor and postural instability, Parkinson's disease also involves gastrointestinal disturbances, which may affect the oral absorption behavior. These disturbances include hypersalivation as a consequence of inefficient and infrequent swallowing, impaired gastric motility, and prolonged colonic transit time. However, several physiological factors which are critical for oral absorption, such as the volume, pH, and composition of gastrointestinal fluids, are still largely unexplored in this patient population.

These knowledge gaps, in conjunction with the poor availability of clinical PK data, result in substantial limitations in the development and validation of in vitro and in silico tools representative of special populations, which is a critical gap in the development of new medicines. In fact, the area of population-specific biopharmaceutics has been identified as a major current challenge in oral absorption research [57].

2.2 *Impact of Co-medication on the Safety and Efficacy of Drugs*

The high prevalence of comorbidities implies that older patients are subject to frequent co-medications. This condition increases the risk for drug-drug interactions at various levels of drug absorption and disposition and has been identified as a major challenge in pharmacological treatment of older patients [58–60]. While there is a broad range of oral medications which could potentially lead to drug-drug interactions at the level of oral absorption, some medications are particularly frequent in this patient population and thus need careful attention. These medications include stomach acid-reducing agents. These compounds represent a high risk for potential drug-drug interactions especially with ionizable compounds exhibiting pH-dependent solubility and absorption behavior. Calcium supplementation is another frequent comedication to prevent or treat bone loss, which shows a trend toward increased use in the older population [61]. Calcium ions, similar to other metal ions administered as food supplements, can form poorly soluble complexes with fluoroquinolones (e.g., ciprofloxacin), penicillamines, and tetracyclines upon concomitant oral administration. As a consequence, oral absorption of these compounds is reduced and may lead to treatment failure.

Motility disorders in older patients are often treated with prokinetic drugs, which increase the gastric transit (e.g., metoclopramide). The reduced time of contact between drug and mucosal area of absorption results in a decrease of drug absorption. This effect could be critical for controlled-release dosage forms or decrease the absorption of compounds exhibiting an absorption window (e.g., digoxin [62]). Finally, drug-drug interactions may occur at the level of drug transporters (e.g., P-glycoprotein (P-gp, MDR1) or metabolizing enzymes in the intestinal membrane. Drug-drug interactions at P-gp level might become clinically relevant when drugs with a low therapeutic index (e.g., digoxin, theophylline, anticancer drugs) are co-administered with macrolides (e.g., erythromycin, clarithromycin), proton-pump inhibitors (e.g., omeprazole), or anti-arrhythmic drugs (e.g., amiodarone, verapamil, or diltiazem). Several drugs that are transported by P-gp are also metabolized by cytochrome P450 3A4 (e.g., cyclosporine, antiepileptic drugs, antidepressant, fluoroquinolones, quinidine, and ranitidine), which can confound interpretation of interactions.

3 Patient-centric Pharmaceutical Drug Product Design

The first definition of the term "patient centric pharmaceutical drug product design" appeared in 2016 in a publication by Stegemann et al. [63]. They defined it as "the process of identifying the comprehensive needs of individuals or the target patient population and utilizing the identified needs to design pharmaceutical products that provide the best overall benefit to risk profile for the target population over the intended duration of treatment." This includes involvement of the patients during the conceptual phase of the development as well as the re-evaluation of the product design during the marketing phase in order to identify potential product enhancements during the product life cycle [63].

Being aware of patient needs, skills, and capabilities helps design better products and reduce healthcare costs. Overall, it is clear that the health of older adults is generally affected by a gradual decline of motor function resulting, for example, in swallowing difficulties, trembling hands, and reduced suppleness. Furthermore, they might suffer from impaired cognition (e.g., dementia), impaired vision, and hearing loss. These aspects can lead to an increased risk of medication errors and also poor adherence [64], which has a direct impact on costs for patients and the healthcare system. As shown by the IMS Institute for Healthcare Informatics [65], 105 out of the 200 billion USD avoidable costs in the US healthcare system are costs created by medication non-adherence in six different chronic diseases, i.e., congestive heart failure, human immunodeficiency virus infection, osteoporosis, hypertension, diabetes, and hypercholesterolemia.

The importance of a patient centric pharmaceutical drug product design and its impact on medication adherence was recently presented by Menditto et al. [64]. A

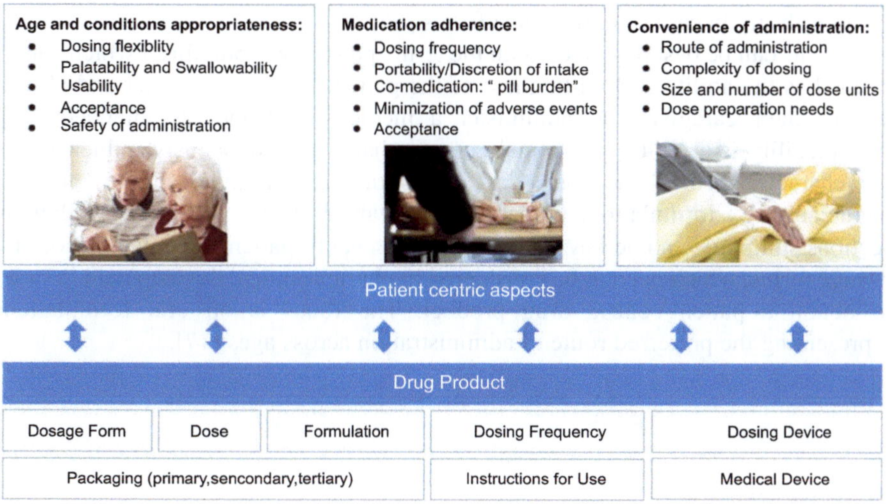

Fig. 2 Interconnection between the drug product. (Elements taken from [64]) and patient-centric aspects

patient-centric pharmaceutical drug product design offers the opportunity to meet older patients' needs and preferences in order to reduce the risk of medication errors and poor adherence. Considering all aspects of the drug product (see Fig. 2 top), Menditto et al. came to the conclusion that a medicinal product for older adults should be easy to swallow, easy to handle, and easy to dose and should make use of a dosing device if needed and be delivered in a packaging that is easy to open. A proper design of the packaging, labelling, and the leaflet ensures that drug product information is easy to read and understandable. Furthermore, it should avoid excipient overload. In addition, the use of fixed-dose combinations should be considered as one solution to reduce the "pill burden." The following subsections focus on three patient-centric aspects, i.e., age and condition appropriateness, medication adherence, and convenience of administration (i.e., safe and effective real-life administration), and provide a more detailed view on the interconnection between the drug product and patient centricity.

3.1 Age and Condition Appropriateness of Dosage Forms

Compared to children and adults of younger or middle age, older people constitute a very heterogeneous group of patients as relevant differences with respect to physical, physiological, and/or cognitive impairments may start at a different chronological age [17]. Thus, contrary to the pediatric population where the appropriateness of dosage forms can often be linked to the chronological age, this is not the case for the older population. Dosage form selection should be based on the specific needs rather than the chronological age and must consider aspects such as medication identification and memorability, medication handling, as well as swallowability in case of oral dosage forms [66].

All this can be best described with the term "patient acceptability," which can be defined as the ability and willingness of the patient to self-administer and also of any of their caregivers to administer a medicinal product as intended [17]. Acceptability is a multidimensional concept that is simultaneously influenced by the characteristics of the user (e.g., demographic, caregiver, setting) and the medicine (e.g., formulation, preparation, container, device [67]). The following sections summarize some key findings with respect to patient acceptability of oral dosage forms, which need to be considered in pharmaceutical development in order to obtain a patient-centric drug product. The focus is on oral formulations representing the preferred route of administration across ages [17].

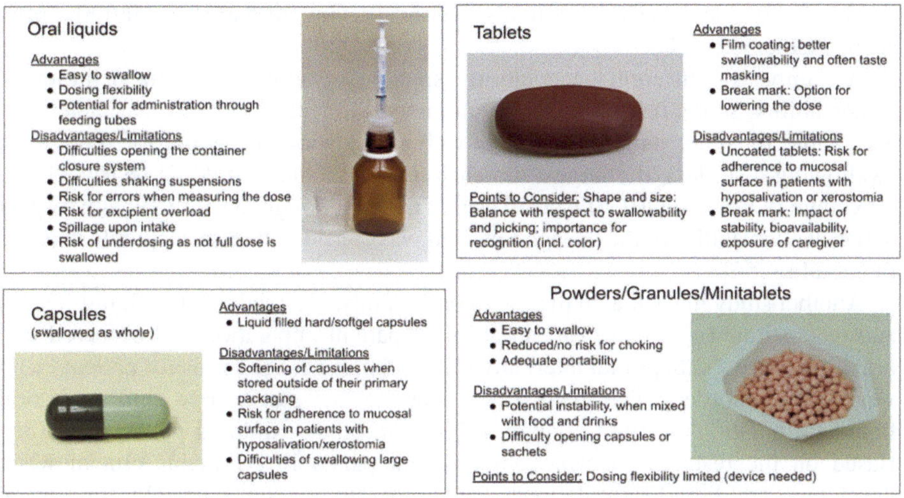

Fig. 3 Advantages, disadvantages, and points to consider when developing "classical" oral dosage forms. (Information taken from [17, 66])

3.1.1 Classical Oral Dosage Forms: Liquids, Granules, Tablets, and Capsules

Acceptability of the dosage form by the end user is crucial as any type of modification to the drug product (e.g., crushing and splitting of tablets; storage in multi-compartment compliance aids/multidose dispensing systems) can alter the bioavailability of the therapeutic agent and/or stability of the dosage form so that a safe and efficacious treatment is no longer guaranteed. Unfortunately, data on acceptability of dosage forms in the older population are still sparse, and often only one factor, such as swallowability, is considered. Figure 3 summarizes the advantages and disadvantages of the classical oral dosage forms from a patient's convenience perspective.

The impact of capsules and tablets shape and size on the swallowability and/or acceptability in older patients was investigated in several recent studies to get a better understanding of the formulation properties required for a more patient-centric design. Overgaard et al. investigated the impact of the shape, size, and color of solid dosage forms on swallowability and patient preference [68]. Although the number of people aged 65 and above was quite limited, the study clearly showed that film-coated tablets were easier to swallow compared to uncoated tablets and that the difficulty to swallow dosage forms increased with size. A similar finding was obtained by Schiele et al. who investigated the prevalence and causes of swallowing difficulties of solid oral dosage forms in a general practice population [69]. This questionnaire-based study showed that swallowing difficulties are often related to the size (74.6%), surface (70.5%), shape (43.5%), and flavor (22.1%) of the dosage form. While a large number of patients had no preference with respect to the dosage form and its shape, patients with swallowing difficulties preferred round-shaped tablets (34.1%).

A number of different acceptability studies was done using the ClinSearch Acceptability Score Test (CAST). In summary, the results showed that tablets are acceptable for older patients without swallowing disorders, while for those with swallowing disorders, the largest dimension of the tablet should be restricted to a maximum of 6.5 mm [70]. The small size of the tablet might be optimal with respect to the swallowability of the dosage form but was considered less acceptable in terms of handling [66].

Another study investigated the use of oral liquids, i.e., ready-to-use liquids (35%) and reconstituted liquids (65%), in an older patient population and observed that oral liquids are a suboptimal alternative to solid oral dosage forms for patients with swallowing disorders. The main reason was the fact that a large number of oral liquid formulations did not contain a flavor and thus exhibited poor palatability. Based on the result, it became obvious that palatability remains crucial when developing oral liquid formulations for the older population, especially for women [71]. Furthermore, the impact of the viscosity of liquids in dysphagia patients was tested using videofluoroscopic swallowing studies. The results clearly showed that a viscosity increase reduced the fraction of patients exhibiting aspiration (15.8% for thin, 8.9% for thick, and 5.8% for ultrathick liquids delivered with a spoon) [72].

3.1.2 Modern Oral Dosage Forms

Oral Films

Oral films are a convenient dosage form as the film can be taken by the patient without any water. It dissolves rapidly and therefore, similar to the orally disintegrating tablets, no monolithic dosage form has to be swallowed. This is advantageous for patients with swallowing difficulties. The research on this dosage form in older patients is still limited. From the nature of the dosage form itself, it can be concluded that there are some limitations when it comes to handling the thin films. For geriatric patients with dexterity issues due to age or their condition, the small and moisture-sensitive nature of oral films presents a challenge [73]. Furthermore, only low doses can be incorporated in oral films, which limits the applicability of this dosage form.

On the other hand, oral films can be advantageous for older adults in emergency situations when drug substances are used that are absorbed via the oral mucosa, such as nitroglycerin. Also molecules such as oxycodon or some triptanes may benefit from such formulation strategy [74].

Orally Disintegrating Tablets

Orally disintegrating tablets (ODTs) are tablets that rapidly disintegrate in the oral cavity without the need for additional water intake, making it a convenient formulation for oral medications. Another benefit is the faster onset of action compared to conventional dosage forms [75]. Additionally, as this dosage form does not need to be swallowed as an intact tablet, it presents an interesting alternative for patients with swallowing difficulties [67]. It was shown that older adults with swallowing disorders were more likely to take the complete dose when administered as ODTs compared to powder for oral solutions (PfoS). Nevertheless, PfoS were still more favored by this population.

In older people with dysphagia, ODTs can improve the compliance [76], and research has shown that ODTs are the preferred dosage form in a study where patients were offered an ODT, a nasal spray, and a conventional tablet [77]. Nevertheless, the evidence on acceptability or preference of ODT as a formulation option in older people is rather limited. An important aspect is how older patients are instructed on the use of an ODT. ODTs look like conventional tablets, and therefore an older patient might assume it has to be swallowed accordingly. Furthermore, older people might develop a dry mouth (xerostomia), and their residual amount of saliva might be limited to 0.05 to 0.1 ml [78]. Hence, instructions for older patients, e.g., taking the ODT with a sip of water, are crucial to achieve a successful therapy with ODTs.

Multiparticulates Including Minitablets

Multiparticulates and minitablets present similar advantages in terms of ease of swallowing, as there is no monolithic dosage form to be swallowed. Multiparticulates are defined as small solid multi-unit dosage forms such as granules, beads, or pellets. The term minitablets describes small compressed tablets with a diameter of 1–3 mm [79]. These dosage forms are smaller than conventional dosage forms such as tablets and capsules and therefore represent valuable alternatives. Minitablets/ multiparticulates can be formulated with modified-release features [80] and can be swallowed easily. Additionally, they can be administered with a vehicle such as soft food or water. For some patient populations (e.g., visually impaired patients), sprinkling multiparticulates or minitablets over food can be challenging in terms of handling and verifying that the complete dose was administered. When the multiparticulates/minitablets are very small, there is a risk of aspiration or choking, which is in contrast to the improvement of swallowing this non-monolithic dosage form [73]. Also, visual impairment has to be taken into account, as multiparticulates are small, and therefore older patients might have difficulties in visually controlling whether the complete dose was taken. Multiparticulates and minitablets are packed in stickpacks/sachets or sprinkle capsules or as bulk in bottles to be used with a dosing spoon. These packaging configurations might be non-intuitive for older patients and therefore prone to erroneous handling. Moreover, child-proof features might be a barrier for older patients suffering from dexterity problems or illnesses such as arthritis [81].

3D-Printed Dosage Forms

Three-dimensional (3D) printing offers the possibility to customize the dosage form to the requirements of the individual end user and to manufacture the personalized printlets (3D-printed drug products) on demand [82]. There are a number of advantages for the use of 3D-printed medicines in older patients such as adaptation of the dose and a combination of different drugs in a single unit based on the patient's individual need. Furthermore, this versatile technology offers some alternative options for patients with visual impairment, swallowing difficulties, or dexterity disabilities.

The acceptability of different geometries of printlets with respect to picking and swallowability was tested in an open-label, randomized, exploratory pilot study with 50 participants. This study surprisingly showed that the torus, as a novel 3D-printed shape, was well accepted in the study population aged 18–45 years and might therefore present an interesting alternative for people aged 65 and above [83].

In another study in polypharmacy patients aged 29–80 years (median 66.5 years), the perception and preference of 3D-printed medicine were explored. Based on a visual assessment, these patients preferred the heart-, the almond-, and to some extent the oval-shaped dosage forms. The sphere that was presented to the participants was disliked, as it is extremely difficult to pick and it might get lost if dropped on the floor. Despite the fact that most of the participants regarded embossing as irrelevant for themselves, they mentioned that embossing such as a sun or a moon could be more suitable for older people to remember the time of administration during the day [84].

There are a number of different research articles showcasing the benefits of using 3D-printed dosage forms in order to reduce the "pill burden" for patients taking a number of different drugs. In the first example [85], nifedipine, captopril, and glipizide were combined in a polypill for the treatment of type II diabetes as well as type II diabetes associated with arterial hypertension. The individual compartments of the polypill were designed to enable a zero-order-release profile for captopril by restricting the drug release through osmosis and ensured a first-order-release kinetics via diffusion for the two other drugs. In the second example [86], five different drugs with two independent release profiles were combined in a polypill for the treatment of cardiovascular pathologies. The polypill consists of three sustained-release compartments for atenolol, pravastatin, and ramipril, and two immediate release compartments for acetylsalicylic acid and hydrochlorothiazide. The same 3D printing technology, i.e., semisolid extrusion, was also used for the manufacturing of 3D-printed tablets with zero-order-release kinetics containing compounds with aqueous solubilities ranging from 9.45 μg/mL to 8.34 mg/mL and drug loads of 3 to 43% [87]. With dual-fused deposition modelling, another 3D printing technology was used to manufacture enalapril maleate and

Determinants of Non-Adherence				
Socioeconomic factor	**Health-care and system related factors**	**Therapy-related factors**	**Patient-related factors**	**Condition related factors**
-Family and social support -Employment status -Cost of drugs and/or treatment	-Access to healthcare -Unclear information about drug administration -Poor provider-patient relationship	-Complexity of therapeutic regimen -Duration -Past failures -Therapeutic changes -Adverse events	-Forgetfulness -Beliefs -Knowledge and/or skills -Misunderstanding instructions for treatment	-Asymptomatic nature of the disease -Disease severity

Fig. 4 Determinants of non-adherence. (Based on information from [64])

hydrochlorothiazide FDCs with a similar release profile for both compounds, despite their different aqueous solubility and independent of the drug load [88]. All these examples illustrate that 3D printing provides superior dosing flexibility and may offer a more patient-centric solution compared to the traditional manufacturing technologies.

3.2 Medication Adherence

Medication adherence is defined as "an active, cooperative and voluntary participation of the patient in following recommendations from a health care provider" [89]. Research has shown that multiple factors have an impact on medication adherence and non-adherence (Fig. 4) and that older patients dealing with polypharmacy are at greatest risk for non-adherence.

Non-adherence often leads to increased costs due to hospitalizations, nursing home admissions, and premature deaths and could be avoided by optimizing the drug products in terms of dosing frequency, "pill burden," acceptability, portability, and discretion of intake as well as minimization of adverse events, when designing products for geriatric patients and older adults.

Two of these aspects, i.e., the reduction in dosing frequency and the use of fixed-dose combination products, will be described in detail below.

3.2.1 Reduction in Dosing Frequency

In order to obtain the pharmacological activity of the drug and achieve the desired therapeutic effect, certain drug levels need to be achieved and maintained over a period of time in the tissue. Fast metabolism and elimination of the drug, e.g., a short half-life, often require an increase in the dosing frequency and thus interfere with the day-to-day activities of the patient or his/her participation in social events. In particular, in cases where the disease is asymptomatic in nature, this can lead to non-adherence and, as a consequence later on, to a worsening of the disease. A

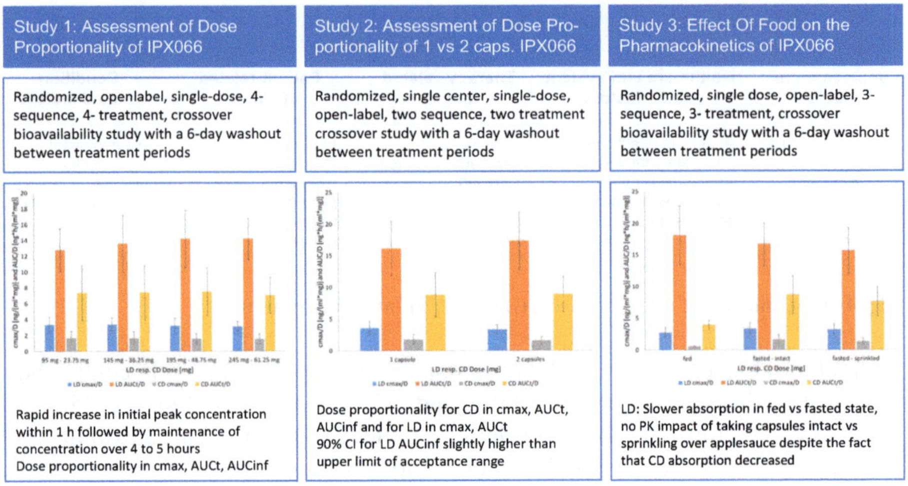

Fig. 5 Pharmacokinetics of Rytary in dependence of dose and food conditions. (Data taken from [92])

reduction in the dosing frequency by using extended-release oral dosage forms, which release the drug for a prolonged period of time, or the use of long-acting parenteral dosage forms contributes to better adherence. The latter also helps reduce costs for the healthcare system, as the administration via the parenteral route often requires the involvement of a healthcare professional.

The benefit of extended-release dosage forms will be explained exemplarily for the treatment of Parkinson's disease (PD). In the population above 60 years of age, approx. 1% of the population is affected by PD [90]. The gold standard therapy for the disease is the administration of levodopa as replacement of dopamine in the brain in combination with a peripheral L-amino acid decarboxylase (AADC) inhibitor in order to reduce the peripheral dopaminergic side effects such as nausea and vomiting and to increase the amount of levodopa taken up in the brain. Carbidopa or benserazide are used as AADC inhibitors. [91] Despite the ease of administration via the oral route, there are some major disadvantages for levodopa. Due to the short half-life of the compound, immediate-release (IR) formulations need to be taken every 4 h to maintain reasonable plasma levels and result in marked fluctuations in the levodopa plasma concentrations [92]. In order to reduce the dosing frequency, a number of different extended-release (ER) formulations of levodopa were developed and clinically tested, and at least one of them was approved in recent years. Rytary® is a capsule formulation that contains one-third IR and two-thirds ER beads of levodopa/carbidopa. The IR beads enable a fast onset of action, while the ER beads maintain sufficient plasma levels over 4–5 h [93]. The extended-release rates are obtained via film coating using methacrylic acid copolymers [94]. The pharmacokinetics of this formulation was tested in three different studies in healthy volunteers [92]. The study design as well as some high-level results of the study are shown in Fig. 5. Overall, the formulation showed dose-proportional pharmacokinetics in fasted states, and no impact on pharmacokinetics was observed when taking the capsule intact versus sprinkled over applesauce. Administration of the drug product after a high-fat, high-calorie meal delayed absorption by 1–2 h, slightly reduced the peak plasma concentration (Cmax), and increased the area under the plasma concentration vs. time curve (AUC). This promising formulation was further investigated in a Phase 3 clinical trial. Based on the positive results, i.e., a reduction in daily OFF time by, on average, an extra −1.17 h with the ER formulation given a mean of 3.6 times per day compared to IR carbidopa-levodopa formulations [95], the ER formulation was approved by FDA in 2015. Additional efforts are underway in order to further reduce the dosing frequency to every 8 h using a new investigational ER formulation of carbidopa-levodopa (CD-LD), and first clinical results look promising [96].

Another formulation, which was investigated in recent years in a number of clinical studies, is the "carbidopa–levodopa accordion pill." This is a gastroretentive slow-release levodopa formulation containing a multilayer planar structure of biodegradable films that are folded in an accordion shape but displayed in standard-size capsules. After the capsule is administered and dissolved/opened, the layers unfold in the stomach and, due to their size and structure, remain in the stomach. The prolonged gastric retention should result in an increase in drug absorption. While the accordion pill resulted in more stable LD plasma concentrations in both fluctuating and non-fluctuating PD patients and significantly improved standard measures of motor symptoms (daily OFF time, total ON time, and good ON time) versus IR-CD/LD in Phase 2 clinical studies [97], the Phase 3 trial did not meet its primary

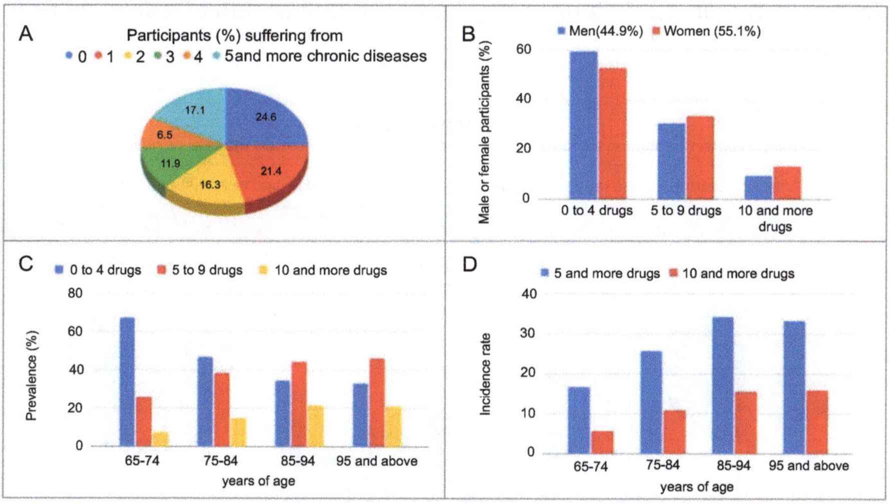

Fig. 6 Aspects of polypharmacy: (**a**) Number of chronic disease(s) in people aged 65 and above, (**b**) polypharmacy in dependence of gender, (**c**) prevalence of polypharmacy and excessive polypharmacy depending on age, (**d**) incidence rate per 100 person-years depending on age. (Data derived from [100])

endpoint [98]. Other oral formulations, including a bilayer tablet (DM-1992), a sustained-release bilayer tablet with the aim to promote colonic absorption (XP21279), as well as a new oral formulation that contains higher amounts of carbidopa (65 or 105 mg instead of 25 mg), were tested in clinical studies, but the development of these formulations seemed to be discontinued.

3.2.2 Polypharmacy

Multimorbidity is associated with the use of multiple medicines (polypharmacy) by a single individual and is quite common in the older population [99, 100]. It usually reflects the presence of various health problems, including chronic diseases, acute conditions such as infections, and symptoms that accumulate with age such as pain [100]. A register-based prospective cohort study in individuals aged ≥ 65 years has shown that individuals were exposed on average to 4.6 drugs (SD 4.0), 44% used 5 or more drugs, and 11.7% used 10 or more drugs. Furthermore, they showed that the incidence rate of polypharmacy increased with age. A summary of their findings is presented in Fig. 6.

Polypharmacy is not only related to an increased risk for adverse drug-related events in older adults (see Sect. 2.2, "Impact of Co-medication on the Safety and Efficacy of Drugs," for further details) but often leads to complex medication schemes and poor medication adherence in the concerned age group. Medication adherence can be improved by using fixed-dose combinations (FDCs) of two or more drugs targeting the same indication. However, mass production of FDCs comes with its own limitations in terms of dosing flexibility. For example, the combination of two drugs, each one being approved in three different dose strengths,

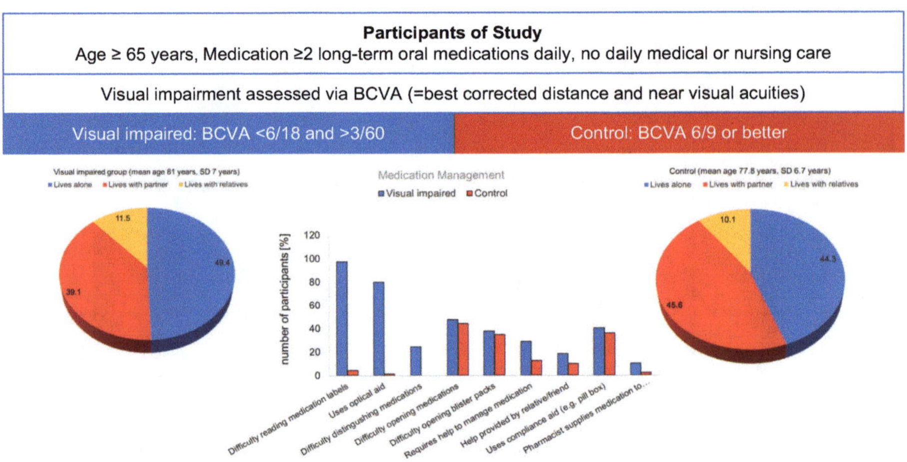

Fig. 7 Results from a case-control study investigating the impact of visual impairment on medication self-management [101]

results in nine possible combinations, and 3 drugs with three dose strengths result in 27 combinations. In these cases, 3D printing could provide a potential alternative by enabling the manufacturing of personalized oral medicines in a community or hospital pharmacy setting. Recent examples from literature, such as Khaled et al. [85, 86], Sadia et al. [88], and Fang et al. [87], clearly show the benefit of this technology with respect to dose individualization. More detailed information on the use of 3D printing was provided in Sect. 3.1.2.4.

3.2.3 Visual Impairment

A case-control study investigated the medication adherence and self-management in visual-impaired patients and those without visual impairment. Despite the fact that the study did not find a difference in medication adherence between both groups, it clearly showed that almost all patients with visual impairment (~97%) have difficulties reading medication labels and around 24% are unable to distinguish their medication. Furthermore, they need more support to manage their medication compared to the control group (see Fig. 7) [101].

Over the recent years, a number of different medication management systems, such as the MedVision [102] or a Braille-based pillbox [103] with audio reminder, were developed, in order to improve medication adherence of visually impaired people, but they still rely on assistance from a family member, friend, the pharmacist, or general practitioner to feed the mobile applications with required information and the medication boxes with the medicine.

Despite these advances and the fact that Braille "letters" are included on the packaging, a proper medicine identification is no longer possible when the patient has removed the medicine from the original container or the pillbox. The patient thus becomes increasingly dependent on others and may feel insecure with regard to its medication management, which increases the risk for non-adherence particularly in visually impaired older adults. A study performed by Awad et al. [104] nicely showed that Braille or Moon patterns could be incorporated in orally disintegrating printlets using selective laser sintering, a 3D printing technology with high resolution, without altering the properties of the dosage form. The recognition of the pattern was verified by a blind person. Furthermore, due to the rapid disintegration of the dosage form in the mouth without the need for water co-administration, these printlets encourage self-administration and thus improve medication adherence.

3.3 *From Clinical Studies to Real-Life Administration*

Within drug product development, the importance given to drug efficacy to support regulatory approval and prescription of the medicines by healthcare professionals seems to be one of the main reasons for the underestimation of patient centric aspects [105]. Instead, patient usability and administration are usually considered to be sufficiently addressed by the user instructions in the package leaflet. However, this gap between efficacy, "how a drug performs under conditions of clinical trials," and effectiveness, "how a drug performs under everyday conditions," has been more and more highlighted by drug regulatory agencies. Improving drug effectiveness in real life seems to have a good value for patient safety, patient adherence, and thus the benefits of the drug product company [106]. A better understanding of patients' needs may improve the interface between patients and drug products, contributing to easier administrations and ultimately to effective therapies.

3.3.1 Usability/Human Factors Engineering

Today, the development of a drug product (where no device constituent part is either integrated or co-packed) does not require pharmaceutical companies to perform usability/human factors activities (such as human factors validation studies), as there are no regulatory usability requirements in place. Nevertheless, the development of any drug, whether it's part of a drug-device combination or not, may benefit from applying usability engineering principles. Usability in this particular context is described as "the product characteristics and attributes of a pharmaceutical product that enables the patient and/or caregiver in its personal environment and life situation to use the pharmaceutical drug product as intended" [105]. Older patients' use difficulties and problems with drug products may be the result of a dysfunctioning interaction between the user and the drug product. The usability/human factors engineering may enable a more patient-centered design of drug products for a safer, easier, and better patient use experience. For example, during the development of a drug-device combination product intended for older patients' use, the intended use, user population, and use environments are typically carefully considered and defined. In order to ensure a drug-device combination product safety and effectiveness, older patients would be involved in iterative usability/human factors evaluations along with other user groups (e.g., caregiver or healthcare professionals) who might support them during the administration of a drug. The following sections are intended to highlight aspects of the usability/human factors engineering process for designing medical devices/drug-device combination products (including their packaging, product labels, instructions for use, etc.) according to the IEC 62366-1 standard and FDA guidelines that may support the development of a more patient centric drug product [107, 108].

3.3.2 Older Patient Population as Users

The patients play an important role in achieving the desired therapeutic outcomes, as they are frequently the ones who manage their medication. To facilitate drug administration and overcome medication issues, the key is to identify and understand older patients' needs and preferences related to product design [106].

Careful consideration should be given to the intended user profile with respect to the perception, cognitive, physical capabilities, psychological and socioeconomic environment, age, and general aptitude of the user. Within the population of older users, general reduced abilities should be expected in terms of:

- *Perception.* How do the older patients perceive the information? How long can they focus? Older people can suffer from reduced or impaired visual abilities, hearing, or even altered sense of smell and taste. This can result in difficulties in reading the product label or use instructions, in handling preparations or opening containers, and in missing instructions/explanations [17]. For example, 20% of the people aged 80 or above have considerable yellowing of the vision that leads to misjudgment and confusion of colors and thus difficulties in reading the dosing graduations to dose oral solutions [109].
- *Cognitive state.* Do the older patients have the required reasoning maturity to understand how to proceed? Older patients can experience reduced or impaired cognition, mental capabilities, forgetfulness, and especially when dementia diseases (e.g., Alzheimer's) are more prevalent. This can result in difficulties remembering when and how to take a medicine, understanding instructions, experiencing confusion, resisting help by others, etc. Older adults can also experience reduced health literacy [17].
- *Physical state.* Do older patients have the physical capabilities to handle the drug product and its packaging? Older adults commonly experience reduced or impaired manual and finger dexterity (e.g. Parkinson's disease), grip strength, hand stiffness, etc. leading to difficulties in picking tablets from the container or pushing tablets through a blister. They can also experience reduced control of mouth and throat muscles, causing swallowing difficulties or even choking if there is dysphagia [17].
- *Emotional/psychological state.* The impact of their medical condition on their ability to use the drug product safely and effectively should also be considered. Older patients often receive more than one medicinal product each day [64]. The increasing age and multimorbidity of the patients correlates with a higher incidence for polypharmacy, and both significantly increase the burden for patients and their caregivers to manage the growing complexity of the medications as recommended [105]. In addition, the experience older patients make in handling and dosing their medication has a strong impact on their compliance/adherence to treatments [64]. It is important to consider that in contrast with pediatrics, getting older is not psychologically easy when it means accepting that one's health conditions are decreasing. Often older people would hide their conditions from their relatives or even react negatively toward medicines [110].

Table 1 Examples of use errors and risks in older adult users

Older adults characteristics	Potential cause for use error	Potential use error	Hazardous situation	Potential harm
Perception	User has visual impairment (e.g., does not see the dosing mark on the syringe or cannot read the information/instructions)	User does not set the dose correctly	Underdosing	Disease progression (therapy not effective)
		User takes an expired pill	Ingestion of degraded drug product	Adverse events
Cognitive state	User does not have the patience to prepare drug product for administration (limited attention span)	User does not administer the required amount of solution	Underdosing	Disease progression (therapy not effective)
	User does not have the cognitive skills to understand which dose to take (e.g., forgot and has different treatments)	User takes the wrong dose	Underdosing/overdosing	Molecule dependent
Physical state	User has difficulties swallowing tablets (e.g., dysphagia)	User tries to swallow tablet	Choking	Irritation, pain, suffocation
	User does not have the strength to self-administer the solution	User is not capable of pushing the plunger to administer the solution	Underdosing	Disease progression (therapy not effective)
Emotional/psychological state	User does not see the immediate benefit of the therapy (e.g., in chronic disease)	User does not take the drug regularly	Missed doses	Disease progression (therapy not effective)
	User has a negative feeling toward the drug product (e.g., has a better handling experience with another type of drug product) or does not want to be helped	User rejects the treatment	No dose	Disease progression (no therapy)

3.3.3 Potential Use Risks and Design Considerations

It is recommended identifying potential use errors that could occur in real life when administering in older patients in order to ensure that these use errors are adequately controlled. Table 1 gives examples of use errors and risks to be considered when designing a drug product for older adults (as patient users).

A drug product should be easy to handle and easy to administer in order to be used by older patients. This section provides examples of drug product design considerations.

Design Should Consider Older Adult Patient Senses Limitations and Preferences Packaging, labeling and leaflet are typical parts of the final drug product. Sometimes it is in addition combined with one or more delivery devices. Any artworks (i.e., the layout, fonts, illustrations, symbols, etc.) should be designed to be readable by the older patient who has reduced or impaired visual acuity. For example, not colored standard fonts, high dark/light contrast, and point sizes higher than 12 should be preferred. Also, certain dosage forms might be more effective (e.g., tablets or capsules) than others (e.g., oral solutions) in their ability to provide the older adult with an accurate dosing. Tablets or capsules can be seen as a better option when considering that their reduced vision capabilities can lead to risks of underdosing/overdosing in case of preparing for an oral solution administration. Also, other appreciated and helpful alternatives for differentiating doses in older patients are the use of different shapes and lighter colors when designing drug products [110].

Design Should Consider that Older Adult Patients Have Limited Cognitive Capabilities to Administer a Drug Product For example, multi-compartment medication devices can be effective in supporting older patients in their daily regimens and in improving therapy outcomes. They allow the patient and caregiver to organize in advance the medicines to take in a specific day period or day of week, reducing the risk of medication errors and simplifying the actions the patients have to do for medicine administration. In addition, the use of fixed-dose combination drug products is simple, the safest, and effective for administration. Nevertheless, some precautions need to be considered when the dosage form becomes too large and leads to problems in swallowing such as with older patients' suffering from dysphagia [106, 111]. Another example is using packaging that meets with specific needs of older adults such as helping them to remember when to take their medication [64].

The Design Should Consider What Is Physically Tolerable for the Older Adult Patient Multi-particulate formulations are good for personalized treatments and enhancing swallowing experience, but they can be difficult for the older patient to handle (e.g., manual dexterity issues). It could be relevant to provide older patients with a suitable dispensing device to enable self-administration of minitablets. In addition, when considering liquid drug products (i.e., syrups, suspensions, and solutions), they can be very flexible in dosing and can be easily swallowed by the patient. However, the dose accuracy is strongly affected by the type of measuring device used by the patient (e.g., spoon, syringe), and the risk of underdose/overdose is high due to incorrect handling [73]. For example, it is important to ensure the older patient has the required strength and hand/fingers dexterity to push on the oral dispenser plunger in order to self-administer the solution.

Design Should Provide a Positive and Safe User Handling Experience The design of the devices and packaging have an important role in the handling of a drug product. Most of the use problems reported with older people are related to the opening of containers and removal of medicines from blisters [112]. The packaging is seen as "the first experience that a patient has with the product, and sets the tone for the overall product experience." When it requires the older patient to use high force, a tool such as scissors, or ask a caregiver to open a package, this could constitute a real barrier to the acceptability and safety of use of the drug product [110]. In addition, using printed features or color coding/symbols on the packaging has been shown as improving the intuitiveness of the drug product. The design of the devices such as oral dispensers or spoons used for dosing and for administration has a considerable impact on the therapy efficacy and the adherence to the drug product. It is frequent that a poor or inadequate design does lead to inaccurate dosing. Portability is also an important aspect to consider especially for those oral liquids products that are to be used with dosing devices. Considering the older adult patient capabilities and limitations, it is widely recommended to focus on minimizing the use steps for preparation of the medicine, for setting the dose (e.g., proving legible dosing graduations), and for the administration of the drug product [110].

3.3.4 Usability Testing with Older Adult Patients

Usability testing is a well-established method usually conducted to inform the development, by revealing potential shortcomings of the user interface, and to validate its safe and effective use by its end users under intended conditions. Whenever possible, it is valuable to conduct exploratory usability testing (generally with 5–8 representative users) in early stages, for instance, prior to a clinical trial in order to identify use errors with the drug product that might have a negative impact on the clinical study [113]. This type of testing is usually performed in the representative conditions (e.g., at home or simulated environment) to better identify the final user's needs and preferences that might impact on their adherence to the treatment. A usability summative testing (also called human factors validation study) is conducted to demonstrate the safe and effective use of the drug device combination product (with at least 15 representative users per user group) and when possible before Phase 3 clinical trials as this allows time for any changes to optimize the user interface prior to completing the clinical investigation. Training might not always happen in real life or differently than expected; thus, it is meaningful to evaluate the intuitiveness of a drug product without having trained the older adult participants prior to this testing. The instructions for use and drug product labeling should be also part of it in order to evaluate the effectiveness of important warnings and critical information (e.g., to assess whether or not older patients are able to locate, read, and understand the information) [107, 108, 113].

When considering such testing with older patients, the following patterns regarding their behavior and performance can be highlighted [114]:

- Older adults might have vision and hearing impairments and might not bring the necessary reading and hearing aids to the test session.
- Older adults are more likely to commit the same use error multiple times and fail to fully process and comprehend the factors that led to the initial use error, perhaps due to reduced short-term memory.
- Older adults might need more time to process requests (i.e., task instructions) and take action.
- Older adults might blame themselves for any use errors and need reassurance that this is not them being tested but the drug product.
- Older adults might get tired sooner than younger participants, so short testing and breaks might be appreciated.

Usability testing can be used as a patient centric method to identify strengths and weaknesses of a drug product to ensure it meets the older patient population needs and its safe and effective administration.

4 Conclusion and Future Opportunities

The older population is rapidly growing in developed countries and is therefore becoming increasingly important from a socioeconomic perspective. The overall health costs are generated to a large extent by this age group, which is often multimorbid and requires care assistance as well as medical and pharmacological treatment. Such complex clinical conditions can contribute to poor medication adherence, which is known to result in a higher likelihood of hospitalization and admission to nursing homes. Therefore, designing medicines that better fit the patient's needs will not only improve patients' experience but also contribute to cost reduction in the health care sector.

This goal has to be achieved in a holistic manner by applying cross-functional thinking within the pharmaceutical industry, an open dialogue between health authorities and the industry, as well as the involvement of specific patient populations, caregivers, geriatricians and practitioners, hospitals, nursing homes, and insurance companies. The increasing awareness of age- and disease-specific needs as well as the improved understanding of changes in physiology and body function will result in the development of more patient-centered formulations targeted to older adults and the geriatric population. From an industrial perspective, the development of better medicines for this age group is clearly a cross-functional effort involving clinical functions, formulation scientists, human factors, packaging and device engineers, and many others. The development teams should gain a common understanding of the disease as well as specific patient needs and limitations already during the formulation design phase in order to ensure that the dosage form, its packaging, and the device (if needed) all contribute to an optimal patient experience and successful drug therapy.

Acknowledgments Authors Susanne Page, Sabrina Bras Da Costa, Cordula Stillhart, and Leonie Wagner are employees of F. Hoffmann-La Roche AG.

References

1. Dolgin E. Big pharma moves from "blockbusters" to "niche busters". Nat Med. 2010;16:837. https://doi.org/10.1038/nm0810-837a.
2. Stegemann S, Ecker F, Maio M, et al. Geriatric drug therapy: neglecting the inevitable majority. Ageing Res Rev. 2010;9:384–98. https://doi.org/10.1016/j.arr.2010.04.005.
3. Swanlund SL. Successful cardiovascular medication management processes as perceived by community-dwelling adults over age 74. Appl Nurs Res. 2010;23:22–9. https://doi.org/10.1016/j.apnr.2008.03.005.
4. Stewart RB, Cooper JW. Polypharmacy in the aged. Drug Aging. 1994;4:449–61. https://doi.org/10.2165/00002512-199404060-00002.
5. Delafuente JC. Understanding and preventing drug interactions in elderly patients. Crit Rev Oncol Hematol. 2003;48:133–43. https://doi.org/10.1016/j.critrevonc.2003.04.004.
6. EC. Regulation (EC) No 1901/2006 of the European Parliament and of the Council of 12 December 2006 on medicinal products for paediatric use and amending Regulation (EEC) No 1768/92, Directive 2001/20/EC, Directive 2001/83/EC and Regulation (EC) No 726/2004 (Text with EEA relevance); 2006.
7. EMA. Committee for Medicinal Products for Human Use (CHMP), Paediatric Committee (PDCO). Guideline on pharmaceutical development of medicines for paediatric use. EMA/CHMP/QWP/805880/2012 Rev. 2; 2013.
8. Penkov D, Tomasi P, Eichler I, et al. Pediatric medicine development. Ther Innov Regul Sci. 2017;51:360–71. https://doi.org/10.1177/2168479017696265.
9. Timpe C, Dischinger A. Drug delivery strategies for oral pediatric and geriatric products. BioPharma Asia. 2014;42
10. ICH. ICH Topic E 7 studies in support of special populations: geriatrics. Note for guidance on studies in support of special populations: geriatrics (CPMP/ICH/379/95); 1994.
11. Jansen PAF. Age and the process of aging. In: Stegemann S, editor. Developing drug products in an aging society. Springer; 2016. p. 67–86.
12. EMA. EMA geriatric medicines strategy. EMA/CHMP/137793/2011. https://www.ema.europa.eu/en/documents/other/geriatric-medicines-strategy_en.pdf (2011). Accessed 19 Sept 2021
13. CDER. Guidance for industry and review staff target product profile – a strategic development process tool. http://www.ncai-cc.ccf.org/skills/documents/U.S.%20FDA%20Target%20Product%20Profile%20Guidance%20Document%20(2007).pdf (2007). Accessed 2 Oct 2021
14. ICH. ICH guideline Q8 (R2) on pharmaceutical development. EMA/CHMP/ICH/167068/2004; 2017.
15. ICH. ICH guideline Q9 on quality risk management. EMA/CHMP/ICH/24235/2006; 2015.

16. Timpe C, Stegemann S, Barrett A, Mujumdar S. Challenges and opportunities to include patient-centric product design in industrial medicines development to improve therapeutic goals. Brit J Clin Pharmacol. 2020;86:2020–7. https://doi.org/10.1111/bcp.14388.
17. EMA. Reflection paper on the pharmaceutical development of medicines for use in the older population; 2020.
18. Dressman JB, Reppas C. Oral drug absorption – prediction and assessment. 2nd ed. Boca Raton: CRC Press; 2010.
19. Bergström CAS, Holm R, Jørgensen SA, et al. Early pharmaceutical profiling to predict oral drug absorption: current status and unmet needs. Eur J Pharm Sci. 2014;57:173–99. https://doi.org/10.1016/j.ejps.2013.10.015.
20. Stillhart C, Vučićević K, Augustijns P, et al. Impact of gastrointestinal physiology on drug absorption in special populations – an UNGAP review. Eur J Pharm Sci. 2020;147:105280. https://doi.org/10.1016/j.ejps.2020.105280.
21. Mangoni AA, Jackson SHD. Age-related changes in pharmacokinetics and pharmacodynamics: basic principles and practical applications. Brit J Clin Pharmacol. 2004;57:6–14. https://doi.org/10.1046/j.1365-2125.2003.02007.x.
22. Klotz U. Pharmacokinetics and drug metabolism in the elderly. Drug Metab Rev. 2009;41:67–76. https://doi.org/10.1080/03602530902722679.
23. Merchant HA, Liu F, Gul MO, Basit AW. Age-mediated changes in the gastrointestinal tract. Int J Pharm. 2016;512:382–95. https://doi.org/10.1016/j.ijpharm.2016.04.024.
24. Bai JPF, Burckart GJ, Mulberg AE. Literature review of gastrointestinal physiology in the elderly, in pediatric patients, and in patients with gastrointestinal diseases. J Pharm Sci. 2015;105:476–83. https://doi.org/10.1002/jps.24696.
25. Britton E, McLaughlin JT. Ageing and the gut. Proc Nutr Soc. 2013;72:173–7. https://doi.org/10.1017/s0029665112002807.
26. Bhutto A, Morley JE. The clinical significance of gastrointestinal changes with aging. Curr Opin Clin Nutr. 2008;11:651–60. https://doi.org/10.1097/mco.0b013e32830b5d37.
27. Grassi M, Petraccia L, Mennuni G, et al. Changes, functional disorders, and diseases in the gastrointestinal tract of elderly. Nutr Hosp. 2011;26:659–68.
28. Hurwitz A, Brady DA, Schaal SE, et al. Gastric acidity in older adults. JAMA. 1997;278:659–62.
29. Feldman. The mature stomach. Still pumping out acid? JAMA. 1997;278:681–2.
30. Russell TL, Berardi RR, Barnett JL, et al. Upper gastrointestinal pH in seventy-nine healthy, elderly, North American men and women. Pharm Res. 1993;10:187–96. https://doi.org/10.1023/a:1018970323716.
31. Gidal BE. Antiepileptic drug formulation and treatment in the elderly: biopharmaceutical considerations. Int Rev Neurobiol. 2007;81:299–311. https://doi.org/10.1016/s0074-7742(06)81020-2.
32. Rotman SR, Bishop TF. Proton pump inhibitor use in the U.S. Ambulatory setting, 2002–2009. Plos One. 2013;8:e56060. https://doi.org/10.1371/journal.pone.0056060.
33. Wallerstedt SM, Fastbom J, Linke J, Vitols S. Long-term use of proton pump inhibitors and prevalence of disease- and drug-related reasons for gastroprotection – a cross-sectional population-based study. Pharmacoepidemiol Drug Saf. 2017;26:9–16. https://doi.org/10.1002/pds.4135.
34. Einarsson K, Nilsell K, Leijd B, Angelin B. Influence of age on secretion of cholesterol and synthesis of bile acids by the liver. N Engl J Med. 1985;313:277–82.
35. Salemans JMJI, Nagengast FM, Tangerman A, et al. Effect of ageing on postprandial conjugated and unconjugated serum bile acid levels in healthy subjects. Eur J Clin Invest. 1993;23:192–8. https://doi.org/10.1111/j.1365-2362.1993.tb00761.x.
36. Annaert P, Brouwers J, Bijnens A, et al. Ex vivo permeability experiments in excised rat intestinal tissue and in vitro solubility measurements in aspirated human intestinal fluids support age-dependent oral drug absorption. Eur J Pharm Sci. 2010;39:15–22. https://doi.org/10.1016/j.ejps.2009.10.005.
37. Saad R, Semler J, Wilding G, Chey W. The effects of age on regional and whole gut transit times in healthy adults. Gastroenterol. 2010;138:S127.

38. Shimamoto C, Hirata I, Hiraike Y, et al. Evaluation of gastric motor activity in the elderly by electrogastrography and the 13C-acetate breath test. Gerontology. 2002;48:381–6. https://doi.org/10.1159/000065500.
39. O'Mahony D, O'Leary P, Quigley EMM. Aging and intestinal motility. Drug Aging. 2002;19:515–27. https://doi.org/10.2165/00002512-200219070-00005.
40. Bouras EP, Tangalos EG. Chronic constipation in the elderly. Gastroenterol Clin N. 2009;38:463–80. https://doi.org/10.1016/j.gtc.2009.06.001.
41. Mabbott NA. A breakdown in communication? Understanding the effects of aging on the human small intestine epithelium. Clin Sci Lond Engl. 2015;129:529–31. https://doi.org/10.1042/cs20150364.
42. Wilms E, Troost FJ, Elizalde M, et al. Intestinal barrier function is maintained with aging – a comprehensive study in healthy subjects and irritable bowel syndrome patients. Sci Rep. 2020;10:475. https://doi.org/10.1038/s41598-019-57106-2.
43. Holt P, Pascal R, Kotler D. Effect of aging upon small intestinal structure in the Fischer rat. J Gerontol. 1984;39:642–7.
44. Höhn P, Gabbert H, Wagner R. Differentiation and aging of the rat intestinal mucosa. II. Morphological, enzyme histochemical and disc electrophoretic aspects of the aging of the small intestinal mucosa. Mech Ageing Dev. 1978;7:217–26.
45. Keelan M, Walker K, Thomson A. Intestinal morphology, marker enzymes and lipid content of brush border membranes from rabbit jejunum and ileum: effect of aging. Mech Ageing Dev. 1985;31:49–68.
46. Lipski P, Bennett M, Kelly P, OF J. Ageing and duodenal morphometry. J Clin Pathol. 1992;45:450–2.
47. Webster S, Leeming J. The appearance of the small bowel mucosa in old age. Age Ageing. 1975;4:168–74.
48. Corazza G, Frazzoni M, Gatto M, Gasbarrini G. Ageing and small-bowel mucosa: a morphometric study. Gerontology. 1986;32:60–5.
49. Collen M, Abdulian J, Chen Y. Gastroesophageal reflux disease in the elderly: more severe disease that requires aggressive therapy. Am J Gastroenterol. 1995;90:1053–7.
50. Hurwitz A, Ruhl CE, Kimler BF, et al. Gastric function in the elderly: effects on absorption of ketoconazole. J Clin Pharmacol. 2003;43:996–1002. https://doi.org/10.1177/0091270003255645.
51. Jacobs A, Miles PM. Role of gastric secretion in iron absorption. Gut. 1969;10:226–9.
52. Morley J. Anorexia of aging: physiologic and pathologic. Am J Clin Nutr. 1997;66:760–73.
53. Wysokiński A, Sobów T, Kłoszewska I, Kostka T. Mechanisms of the anorexia of aging – a review. Age. 2015;37:81. https://doi.org/10.1007/s11357-015-9821-x.
54. Riordan S, McIver C, Wakefield D, et al. Small intestinal bacterial overgrowth in the symptomatic elderly. Am J Gastroenterol. 1997;92:47–51.
55. Fillenbaum GG, Pieper CF, Cohen HJ, et al. Comorbidity of five chronic health conditions in elderly communityresidents: determinants and impact on mortality. J Gerontol Ser. 2000;55:M84–9. https://doi.org/10.1093/gerona/55.2.m84.
56. Wollmer E, Klein S. A review of patient-specific gastrointestinal parameters as a platform for developing in vitro models for predicting the in vivo performance of oral dosage forms in patients with Parkinson's disease. Int J Pharm. 2017;533:298–314. https://doi.org/10.1016/j.ijpharm.2017.08.126.
57. Vinarov Z, Abrahamsson B, Artursson P, et al. Current challenges and future perspectives in oral absorption research: an opinion of the UNGAP network. Adv Drug Deliver Rev. 2021;171:289–331. https://doi.org/10.1016/j.addr.2021.02.001.
58. Courlet P, Livio F, Guidi M, et al. Polypharmacy, drug-drug interactions and inappropriate drugs: new challenges in the aging population living with HIV. Open Forum Infect Dis. 2019;6:ofz531. https://doi.org/10.1093/ofid/ofz531.
59. Efjestad AS, Ihle-Hansen H, Hjellvik V, Blix HS. Comedication and treatment length in users of acetylcholinesterase inhibitors. Dement Geriatr Cogn Disord Extra. 2017;7:30–40. https://doi.org/10.1159/000454948.

60. Teka F, Teklay G, Ayalew E, Teshome T. Potential drug–drug interactions among elderly patients admitted to medical ward of Ayder Referral Hospital, Northern Ethiopia: a cross sectional study. BMC Res Notes. 2016;9:431. https://doi.org/10.1186/s13104-016-2238-5.
61. Lee S, Teschemaker AR, Daniel M, et al. Calcium and vitamin D use among older adults in U.S.: results from national survey. J Nutr Health Aging. 2016;20:300–5. https://doi.org/10.1007/s12603-015-0614-9.
62. Johnson BF, Bustrack JA, Urbach DR, et al. Effect of metoclopramide on digoxin absorption from tablets and capsules. Clin Pharmacol Ther. 1984;36:724–30. https://doi.org/10.1038/clpt.1984.249.
63. Stegemann S, Ternik RL, Onder G, et al. Defining patient centric pharmaceutical drug product design. Aaps J. 2016;18:1047–55. https://doi.org/10.1208/s12248-016-9938-6.
64. Menditto E, Orlando V, Rosa GD, et al. Patient centric pharmaceutical drug product design – the impact on medication adherence. Pharmaceutics. 2020;12:44. https://doi.org/10.3390/pharmaceutics12010044.
65. Informatics II for H. Avoidable costs in U.S. Healthcare – the $200 billion opportunity from using medicines more responsibly; 2013.
66. Shariff Z, Kirby D, Missaghi S, et al. Patient-centric medicine design: key characteristics of oral solid dosage forms that improve adherence and acceptance in older people. file:///C:/Users/hummleh1/Downloads/pharmaceutics-12-00905.pdf (2020). Accessed 22 Jun 2021.
67. Ruiz F, Vallet T, Wojcicki AD, et al. Dosage form suitability in vulnerable populations: a focus on paracetamol acceptability from infants to centenarians. Plos One. 2019;14:e0221261. https://doi.org/10.1371/journal.pone.0221261.
68. Overgaard ABA, Møller-Sonnergaard J, Christrup LL, et al. Patients' evaluation of shape, size and colour of solid dosage forms. Pharm World Sci. 2001;23:185–8. https://doi.org/10.1023/a:1012050931018.
69. Schiele JT, Quinzler R, Klimm H-D, et al. Difficulties swallowing solid oral dosage forms in a general practice population: prevalence, causes, and relationship to dosage forms. Eur J Clin Pharmacol. 2013;69:937–48. https://doi.org/10.1007/s00228-012-1417-0.
70. Vallet T, Michelon H, Orlu M, et al. Acceptability in the older population: the importance of an appropriate tablet size. Pharmaceutics. 2020;12:746. https://doi.org/10.3390/pharmaceutics12080746.
71. Belissa E, Vallet T, Laribe-Caget S, et al. Acceptability of oral liquid pharmaceutical products in older adults: palatability and swallowability issues. Bmc Geriatr. 2019;19:344. https://doi.org/10.1186/s12877-019-1337-2.
72. Kuhlemeier KV, Palmer JB, Rosenberg D. Effect of liquid bolus consistency and delivery method on aspiration and pharyngeal retention in dysphagia patients. Dysphagia. 2001;16:119–22. https://doi.org/10.1007/s004550011003.
73. Walsh J, Ranmal SR, Ernest TB, Liu F. Patient acceptability, safety and access: a balancing act for selecting age-appropriate oral dosage forms for paediatric and geriatric populations. Int J Pharm. 2018;536:547–62. https://doi.org/10.1016/j.ijpharm.2017.07.017.
74. Mahboob MBH, Riaz T, Jamshaid M, et al. Oral films: a comprehensive review. Int Curr Pharm J. 2016;5:111–7. https://doi.org/10.3329/icpj.v5i12.30413.
75. Debruyne FMJ, Gittelman M, Sperling H, et al. Time to onset of action of vardenafil: a retrospective analysis of the pivotal trials for the orodispersible and film-coated tablet formulations. J Sex Med. 2011;8:2912–23. https://doi.org/10.1111/j.1743-6109.2011.02462.x.
76. Stegemann S, Gosch M, Breitkreutz J. Swallowing dysfunction and dysphagia is an unrecognized challenge for oral drug therapy. Int J Pharm. 2012;430:197–206. https://doi.org/10.1016/j.ijpharm.2012.04.022.
77. Blanco MÁ, Prieto M, Mearin F, et al. Evaluación de la preferencia de los pacientes con enfermedad por reflujo gastroesofágico y disfagia por el tratamiento con comprimidos bucodispersables de lansoprazol. Gastroenterología y Hepatología. 2009;32:542–8. https://doi.org/10.1016/j.gastrohep.2009.05.004.

78. Malallah OS, Garcia CMA, Proctor GB, et al. Buccal drug delivery technologies for patient-centred treatment of radiation-induced xerostomia (dry mouth). Int J Pharm. 2018;541:157–66. https://doi.org/10.1016/j.ijpharm.2018.02.004.
79. Harris D, Hermans E, Klein S, et al. Age-appropriate solid oral formulations for pediatric applications with a focus on multiparticulates and minitablets: Summary of September 2019 EuPFI workshop. Eur J Pharm Biopharm. 2020;153:222–5. https://doi.org/10.1016/j.ejpb.2020.06.012.
80. Tawfeek HM, Roberts M, Hamd MAE, et al. Glibenclamide Mini-tablets with an Enhanced Pharmacokinetic and Pharmacodynamic Performance. Aaps Pharmscitech. 2018;19:2948–60. https://doi.org/10.1208/s12249-018-1108-y.
81. Haack D, Koeberle M. How advances in pharmaceutical packaging are better meeting patients' needs. Drug Dev Deliv. 2017;17:20–4.
82. Seoane-Viaño I, Trenfield SJ, Basit AW, Goyanes Á. Translating 3D printed pharmaceuticals: from hype to real-world clinical applications. Adv Drug Deliver Rev. 2021;174:553–75. https://doi.org/10.1016/j.addr.2021.05.003.
83. Goyanes A, Scarpa M, Kamlow M, et al. Patient acceptability of 3D printed medicines. Int J Pharmaceut. 2017;530:71–8. https://doi.org/10.1016/j.ijpharm.2017.07.064.
84. Fastø MM, Genina N, Kaae S, Sporrong SK. Perceptions, preferences and acceptability of patient designed 3D printed medicine by polypharmacy patients: a pilot study. Int J Clin Pharm-net. 2019;41:1290–8. https://doi.org/10.1007/s11096-019-00892-6.
85. Khaled SA, Burley JC, Alexander MR, et al. 3D printing of tablets containing multiple drugs with defined release profiles. Int J Pharmaceut. 2015;494:643–50. https://doi.org/10.1016/j.ijpharm.2015.07.067.
86. Khaled SA, Burley JC, Alexander MR, et al. 3D printing of five-in-one dose combination polypill with defined immediate and sustained release profiles. J Control Release. 2015;217:308–14. https://doi.org/10.1016/j.jconrel.2015.09.028.
87. Fang D, Yang Y, Cui M, et al. Three-dimensional (3D)–printed zero-order released platform: a novel method of personalized dosage form design and manufacturing. Aaps Pharmscitech. 2021;22:37. https://doi.org/10.1208/s12249-020-01886-8.
88. Sadia M, Isreb A, Abbadi I, et al. From 'fixed dose combinations' to 'a dynamic dose combiner': 3D printed bi-layer antihypertensive tablets. Eur J Pharm Sci. 2018;123:484–94. https://doi.org/10.1016/j.ejps.2018.07.045.
89. Vrijens B, Geest SD, Hughes DA, et al. A new taxonomy for describing and defining adherence to medications. Brit J Clin Pharmacol. 2012;73:691–705. https://doi.org/10.1111/j.1365-2125.2012.04167.x.
90. Tysnes O-B, Storstein A. Epidemiology of parkinson's disease. J Neural Transm. 2017;124:901–5. https://doi.org/10.1007/s00702-017-1686-y.
91. Freitas ME, Ruiz-Lopez M, Fox SH. Novel levodopa formulations for parkinson's disease. CNS Drugs. 2016;30:1079–95. https://doi.org/10.1007/s40263-016-0386-8.
92. Yao H-M, Hsu A, Gupta S, Modi NB. Clinical pharmacokinetics of IPX066. Clin Neuropharmacol. 2016;39:10–7. https://doi.org/10.1097/wnf.0000000000000126.
93. Amneal. How Rytary is thought to work. https://rytary.com/how-rytary-is-thought-to-work (2021). Accessed 1 Jun 2021
94. FDA. CENTER FOR DRUG EVALUATION AND RESEARCH APPLICATION NUMBER: 203312Orig1s000 LABELING. https://www.accessdata.fda.gov/drugsatfda_docs/nda/2015/203312Orig1s000LBL.pdf (n.d.). Accessed 3 Oct 2021
95. Hauser RA, Hsu A, Kell S, et al. Extended-release carbidopa-levodopa (IPX066) compared with immediate-release carbidopa-levodopa in patients with Parkinson's disease and motor fluctuations: a phase 3 randomised, double-blind trial. Lancet Neurol. 2013;12:346–56. https://doi.org/10.1016/s1474-4422(13)70025-5.
96. Modi NB, Mittur A, Rubens R, et al. Single-dose pharmacokinetics and pharmacodynamics of IPX203 in patients with advanced parkinson disease. Clin Neuropharmacol. 2019;42:4–8. https://doi.org/10.1097/wnf.0000000000000314.

97. LeWitt PA, Giladi N, Navon N. Pharmacokinetics and efficacy of a novel formulation of carbidopa-levodopa (Accordion Pill®) in Parkinson's disease. Parkinsonism Relat D. 2019;65:131–8. https://doi.org/10.1016/j.parkreldis.2019.05.032.
98. Pharma I. Intec pharma reports top-line phase 3 trial results of accordion pill-carbidopa/levodopa in advanced parkinson's disease patients. https://www.biospace.com/article/releases/intec-pharma-reports-top-line-phase-3-trial-results-of-accordion-pill-carbidopa-levodopa-in-advanced-parkinson-s-disease-patients/ (2019). Accessed 3 Oct 2021.
99. Masnoon N, Shakib S, Kalisch-Ellett L, Caughey GE. What is polypharmacy? A systematic review of definitions. Bmc Geriatr. 2017;17:230. https://doi.org/10.1186/s12877-017-0621-2.
100. Morin L, Johnell K, Laroche M-L, et al. The epidemiology of polypharmacy in older adults: register-based prospective cohort study. Clin Epidemiol. 2018;10:289–98. https://doi.org/10.2147/clep.s153458.
101. McCann RM, Jackson AJ, Stevenson M, et al. Help needed in medication self-management for people with visual impairment: case–control study. Br J Gen Pract. 2012;62:e530–7. https://doi.org/10.3399/bjgp12x653570.
102. Farhadyar K, Safdari R, Behpajooh A, Nematollahi I. Assistive medication management system for users with visual impairment. Stud Health Technol. 2018;249:53–60.
103. Villanueva ELD, Tarampi JDG, Cayetano AJM, Linsangan NB. Braille-based pillbox for visually impaired with audio reminder. In: 2020 4rd Int conf electr telecommun comput eng elticom; 2020. pp. 29–34. https://doi.org/10.1109/elticom50775.2020.9230519.
104. Awad A, Yao A, Trenfield SJ, et al. 3D printed tablets (printlets) with braille and moon patterns for visually impaired patients. Pharm. 2020;12:172. https://doi.org/10.3390/pharmaceutics12020172.
105. Drumond N, van Riet-Nales DA, Karapinar-Çarkit F, Stegemann S. Patients' appropriateness, acceptability, usability and preferences for pharmaceutical preparations: results from a literature review on clinical evidence. Int J Pharmaceut. 2017;521:294–305. https://doi.org/10.1016/j.ijpharm.2017.02.029.
106. Eichler H-G, Abadie E, Breckenridge A, et al. Bridging the efficacy–effectiveness gap: a regulator's perspective on addressing variability of drug response. Nat Rev Drug Discov. 2011;10:495–506. https://doi.org/10.1038/nrd3501.
107. FDA. Human factors studies and related clinical study considerations in combination product design and development; 2016.
108. Devices I 210 Q management and corresponding general aspects for medical. IEC 62366-1:2015 Medical devices – Part 1: application of usability engineering to medical devices; 2015.
109. Ishihara K, Ishihara S, Nagamachi M, et al. Age-related decline in color perception and difficulties with daily activities–measurement, questionnaire, optical and computer-graphics simulation studies. Int J Ind Ergon. 2001;28:153–63.
110. Ternik RL. Oral drug product use in the elderly patient population. In: Stegemann S, editor. Developing drug products in an aging society. Cham.: Springer; 2016. p. 225–46.
111. EMA. Guideline on clinical development of fixed combination medicinal products. EMA/CHMP/158268/2017. https://www.ema.europa.eu/en/documents/scientific-guideline/guideline-clinical-development-fixed-combination-medicinal-products-revision-2_en.pdf (2017). Accessed 3 Oct 2021.
112. Notenboom K, Leufkens HG, Vromans H, Bouvy ML. Learning from patients: Identifying design features of medicines that cause medication use problems. Int J Pharmaceut. 2017;517:128–34. https://doi.org/10.1016/j.ijpharm.2016.12.004.
113. Lemke MR, Broky DB. Human factors regulations and standards in combination product development: IEC 62366 and FDA guidance documents. In: Development of biopharmaceutical drug-device products. Cham.: Springer; 2020. p. 741–66.
114. Wiklund ME, Kendler J, Strochlic AY. Usability testing of medical devices. 1st ed. Boca Raton: CRC Press; 2011.

[illegible]

Development of Appropriate Medicines for Older Patients: An Industrial Perspect ive

Kevin Hughes and Charlotte Miller

Contents

Abstract In this chapter, we review the regulatory context behind the development of medicines for the older adult population, how to define the needs of this population and then design a drug product that is appropriate. This patient-centric approach is explored in some detail and discusses various approaches to align patient needs and capabilities with the final drug product. Many factors need to be considered in

---

K. Hughes (✉) · C. Miller
Colorcon Ltd, Dartford, Kent, UK
e-mail: khughes@colorcon.com; cmiller@colorcon.com

M. Orlu, F. Liu (eds.), *Pharmaceutical Formulations for Older Patients*, AAPS Advances in the Pharmaceutical Sciences Series 51,
https://doi.org/10.1007/978-3-031-35811-1_5

dosage form design such as selecting appropriate excipients, taking into consideration cumulative effects, excipient sensitivity and potential pharmacokinetic impact; designing a product that can be easily identified, handled and swallowed by the patient is critical; and size, shape, colour and packaging need to be carefully selected. Another important area that is covered is the need to assess the effectiveness of these design features and ensure they meet patient needs in real-world simulations.

Keywords Older adult · Patient-centric · Dosage form design · Colour · Human factors

1 Introduction

In addressing this topic, it is necessary to consider the needs of the target population, and if they are being addressed by drug products already on the market, the very fact that a book like this is required indicates that this is not necessarily the case. So, before we explore the development of a medicine for older adults, it is necessary to understand the regulatory context that will influence and define the steps that need to be taken, an obvious parallel being the legislation that was required to drive the development of paediatric medicines.

Medicinal drug products are consumed by the entire human population at some point in their lifetime, and the format of the drug product they receive is likely to vary depending on the age or capability of the patient when they receive it. In general, the format of the medicinal drug product will be the same for the majority of the patient population, although this does not necessarily mean that it is an appropriate format for the patient. Two "special populations" of patients that require drug products aligned to their needs are paediatric patients and older adults.

The EMA defines "drug product" as "a pharmaceutical preparation in its container closure system, together with any measuring and administration device and the authorized written user instructions.

(SmPC section 6, package leaflet, product label)". It further elaborates that "Pharmaceutical preparation" is "a formulation in a particular strength/ concentration" or in case of formulations for single use, the labelled container contents, and a formulation as being "the composition of a particular dosage form". It is commonly understood that "formulation" also entails the product appearance although this is not singled out. So, when we are talking about pharmaceutical drug products in this chapter, we are referring to the composition, dosage form, route of administration, dosing frequency, packaging, measuring or administration device and the user instruction of the medicine.

1.1 Paediatric Population

This segment of the population is very different to the adult population, not only in terms of size and capability but also in terms of developmental physiology, and this should be considered during drug product development. Medicines for paediatric patients have been the focus of a lot of attention over the last 20 years, and the lack of appropriate dosage forms for this population has long been a concern for clinicians and caregivers, leading to the manipulation of adult dosage forms (that are not suitable for paediatric patients) to ensure the correct dose is administered or to encourage patient compliance. This need culminated in regulations in the USA [1] and the EU [2], requiring drug development companies to actively investigate the suitability of their active pharmaceutical ingredients (APIs) for the paediatric population. Now paediatric investigation plans (PIPs) or equivalents are firmly embedded in the drug development process, and paediatric patients now have a wider choice of age-appropriate medicines available to them.

1.2 Older Adult Population

This is another section of the population that has different levels of capabilities and needs compared to the general adult population. It is also the segment of the population that is growing the fastest [3] and the main user of medicines. Consideration should be taken to ensure that the medicines they need are appropriate for their needs and encourage patient adherence.

In Europe, the European Medicines Agency (EMA) plays an important role in ensuring that the needs of older patients are taken into account in the development, approval and use of medicines in the European Union. The EMA has developed a Geriatric Medicines Strategy [4], which provides a roadmap to 2015 and aims to consider the needs of the older population. The key features of the strategy are:

Medicine development

- Ensure older adults are included in clinical trials.
- Examine potential tools used to investigate effect and safety in frail patients.
- Include safety and efficacy requirements in EMA and CHMP (Committee for Medicinal Products for Human Use) guidelines as routine.

Medicines evaluation

- Ensure older adults are represented in data collection.
- Assessment reports will specifically include a section on experience in geriatric population.
- Consider older adults specifically as part of post-authorization commitments.

Within this strategy, a Geriatric Expert Group (GEG) has been set up to advise the EMA and CHMP on issues related to the elderly and to provide scientific evidence to developers.

1.2.1 Patient Capability and Acceptability

With people living longer than ever before and life span increasing year on year, it is inevitable that medication uptake will continue to increase. With that said, it has never been more important to ensure medications are designed with a patient-centric approach. Over the past few years, reducing medication errors and improving patient adherence has become a hot topic, with global regulatory authorities issuing guidance to drug developers. Any barrier that prevents a patient from taking medication must be reduced or eliminated to ensure medication adherence.

Approximately 50% of general population is estimated to take their medicines as instructed [5]. Understanding the patient's capability to take a medication is critical and must be considered when designing solid oral dosage forms. This is important for all medicines but especially crucial when developing medicines for the elderly population. The EMA reflection paper on the pharmaceutical development of medicines for use in the older population [6] defines patient acceptability as the ability and willingness of a patient to self-administer or of any lay or professional caregiver to administer a medicinal product as intended. If patient acceptability is not considered, it is likely to have a significant impact on adherence. Several product characteristics influence patient acceptability in older people and include:

- Route of administration.
- Type of dosage form.
- Site of application or administration.
- Appearance.
- Swallowability.
- Palatability.
- Patient perception.
- Dosing regimen and complexity of dosing instructions.
- Shelf life and storage conditions.
- Ease of use.

To ensure sufficient acceptability, this means that a company should identify the patient needs, evaluate if the product is accepted by the patients for which it has been designed and justify that the achieved level of patient acceptability corresponds with the level of risk involved. The expectation from the authorities is that evidence of this activity should be included in the dossier submitted for marketing authorization.

1.2.2 Swallowability

It is well documented that oral medications continue to be the most popular administration route, due to their ease of use and manufacture. Although solid oral dosage forms are convenient for the patient, they do pose challenges, particularly for the elderly and those suffering from illnesses that inhibit swallowing (dysphagia is the medical term relating to swallowing difficulties).

Firstly, it is helpful to understand the functional processes involved in the swallowing process. Sven Stegeman et al. summarize this very well in their review. Swallowing dysfunction and dysphagia is an unrecognized challenge for oral drug therapy [7]. The process can be broken down simply into three steps, the oral stage, the pharyngeal stage and the oesophageal stage.

During the oral stage, food is chewed and mixed with saliva to form a bolus (ball), and the tongue pushes the bolus toward the pharynx-stimulating receptors, which start to trigger the swallowing response. The pharyngeal stage initiates the swallowing process by stopping the breathing, raising the larynx and preparing the bolus to travel. Peristaltic muscle movement transports the bolus into the oesophagus through the cricopharyngeal sphincter. In the oesophageal stage, the bolus moves downward through the lower oesophageal sphincter by peristalsis into the stomach.

As we age, our swallowing ability declines. This may be related to anatomical, physiological, motoric or sensory changes, but very often disease plays a role in inhibiting swallowability.

1.2.3 Disease State

Several diseases have been identified, which cause dysphagia. These include but are not limited to Alzheimer's disease, Parkinson's disease, multiple sclerosis, osteoarthritis, diabetes and some cancers [8]. Since many of these diseases are disproportionately inflicted on the elderly population, it is important to evaluate the disease state, as well as considering the condition being treated.

Age-related vision changes naturally occur as people age, and accordingly older patients are far more likely to suffer from impaired vision. Therefore, consideration must be given to these patients to minimize the risk of medication error. Clear labelling and the use of braille format should be used where possible.

Using unique shapes, colours and logos adds another level of protection [9]. If the product is specially targeted to patients that are likely to have visual impairment, risk assessments should be used to understand the risk.

1.3 The EMA Strategy

One of the key aspects of the EMA vision for older adult medicine is to ensure that medicines used by older adults are of high quality, well studied and evaluated throughout the product life cycle to ensure suitability for this population.

This strategy determined that within the existing regulatory framework [10, 11], there was already sufficient provision to ensure that "In all cases, the product should be designed to meet patients' needs and the intended product performance" [12], i.e. the European Commission do not consider that a specific regulation is necessary to drive the development of appropriate medicines for older adults as this requirement is already there. In order to achieve this goal, it is necessary to ensure that the regulatory framework includes considerations for older adults as it is reviewed and updated, so the EMA and GEG actively commented on guidelines, concept papers and reflection papers, to ensure that where older adults make up a significant portion of the target patient population, the safety and efficacy requirements of this age group are addressed. In the period from 2011 to 2013 [13], 28 guidelines were identified as relevant to the older adult population. These were reviewed, and in 93% of cases, comments were submitted and taken into consideration before adoption of the final document versions. Not all comments were adopted in the final versions, and it was recognized that it was necessary for the GEG activity to take place earlier in the document review process.

The final phase of the EMA strategy was the development of a reflection paper [6], with the first draft being published for consultation in 2017. Input from industry was encouraged through presentation of the paper at two conferences: The APV Graz meeting Medicines for Older Adults: Getting Prepared for the Scientific and Regulatory Evolution (March 2017) and the DIA Basel Chemistry Manufacturing Controls (CMC) workshop on Medical Errors Due to Product Design and Development (June 2018). Stakeholder feedback revealed a common agreement on the importance of adapting drug development and pharmaceutical product design to the needs of older patients to support medication safety and efficacy [14].

The authors of the paper were keen to learn lessons from the regulations and guidance around the provision of paediatric medicines, avoid repetition of data already included in other regulations and follow an overall patient centric approach. A gap analysis was carried out on how existing marketing authorizations were not fully meeting the needs of elderly patients [6], which identified that tablet breaking, enteral feeding tubes and packaging were three of the main issues with existing drug products not meeting the needs of elderly patients.

1.4 The FDA Strategy

Since the 1980s, the FDA has long established guidance on labelling of medicines for older adults to ensure that there is adequate information provided to healthcare practitioners [15]. This also encourages the thorough examination of age-related differences in efficacy, dose response and safety. There is a requirement for new drug applications (NDA) [16] and a recommendation for biologics application holders (BLA) [17] to present their applications' effectiveness and safety data based on age. In 2012, the FDA announced [18] the availability of a new guidance [19] prepared under the auspices of the International Conference on Harmonisation (ICH) of Technical Requirements for Registration of Pharmaceuticals for Human Use. The Q&A guidance addresses the need to design clinical trials for drugs, which are likely to have a significant use in the elderly, in such a way that the needs and capabilities of the target patient population are addressed. The Q&A provides guidance on the use of older adult data to adequately characterize and represent the safety and efficacy of a drug in a marketing application including data collected after launch. Particular emphasis is placed on the increasing size of the geriatric population and how geriatric patients can respond differently to drug therapies compared to younger patients, particularly after the age of 75. Despite this, there is still a long history of the exclusion of older adults from clinical trials. To address this, the National Institutes of Health (NIH) instituted the Inclusion Across the Lifespan policy requiring the inclusion of older adults in clinical trials [20]. There is still progress to be made in this area, for example, Helfand et al. showed that older adults are likely to be excluded from more than 50% of COVID-19 clinical trials and 100% of vaccine trials [21]. It is recognized that some clinical trial exclusions will be essential to protect older adults from severe or uncontrolled comorbidities; however, a general exclusion based on age limits the ability to evaluate the efficacy, dose and adverse effects of intended treatments in the older adult population that is the focus of the medication and is the one that is at greatest risk of serious illness from the COVID-19 coronavirus.

2 The Principle of Patient Centricity

Designing a drug product to meet the needs of the older population requires an understanding of both the requirements of that population and the characteristics of the proposed drug product and how the two interact. Historically, this has not always been taken into consideration during drug product development, where time and cost pressures may steer the formulators to tried and tested traditional formulations, which may not always meet the patient's need.

The provision of healthcare is heavily influenced by the need to control costs, and this is even more critical in the context of the COVID-19 coronavirus pandemic, which started in 2020, and has put great strain on healthcare provision around the world. Unfortunately, the cost-saving measures that are adopted do not necessarily provide improvements in the level of care to patients [22]. The focus on cost control may deliver healthcare solutions that do not always fit the needs of the patients [23] and ultimately could result in additional costs where the patients are not comfortable with the options available. One example of this is the need for compounding of unsuitable solid oral dosage forms to ensure compliance in patients with dysphagia. An estimate of the costs related to patients' non-adherence to prescribed therapies is approximately $100 billion per annum [24]. Drumond identifies that to address these issues and reduce healthcare costs, fundamental changes are needed in how healthcare provision is administered to patients, which could be achieved by the implementation of patient-centric healthcare systems [25]. The key is to find the balance between the two, where innovation and affordability come together to deliver patient-centric drug products.

A patient-centric approach involves the matching up of the needs and abilities of the patient with the appropriate therapy or drug product, and this approach is reflected in the general strategies of both the FDA and EMA and described in the EMA reflection paper. It sounds simple, but a deep understanding of the patient's capabilities, a focussed patient-centric drug development approach and buy-in from healthcare providers are necessary to deliver a truly patient-centric pharmaceutical drug product. The resulting drug product should contain features, which intuitively leads the patient to use it effectively and as intended, thereby preventing adherence issues or administration errors.

However, whilst "patient-centric" is increasingly used in the scientific literature in a wide variety of contexts, a harmonized definition of patient-centric pharmaceutical drug product design has not yet been established. In a White Paper from Stegemann et al. from 2016, this concept is explored in detail and presents a first draft definition, resulting from discussions held by an interdisciplinary AAPS focus group of experts [26]. The older adult population in general can be broken down further into special populations, which may have different skills and capabilities and need to be taken into account during development, for example, the very old, frail and multi-morbid, the long-term cancer survivor and the cognitively impaired (dementia). The ability of patients to adhere to a recommended therapy is likely to require a specific set of skills and capabilities, and if this ability is lacking, then patients are likely to have difficulty in taking the medicine, resulting in inappropriate drug use, improper administration and poor adherence (which may culminate in the drug product simply not being taken). In many cases, patients will rely on support from caregivers to manage their medications, but even here, the management becomes more demanding as the number of drug products, dosage forms and dosing regimens increases. It is important to obtain feedback from these patients and caregivers on how they interact with their medication, in order to ensure that the dosage form design is appropriate.

2.1 *Definitions*

The Stegeman White Paper starts by setting out a set of definitions of terms, which are commonly used but do not necessarily mean the same thing. In addition, a full-sentence definition is set out to define what *patient-centric pharmaceutical drug product design* means and to distinguish it from *patient-centric* and *pharmaceutical drug product design*:

Pharmaceutical drug product design: The design of the comprehensive presentation of the therapeutic entity to the end user (patient/caregiver/health care provider) including the type of dosage form; formulation; dose; dosing frequency; primary, secondary and tertiary packaging; medical device; dosing devices; instructions for use (as in the SmPC/PIL/product label); and other authority-approved patient support tools and programs.

Patient centricity: The recognition of the needs of an individual patient or distinct patient populations and their specific needs as the focal point in the overall design of a medicine including the targeted patients' physiological, physical, psychological and social characteristics.

Patient-centric pharmaceutical drug product design: The process of identifying the comprehensive needs of individuals or the target patient population and utilizing the identified needs to design pharmaceutical drug products that provide the best overall benefit to risk profile for that target patient population over the intended duration of treatment.

When taken into account, the definitions above ensure that the drug product-patient/caregiver interaction is given the appropriate level of attention and will enable the resulting medication to be taken intuitively as intended and prevent medication non-adherence and medication errors. Table 1 lists the main product and patient characteristics, which need to be considered for a patient-centric pharmaceutical drug product design process. Of particular importance is how these two sets of characteristics interact and impact the acceptability of the drug product to the patient.

2.2 *Patient and Drug Product Interactions*

2.3 *Considerations for Patient-Centric Drug Product Design*

The Stegeman paper also highlights that patient-centric pharmaceutical product design cannot solely be based on theoretical data but must examine the patient/caregiver response to the resulting medication. This type of testing is highlighted in several FDA guidance documents and requires the drug product design to be evaluated with the target patient population in their own individual circumstances (i.e. hospital environment with supervision or an unsupervised home environment) [27, 28]. This can be achieved in several different ways, such as conducting usability studies in a domestic setting or in schools, nurseries, etc. The FDA uses the "patient-centric" terminology in conjunction with "patient-reported outcomes", where the intention is to obtain patient feedback on the benefit to risk ratio of a new drug substance under development and the "value" to the patient [29]. Having the patient directly involved in the drug development process and providing feedback on the product design from interacting with it in their own environment (Human Factors Trials) will help identify and risk assess potential sources of non-adherence and medication errors and thereby enable a truly patient-centric medication.

The Stegeman white paper advocates the documentation and examination of the major product- and patient-related characteristics (Table 1) to understand and anticipate how they will affect the patient. Utilization of this approach will allow the formulator to identify specific design drivers derived from a deeper understanding of the target patient population than would perhaps traditionally be the case. As is the case in most formulations, the relationship between the drivers and the patient needs is unlikely to be a direct match; multiple drivers may affect each need and vice versa (Table 2). When this conflict among design elements occurs, decisions

Table 1 Summary of major drug product-related and patient-related characteristics for consideration in a patient-centric pharmaceutical drug product design and development process

Product-related characteristics	Patient-related characteristics
Type of drug substance (drug profile, desired effects, side effects, therapeutic Window, mechanism of action) Dose to therapeutic effect (required dose range for different patient populations; Need for small or larger dosing increments, dose flexibility) Route of administration (oral, nasal, rectal, parenteral, etc.) Type of dosage form (tablet, capsule, solution, injection, etc.) Formulation characteristics (immediate release, modified release, tablet Size/shape/colour/strength/concentration/ excipient composition, etc.) Product strength/concentration Packaging (primary, secondary, tertiary) Dosing and administration devices (oral syringe, inhalator, etc.) Product recognition, identification (appearance unpacked product and in its primary, Secondary or tertiary packaging name, name, etc.) Instructions for use (SmPC, PIL, labelling, others) Specific dosing requirements (e.g. before meal, not together with drug x, etc.) Product stability Storage conditions In-use shelf life and storage conditions Preparation steps to administration Compatibility with diluents and other medicines where appropriate Compatibility with food or drink Industry-verified manipulations of the dosage form	Developmental stage/patient age Newborn, toddler, elderly Organ and body functions (kidney, liver, lung, etc.) Social-emotional development Way of living (alone, with spouse, with parents) Access to caregivers PK/PD and physiological changes (e.g. receptor density, kidney clearance, etc.) Visual impairment Near acuity Visual field Blindness Loss of hearing Swallowing impairment Safe solid dose swallowing Dysphagia Poor hand sensitivity Control of movement Control of strength Sensing surfaces or temperature Control of eye-hand coordination Motoric impairment Grip strength Pinch strength Manual dexterity Arm mobility (e.g. lift above head) Difficulties walking Bedridden Cognitive impairment Mild dementia, memory loss Information processing speed (hearing, understanding) Health literacy Dentition Psychological issues Adherence Negative perception Depressive disorders Disease state, comorbidity Experience of disease Disease cluster Disabilities End point of disease progression

Reproduced unchanged from Stegemann et al. [26] (Creative Commons — Attribution 4.0 International — CC BY 4.0)

Table 2 Summary of major items considered as design drivers, their related design inputs and respective patient needs

Design drivers	Design inputs	Design outputs
Characteristics disease/condition	Disease-specific expression Multi- and comorbidity Frailty Disease severity/burden Disease stage	Individual drug/drug combination Individual drug dose accuracy Dose range Disease-specific disabilities
Characteristics drug substance/physiology	Developmental stage (maturation, Declining body functions) Oro-oesophageal and GI transit Permeability Fat/water ratio Drug metabolism and clearance Homeostasis Reserves	Flexible dose adjustment Appropriate dosage form Excipient safety/total amount of excipients Low adverse drug reactions Low intake/administration frequency
Characteristics drug therapy	Need for different types of dosage forms Availability of combination products	Reimbursement Dose tracking
Characteristics drug product	Multiple and polypharmacy Therapeutic complexity Prescription guidelines Different dosage forms Possibility of product modifications (manipulations) Range of trademarks	Simplified regimen Appropriate dosage form Drug product identification Drug product recall Dosing frequency Dosing moments Dispensing, substitution and re-substitutions Reimbursement Dose tracking
Patient characteristics	Age, gender, socio-emotional development Mobility (travel) Perceived wellbeing Functional limitations (motoric, sensory, Cognitive) Health literacy (disease/ therapy Understanding) Dehydration/malnutrition Motivation psychological traits Remaining life time Living alone or with others Daily occupation (work, school) Social support and interaction Stress resistance	Usability/ergonomics Self-explaining/intuitive use Drug product information Product identification Swallowability Palatability (taste, smell, texture) Reminder Dosing frequency/moments Least number of units/drug products Feedback/communication/motivation

(continued)

Table 2 (continued)

Design drivers	Design inputs	Design outputs
Medication management (adherence and administration)	Intended and non-intended non-adherence Therapy simplification Pill boxes and compliance aids Hoarding Environment where the medication needs to be prepared and taken Access/cost Co-medications and changing (generic) prescriptions	Identification/differentiation outside packaging Dosing frequency/moments In-use stability (e.g., external "pill box," airport scanning) Ease of storage Convenience of use (e.g., specific requirements like before breakfast) Use discretion Portability Food effects Refill reminders Harmonized labelling, naming elements, product elements (e.g. packaging) or cue tags
Usability (handling, storage And disposal)	Drug product appearance Drug product shelf life Drug product storage Drug product disposal Drug product packaging Dose measurement and preparation Dosing frequency Need for administration device Ability to self-administer with ease Need for help from caregiver Learned usage/experience	Handling issues (e.g., round tablets roll off the table) Ergonomics Formulations enabling (easier) self-administration Refrigeration requirements Stability during use period Minimize waste (dose form, packaging) Mechanical stress stability of the product "Predicted usage" (modification by patient)

and compromises will be necessary, and these should be evaluated on a risk-benefit approach for the formulation as a whole. Several prototype variations can be prepared and evaluated in Human Factor style trials that examine the drug product and patient/caregiver interactions in the relevant environment, which provides the essential feedback loop that allows the formulation to be optimized for the patient. This stage will take time, but a thorough understanding of the product/patient characteristics and the resulting design drivers allows a precise patient-centric target product profile (PCTPP) to be defined, which should mean that the prototype formulations will be closer to optimal for the first human factors trials, than if this evaluation had not been undertaken.

2.4 Human Factor Trials

Human factor trials are more well-known from the development of medical devices, where they are used to evaluate the interaction of the patient with the medical device. The purpose and intention of human factor engineering trials used for medical device development are aligned with the information required for patient-centric pharmaceutical drug product design, and the trials can easily be adapted to give valuable insight to this development process. Human factor trials and clinical studies have different objectives, and for this reason, inclusion of a human factors study in a clinical trial setting is not usually recommended (Fig. 1).

Human factors engineering (HFE) is the terminology used in the USA (also known as usability engineering (UE) in other countries). This type of approach has the advantage of generating more patient centric information than a traditional clinical trial, which may only return subjective feedback:

- Observed performance assessment of users interacting with the drug product in relevant environments.
- Full assessment of labelling comprehension.
- Identification of product-related medication errors or potential issues that could affect drug product safety or effectiveness.
- Interview to establish root cause for use errors.
- Greater ability to stratify distinct user groups (i.e. by ability or frailty of patients or healthcare role nurses, pharmacists, etc.)
- Helps define end use training requirements.
- Helps validate that PCTPP is met.

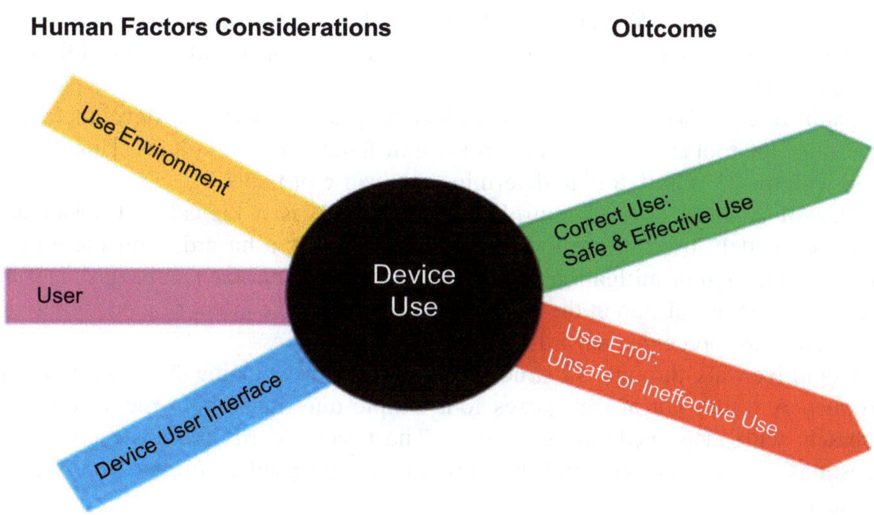

Fig. 1 Interactions among HFE/UE considerations in either safe and effective use or unsafe and ineffective use. (Reprinted from "Applying Human Factors and Usability Engineering to Medical Devices. Guidance for Industry. Food and Drug Administration Staff", 2016 [27])

The EMA [30], FDA [28] and industry [31] provide some recommendations and discussion on how human factors testing can be carried out using tools such as:

- Failure mode effects analysis (FMEA).
- Simulated use testing.
- Fault tree-analysis (FTA).
- Hazard analysis.
- Cognitive walk-through.
- Perception-cognition-action (PCA).

For drug products delivered via a medical device, the relevant harmonized standards are EN 14971 Application of Risk Management to Medical Devices and ISO/IEC 62366: Medical Devices – Application of Usability Engineering to Medical Devices. Methods and standards are also available for the assessment of packaging: Technical Standards for testing the ease of opening of consumer packaging (CEN/TS 15945) and Child Resistance (ISO 8317:2015).

The Institute for Safe Medication Practices (ISMP) has issued guidance on the principles of conducting failure mode effect analysis (FMEA) [32]. The principle is to analyse all potential sources of medication error before they occur and the situations in which they may occur and potential causes. These are then rated for effect on the patient, severity and probability of occurrence and are scored accordingly. Once these failure modes are identified, measures can be put in place to eliminate or mitigate the failure.

Simulated use testing is the evaluation of a product or medical device in a laboratory where the test methodology is used to replicate the patient interaction with the device or product. A useful comparison of actual vs simulated use in human factors testing can be found on the Chartered Institute of Ergonomics and Human Factors website [33]. Written in 2014, this article already alludes to the increasing trend of regulators requesting human factors data based on actual use, in addition to simulated use.

Fault tree analysis (FTA) is a deductive process whereby developers can determine the root cause or human errors for different types of failures. This is often used with the "5 Whys" tool to determine root cause of faults.

Hazard analysis critical control point (HACCP) is a logical and systematic approach widely used in the food industry to identify a hazard, with the aim of improving design or mitigation steps to eliminate or minimize that hazard. Hazard analysis is the initial step in the HACCP process, with the intent of identifying and prioritizing the perceived hazards in the product or process.

Cognitive walk-through is a structured approach to evaluating the usability of the product. A tester, not a user, poses four simple questions about the way a user interacts with their medicine or device. The tester records the answer to these questions in their opinion, and the observations are used to further improve the product.

Perception-cognition-action (PCA), which are the main categories of use errors, can be carried out early in the development process. *Perception errors* may involve failures to see visual information or to hear auditory information; *cognition* errors fall into three categories: memory failure, rule-based failure and knowledge-based failure; *action errors* are caused by events such as failure to reach controls, i.e. using the device or equipment incorrectly.

The consumer healthcare industry does have a history of involving consumers and patients in the drug development process through patient testing and consumer surveys [34, 35] and is constantly gathering feedback on products and services to improve the consumer experience. Traditional consumer-centric approaches for pharmaceutical development may need to be adapted and enhanced to ensure that more attention is paid to the specific needs of the patient population. It is widely recognized that quality cannot be tested into a product, so therefore, before development starts, quality by design (QbD) principles establish the requirement for a defined quality target product profile (QTPP) to ensure that full product and process understanding is achieved and a consistent product is delivered with a pre-defined performance. This approach is widely used [36] and endorsed by the regulatory authorities, which set the regulatory precedence for patient-centric dosage forms; it simply needs to be re-enforced and refined to ensure that a patient-centric target product profile (PCTPP) is defined to meet the needs of the target patient population and documented ahead of product development. The authors propose a potential patient centric pharmaceutical drug product design (PCPDPD) pathway (Fig. 2), where all of these factors come together to aid the development of a truly patient-centric drug product.

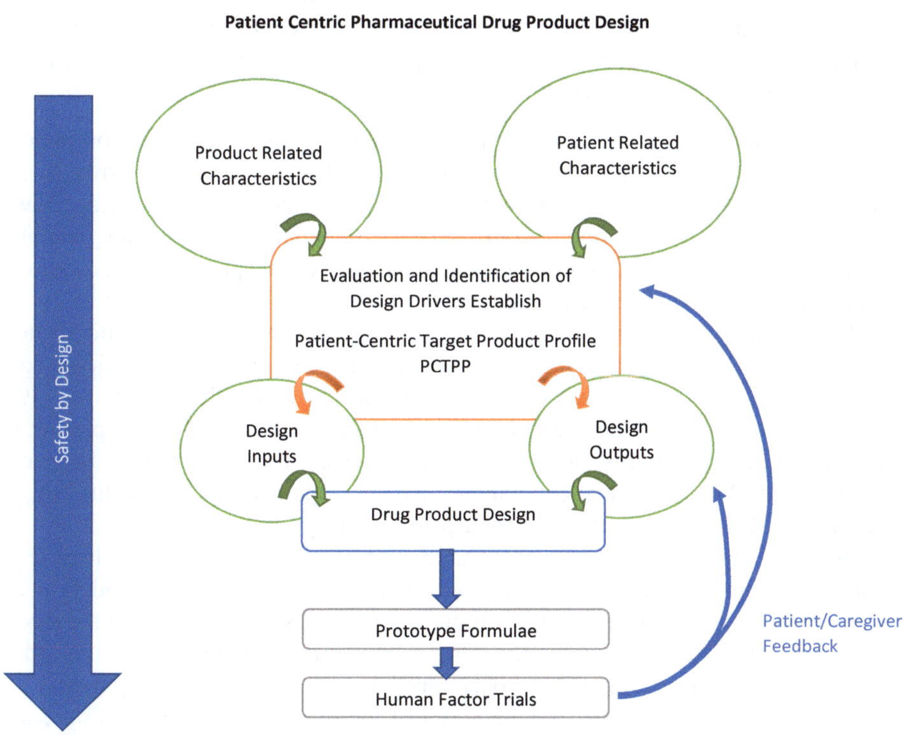

Fig. 2 Patient-centric pharmaceutical drug product design

3 Drug Product Design, Appearance and Identification

So far, we have reviewed the various guidelines and strategies of different agencies and the work carried out by other authors, which has provided the basis for a systematic review of the patients' capabilities and needs. What we will now explore is how to address these different requirements in the drug product design; the expertise of the authors is in solid oral dosage forms, which are also the format that 65–70% of drug products are prescribed to elderly patients [37]. Therefore, the focus will be on solid oral dosage forms, which do appear to be the drug product format that needs most attention to make it truly patient-centric [38]. There are still many other dosage forms that are applicable to older adult patients, which form a vital part of their healthcare provision, e.g., parenterals, liquids, topicals, etc.

The EMA has provided some definitions to help clarify the terminology that is used interchangeably when they do not necessarily mean the same thing:

"Drug product": "a pharmaceutical preparation in its container closure system, together with any measuring and administration device and the authorized written user instructions (SmPC section 6, package leaflet, product label)".

"Pharmaceutical preparation": "a formulation in a particular strength/concentration" or, in case of formulations for single use, the labelled container contents and a formulation as being "the composition of a particular dosage form"; commonly understood that "formulation" also entails the product appearance.

"Pharmaceutical design of a drug (medicine)": "the composition, dosage form, route of administration, dosing frequency, packaging, measuring or administration device and the user instruction of a medicine".

The appearance of a solid oral dosage form has a significant impact on patient adherence. If a tablet looks large and unappealing, this could result in doses being missed, or worst of all, the patient may stop taking the medication altogether. Easy identification is also important, not only for the patient but also for the caregiver and pharmacist. Since tablets can become detached from external packaging, it is critical that tablets themselves are easy to identify.

Unfortunately, there have been, and continue to be, many cases of medication errors. The *British Medical Journal* reports more than 237 million medication errors are made in England every year, resulting in the death of more than 1700 lives [39].

As older patients may need to take multiple medications daily, it is important to incorporate design features into dosage form to ensure that tablets do not become misidentified. Pharmaceutical companies have the responsibility, and ability[29, 30], to use shape, colour and logo to make their product easily identifiable.

Logos, whether debossed or printed onto a tablet, should be clear and legible to enable easy identification. In industry, debossing tends to be the most widely used method due to its ease of use, as the identifying mark is simply applied to the tablet tooling. Identification markings can also be applied by printing, but this adds another manufacturing step to the process and requires specialized equipment.

There are of course limits to the amount of lettering that can be fitted onto the small area of a tablet. From a design perspective, less, as they say, is more. Detailed logos pose challenges during manufacturing and could potentially be positioned in areas that are prone to erosion, resulting in lettering becoming illegible, which would be problematic for older patients with impaired vision.

One advantage of printing onto tablets and capsules is that it enables manufacturers to use more detailed logos, increasing the possibilities to improve identification and eliminate ambiguity. Printing can also be useful for tablets with breaklines since both halves of the tablet could be printed with the same logo to identify each half of the tablet. Applying printed images onto tablets or capsules, for example, a heart for cardiovascular medicines, can also remind the patient what the tablet is intended for.

3.1 *Size*

It's a common misconception that small tablets are preferred by patients, but this isn't always the case, especially with regard to handling. Daanish, M (2014) [40] found that larger tablets were preferred since it made them much easier to handle. A study by Notenboom, K. et al. [41] noted one patient found the tablet so small they had difficulty locating the product in the mouth, making it difficult to know if they had swallowed their tablet.

Size is the main characteristic that impacts swallowability, not just in terms of functionality but also perception. Tablet shape and dimensions must be carefully considered to ensure the dosage form appears as small as possible. The FDA guidance issued in 2015, "Size, Shape, and other Physical Attributes of Generic Tablets and Capsules", recommends keeping the largest dimension below 21 mm, and elongated oblong type shapes are preferred for larger tablets (typically over 500 mg tablet weight) [42]. Computer-aided design software can be helpful for optimizing tablet dimensions, enabling the formulator to choose the design that is perceived to look the smallest.

Schiele et al. reported that round tablets of 8 mm in diameter caused swallowing difficulties in patients, and for oval and oblong shapes, the length of tablet that caused patients difficulties in swallowing was 15 mm and 16 mm, respectively [43].

Through a series of interviews with patients, Stegemann et al. [44] reported that a small white round tablet did not capture attention and required a controlled attention mode to differentiate from other distracting small shapes. Smaller tablets (6 mm round) were described as difficult to see and often led to unintentional non-adherence [45].

3.1.1 Handling

As we age, our fine motor skills decline, and the simple process of picking up and taking a tablet or capsule may become more difficult. Symptoms such as tremors and stiffness can also pose significant challenges, particularly for those suffering with illnesses such as Parkinson's disease [46]. If the patient cannot easily handle the medicine, they may not be able to administer their medication, and treatment stops.

There are specific shapes and design features that can be applied to improve tablet handling. Flat tablets are best avoided since these are difficult to pick up from a flat surface. By adding a curvature or rounded point to the tablet, it sits proud on a flat surface, making the tablet easier to pick up. Known as Tilt-Tabs, GlaxoSmithKline's Parkinson's drug Requip® is an example of this type of design. Images of these tablets can be found easily online.

3.2 Tablet Shape

Tablet shape selection is important from a patient handling perspective, but it can also make the medication more memorable.

When asked, people tend to prefer traditional tablet shapes, such as round, caplet or oval [47], probably because they are familiar. There is no issue with using a conventional shape, but colour and logo become more important to make the tablet identifiable and distinguishable. As well as aiding identification, a unique tablet shape can strengthen branding and make the medication more memorable.

From a manufacturing perspective, sharp points should always be avoided as they have the potential to be weak and susceptible to breakage/erosion during compression, coating and packaging. Additionally, they can make swallowing challenging as even before the tablet is taken, the patient may perceive pointed shapes as difficult to swallow.

Flat sides/edges should also be minimized, as these can also pose manufacturing challenges (particularly during film coating, where tablets can become stuck together (twinning) along the flat edges). Flat edges have also caused challenges during swallowing; a study by Bio-Images Research and Colorcon demonstrated that coated oval/oblong shapes are the easiest to swallow when compared to a caplet-shaped tablet [48]. It is believed that the large flat sides have greater surface area in contact with the oesophagus and can therefore become stuck. In the same study, hard and soft gelatine capsules also became lodged in the oesophagus, typically because of their sticky gelatinous nature, which in turn adheres to the throat.

3.2.1 Tablet Splitting

Breaklines, or score lines, are typically applied to aid splitting of tablets to facilitate swallowing or to modify the dosage. Most patients prefer to receive a tablet of the prescribed dose rather than having to split their tablets, but in reality, this isn't always possible. There are pros and cons associated with tablet splitting, but for older patients, the main challenge may be difficulty in breaking the tablets or breaking them into equal parts.

In practice, it is difficult to ensure 50% of the dose is contained in half a tablet when it is broken by the patient, but each part of the tablet should comply with the content uniformity requirements of the European Pharmacopoeia. Improvement in the provision of information about splitting of tablets is necessary to avoid potential medication errors [49]. The resulting variation in drug uniformity may have serious implications for the patient, including overdose or loss of efficacy due to underdosing. This is especially important for drugs that are dose critical or have a narrow therapeutic index [50]. It is no surprise that after splitting, dose variation is greater for non-scored tablets than scored tablets [51].

From personal experience, manually splitting a tablet typically results in sharp edges or points, and these pieces not only appear daunting but can actually be difficult to swallow. The patient is also more likely to taste the drug since the film coating (if applied) has been compromised and the tablet core exposed. Large round shapes that have been broken can become very difficult to swallow. An oblong shape with breakline may be easier to swallow once broken, since it typically has a smaller surface area at the break point (Fig. 3).

Several breakline designs are possible, and commonly used examples include standard bisect, cut-through bisect and decreasing bisect (Fig. 4). The breakline must not weaken the tablet, causing it to break under the slightest pressure, and must be able to withstand the pressure applied to remove the tablet from the blister pack. It must also be robust enough to undergo further manufacturing steps, such as coating and packaging, but it should not be so hard that patients have difficulty breaking the tablet.

It should be pointed out that splitting of modified release tablets (i.e. delayed release or extended release) is dangerous, as this activity removes the ability of the formulation to control release of the API and can lead to dose dumping from the compromised tablet, which may harm the patient. If pharmaceutical manufacturers want to prevent patients from breaking tablets, the obvious solution is to avoid using

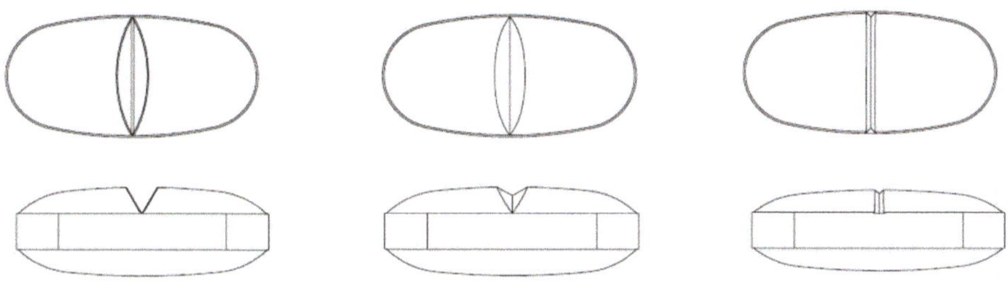

Fig. 3 Commonly used breaklines. (Image: from Colorcon library)

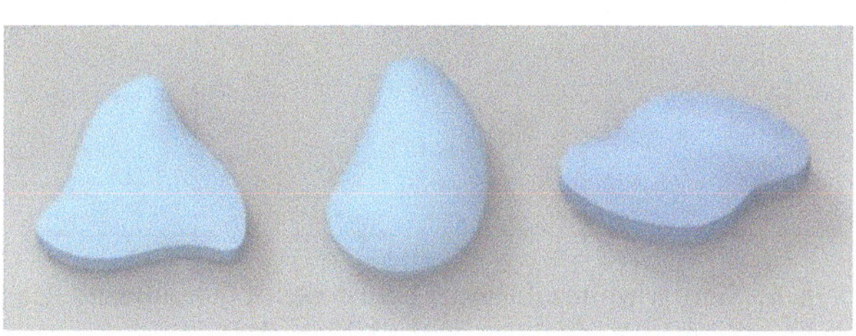

Fig. 4 Examples of asymmetrical tablet shapes. (Image from Colorcon library)

a breakline, but it's surprising how many tablets include a breakline even when breakage is not recommended, for example, on enteric coated tablets. The use of asymmetrical tablet shapes for smaller tablets could be advantageous in this respect, as this makes splitting tablets in half particularly difficult and could act as a deterrent (Fig. 4).

3.3 Colour

The power of colour in marketing and branding has been well documented for many years. It is understood that colour can increase brand recognition by up to 80% [52] and influences our perception, behaviour and expectation. Significant time is often spent on packaging design, especially for over-the-counter products where the patient or consumer is faced with a wide array of options to choose from. Unfortunately, many solid oral dosage forms are still white, and whilst branding is not so important for prescription medications, colour still plays a significant role in patient acceptability and adherence.

Older people typically take multiple medicines daily, and colour can make a tablet memorable and potentially prevent the patient from taking the wrong medication. If all tablets are white, this could easily confuse the patient, whatever their age.

The images below show two examples of very similar tablets. Figure 5 shows tablets with very similar drug names and appearance, whilst Fig. 6 shows two different doses of the same product.

Clobazam 10 mg tablets	Clonazepam 500 microgram tablets
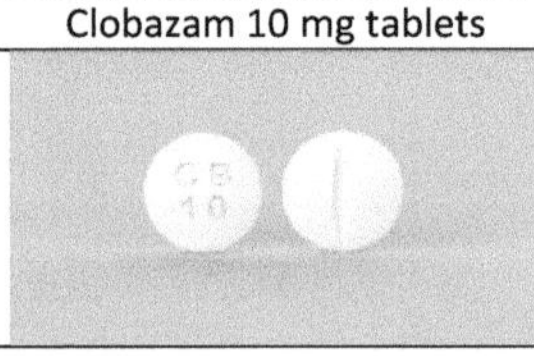	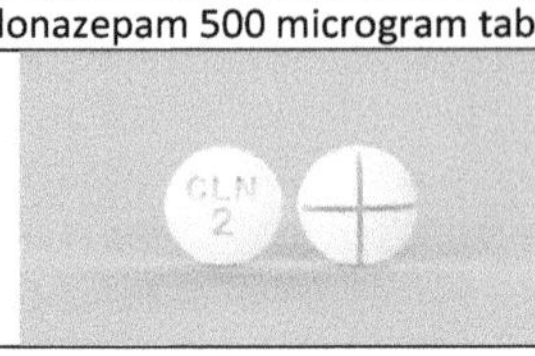

Fig. 5 Similar white round tablets with similar drug names. (Images from Colorcon library are representative of tablets on the UK market)

Amlodipine 5mg	Amlodipine 10 mg
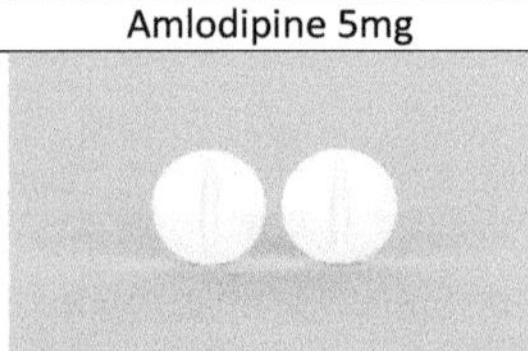	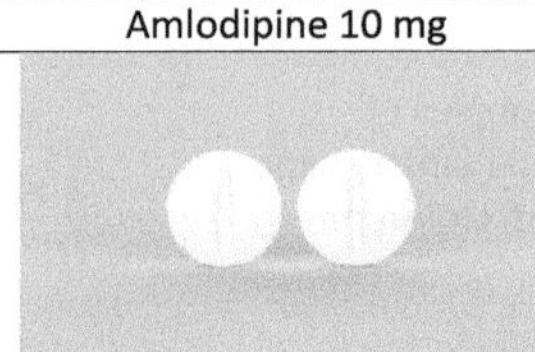

Fig. 6 Similar white tablets for both doses of the same medication. (Images from Colorcon library are representative of tablets on the UK market)

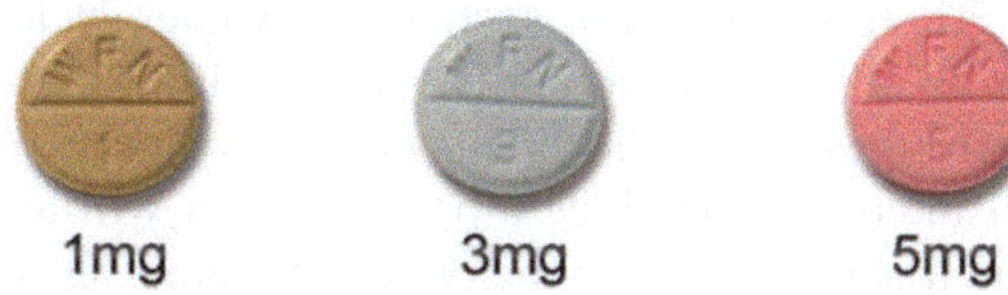

Fig. 7 Differentiation between doses using colour and logo. (Image: European Medicine Agency [30])

Both the FDA and EMA have promoted the importance of colour when designing medicines [28, 30]. Colour can be used to help the pharmacist, patient and/or caregiver easily identify the tablet or capsule. It is also particularly beneficial when used as a differentiator to help distinguish products with multiple doses (see Fig. 7).

The use of colour should also be considered to ensure it does not pose a risk of confusion, especially with other established products. Shariff, Z. et al. [45] reported that healthcare professionals referred to the importance of brighter colours, particularly for elderly patients, due to visual afflictions such as macular degeneration or cataracts.

Anecdotal evidence has shown that some patients do not care what colour their tablet is, as long as it's effective; however, colours can influence a patient's perception and expectation before they even take the tablet or capsule. Wan, X. et al. [53] concluded that differing the colours and shapes of pharmaceutical tablets sets up significantly different expectations, which likely play an important role in the patient experience. A systematic review by de Craen, A. et al. [54] found that red, yellow and orange are most associated with a stimulant effect, whilst blue and green are perceived to have a tranquilizing effect.

Colours are also associated with different tastes [55]. Results of online and in-lab tests have revealed that most people match sweet tastes with red and pink, whilst bitterness is associated with brown-black and violet/purple. Taste masking is discussed in more detail later in this chapter.

After a branded prescription medicine loses market exclusivity, generic prescription drugs become available. Although generic drugs must be bioequivalent to the branded drug, there is no requirement for them to look the same. Innovator companies may also trademark the trade dress (visual appearance) of the medicine, restricting generic companies from copying the colour, shape and size. This in turn can cause confusion for patients. Kesselheim, A. et al. [56] reported such differences between generic products with different physical characteristics may impact patient care by causing confusion, resulting in non-adherence or prescription error. This can be resolved by conducting in use trials, education of the patient and communication at the pharmacy.

It's clear that acceptable differentiation can vary between regulatory authorities and assessors. The best way to ensure the product is well differentiated is to prove safety by conducting in-use testing and risk assessments as detailed earlier in this chapter. Three-dimensional printed composite tablet models are a safe way of conducting human factor trials. The models are exact shape, size and colour replicas of how the final design might look. Patients can handle (but not swallow) the models to access the ease of handling, the perception of size, ease of identification, etc.

There are many cases where companies decide to produce each dose in a different shade of the same colour (i.e. light yellow, mid-yellow and dark yellow). Although each dose is a different colour visually, this approach could be open to misinterpretation since a patient leaflet will usually list the appearance as yellow and it is not known to the patient or caregiver that there are two other yellow doses available. For the same reason, colour must also be considered at the manufacturing site as similar colours can cause confusion during quality control.

3.4 Film Coating

Although some pharmaceutical companies may choose not to coat tablets, this could prove a costly error in the long term. From an aesthetic point of view, uncoated tablets are usually dusty or powdery, have a rough tablet surface and may even be perceived as poor quality. Coating is a relatively easy way to improve the appearance of any tablet and at the same time ensure it is easily identifiable.

Coating a tablet adds minimal cost but, from a manufacturing perspective, reduces the amount of dust present, which in turn decreases cleaning burden and minimizes exposure of the workforce (and ultimately the patient) to the API. Coated tablets also tend to run faster through packaging lines as they are more slippery than uncoated tablets and therefore flow through hoppers and transfer systems more readily. Reject rates may also be lower since the tablets are protected by the coating and are therefore less likely to break.

The process of film coating involves the deposition of a thin, uniform polymer-based formulation onto the surface of a tablet, capsule, powder, granule or pellet. In its simplest form, film coatings typically consist of a polymer, a plasticizer and pigment(s). The polymer provides mechanical strength, the plasticizer improves flexibility of the film and pigments increase opacity, protect from light and add colour to the coating. Other excipients may be added for other purposes, such as to improve slip or decrease tack.

Common polymers used in tablet coatings are as follows:

- *Cellulose ethers*, e.g. hydroxypropyl methylcellulose (HPMC), hydroxypropyl cellulose (HPC),
- *Vinyl polymers*, e.g. polyvinyl alcohol (PVA), polyvinyl pyrrolidone (PVP), polyvinyl alcohol-polyethylene glycol graft copolymer (PVA-PEG copolymer).
- *Acrylic acid polymers*, e.g. methacrylic acid copolymer (MAA).

Texture and mouthfeel play an important role in product acceptability, and tribology is the science of friction and lubrication of surfaces in relative motion. In the pharmaceutical industry, it refers to the study of texture and mouthfeel.

Most people who have taken a tablet can feel the difference between a coated and uncoated tablet. An uncoated tablet may feel dry inside the mouth and stick to the tongue, whereas a coated tablet induces a more slippery sensation. The viscosity, solubility and composition of coatings all play an important role in lubrication [57].

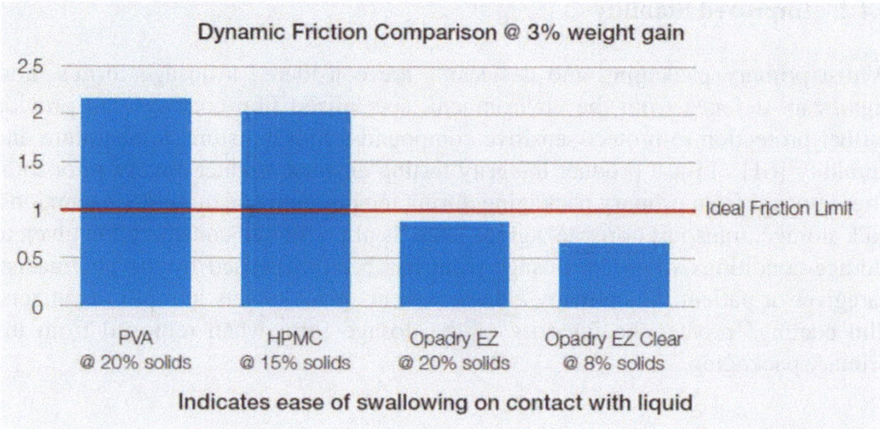

Fig. 8 Dynamic friction comparison of different film coatings. (Image: from Colorcon library)

In recent years, newer coating technologies, such as Opadry® EZ Easy Swallow Coating, have been developed, which increase tablet mobility, reducing the probability of sticking in the throat or oesophagus. When compared to other film coatings (Fig. 8), once wet, the reduced friction on the surface of the Opadry EZ coated tablet results in superior slip movement [58].

Glossy tablets also improve the perception of swallowability [45], providing the illusion that the tablet will slide down the oesophagus far easier than an uncoated or rough tablet. Although aesthetically this may be an illusion, many studies have scientifically demonstrated that coatings enable patients to swallow tablets more easily than uncoated tablets [59, 60]. Uncoated tablets can take 10 minutes or longer to move from the mouth to the stomach. Early research used gamma scintigraphy to measure the influence of film coatings on reducing transit times and demonstrated that the most effective coatings can reduce transit times to around 20–30 s [48].

3.4.1 Taste or Odour Masking

Optimization of taste and mouthfeel can be achieved through the application of a coating. Given the fact that many APIs are bitter or pungent, approaches should be made to mask unpleasant tastes or odours, which could otherwise prevent a patient from taking their medication.

The simplest approach to taste masking is the addition of sweeteners or flavours, but for extremely bitter APIs (e.g. ferrous fumarate, quinine hydrochloride), this may not be feasible since the amount of taste-masking agent will either increase the overall size of the dosage form or erode the therapeutic dose amount in the dosage form. In this case, the best option is to create a physical barrier between the drug substance and taste receptors in the mouth. The barrier can be a capsule shell, or a coating applied to tablet, granule or directly onto the drug substance itself.

3.4.2 Improved Stability

Whilst primary packaging and desiccants are considered a dosage form's most significant defence from the environment, specialized film coatings can provide further protection to protect sensitive compounds from moisture, temperature and humidity [61] . In-use product integrity testing ensures product quality prior to or after removal from primary packaging during temperature and humidity excursions, bulk storage, transport or repackaging. There is often limited consideration given to storage conditions when the dosage form has been dispensed by the pharmacist, caregiver or patient. Since many elderly patients place tablets into pill organizers, film coatings protect the integrity of the dosage form when removed from the primary packaging.

3.4.3 Tablet Crushing

Film coatings are often applied to protect highly unstable drugs, such as nifedipine, which is extremely light sensitive, and therefore crushing a tablet may negatively impact the stability of the drug.

Many patients who struggle to swallow whole tablets may decide to crush their medication before swallowing. In some cases, this may not cause problems, but there are instances where this should be avoided. Patient education is paramount, whether this be provided by the doctor or pharmacist or clearly explained in the patient leaflet or on the label.

Crucially, crushing modified-release or enteric coated tablets could cause the medicine to stop working as intended or result in increased side effects due to overdose. Drug toxicity in the gastro-intestinal tract (GIT) is a common medical problem linked to tablet crushing, leading to minor symptoms such as diarrhoea or more serious consequences including haemorrhage. To prevent GIT injury, formulators use gastro-resistant enteric coatings comprised of polymers, which remain intact in the stomach but dissolve and release the API in the more neutral pH of the small intestine. Film coatings will be discussed in more detail later in this chapter.

3.5 Modified Release

The United States Pharmacopoeia (USP) defines a modified-release (MR) dosage form as "one for which the drug release characteristics of time course and/or location are chosen to accomplish therapeutic or convenience objectives". Extended release (ER) is a type of MR in which the active ingredient is released over a longer period of time in the body, thereby maintaining a therapeutic dose for longer. Other terms for this form of MR include controlled release (CR), prolonged release, sustained release (SR) and long-acting (LA). Additionally, the names of ER marketed products often incorporate a suffix such as CR, XL, LA, SR, XR, retard, etc.

There are numerous advantages to formulating an active as an ER dosage form. A key benefit is a reduction in the frequency of dose administration (maybe as low as once per day), resulting in improved patient compliance. This can be particularly advantageous to elderly patients who are often prescribed complex dosing regimens. Another important feature is the reduction in fluctuations of drug plasma levels, resulting in fewer undesirable side effects and therefore also increasing patient compliance; however, once-per-day tablets are generally larger and may present swallowing difficulties for some patients.

The simplest and most cost-effective (and therefore the most common) form of ER tablet is the hydrophilic matrix. This is simply a physical blend of the active and a hydrophilic polymer, plus additional excipients such as a diluent, a glidants and a lubricant, which are compressed into a tablet form using conventional tabletting processes. The most popular hydrophilic polymer for this application is hypromellose (hydroxypropyl methylcellulose, HPMC), a high-molecular-weight water-swellable polymer produced from cellulose. This is an extensively studied and well-understood pharmaceutical excipient, which is highly stable and non-ionic, has excellent global regulatory acceptance, is easily processed using established tabletting technology (direct compression, wet granulation, or roller compaction) and is readily available. Moreover, multiple grades are available (with different molecular weights and chemical composition), which provides a high level of versatility for achieving the desired duration of drug release and suitability for drugs of different doses and solubilities.

3.6 Capsules

Both hard and soft gelatine capsules are convenient dosage forms in terms of formulation and administration. They can also be coloured and printed to help aid identification, and the principles explored above in this chapter regarding tablets can be applied to capsules as well, but as explained previously, capsules can be more difficult to swallow and have a tendency to become stuck or lodged in the oesophagus [48].

3.7 Excipient Selection

The principles of excipient selection for pharmaceutical drug products are the same regardless of the age of the intended target population; this is based around safety, quality, functionality, cost and regulatory acceptability. The exception for special populations such as young or elderly patients is the ability of the patient to tolerate the excipient. In the case of the elderly patient, they are likely to have had exposure to most, if not all, excipients during their lifetime, so intolerances are likely to be known and most likely to be based around an allergic reaction rather than an intolerance due to the developmental state of the patient (e.g. such as you would have for a paediatric patient). However, what needs to be considered are the changes associated with aging and the impact on pharmacokinetics and pharmacodynamics that will not only have an impact on API absorption but also potentially on how the body deals with excipients. Pharmacokinetic changes include a reduction in renal and hepatic clearance with an increased volume of distribution of lipid-soluble APIs [62], resulting in a prolonged half-life [63]. Pharmacodynamic changes involve altered sensitivity to several classes of drugs such as anticoagulants and cardiovascular and psychotropic drugs, and it is logical to assume that the way the aging body deals with excipients could change as well although there is limited research in this area. Additional consideration should be paid to the potential for there to be a cumulative effect of excipients coming from many different medications (i.e. polypharmacy) and also from dietary supplements. In the European Union, it is estimated that 40% of people over 60 consume dietary supplements [64]; the most commonly used are vitamin and mineral products, fish oils probiotics and herbal products [65]; these are typically consumed in a tablet or capsule format and therefore use the same excipients as those found in drug products. In such cases, the cumulative levels of certain excipients may be detrimental to the patient or impact drug metabolism (i.e. lactose intolerance [66]; high levels of polyols can cause a laxative effect; there is a link between elevated phosphate levels and cardiovascular morbidity [67], degradation of neomycin by sodium carboxymethylcellulose or absorption of nitrazepam in tablet dosage form by colloidal silicon) [68].

3.8 Other Dosage Forms

There are many ways to deliver APIs to elderly patients, and these are discussed in the EMA Reflection Paper [9]; however, these still need to be considered carefully to ensure all aspects are taken into account. At first glance, these dosage forms seem to be more aligned with the needs of the older adult patient [5]; there is no issue with swallowability or taste, and they allow more accurate dosage adjustment, but care still needs to be taken to ensure that they can be used effectively by the patient themselves or their caregiver [37].

A liquid formulation could help overcome the swallowing challenges associated with solid oral dosage forms, and taste masking can be achieved through the use of masking agents, but if the API has a very poor taste, these agents may not be enough to produce a palatable formulation. This may not be the best option for poorly soluble or unstable actives, and additionally liquid formulations usually require increased levels of excipients, particularly preservatives and taste-masking agents/sweeteners, which can add significantly to the cumulative excipient burden in a polypharmacy situation.

Intravenous or intramuscular dosage forms are usually dosed to patients under medical supervision where age and capability of the patient are not a limiting factor; this type of drug product plays a vital role in long-term treatment regimes. Intravenous administration routes can be difficult if the patient is not drinking enough and is dehydrated.

Subcutaneous delivery can be administered by the patient themselves depending upon their capability and is a frequently prescribed form of administration for insulin or anticoagulants, for example. However, in patients with reduced vision and/or manual movement limitations, significant difficulties can arise.

Suppositories are a dosage form that can provide many advantages, but it may not be suitable for patients with restricted mobility.

Inhalers are an effective dosage form, but special training does need to be provided to the patient to ensure that their technique and understanding are effective. There is a danger that technique or dexterity can change over time, so periodic reassessments are recommended.

Transdermal patches, ointments or creams are simple and effective dosage forms that can be administered by the patient or caregiver with minimal issues.

3.9 Polypharmacy

Polypharmacy has been mentioned already in this chapter and will be addressed in detail elsewhere in this book, but for reference, polypharmacy is traditionally defined as the long-term use of five or more drugs [69]. It is estimated that in developed countries, 30–40% of people over 65 years old fall into this definition, with 12% in this group using ten or more different medicines [70].

3.10 Packaging

It's no surprise that packaging often poses challenges for older people. The design of packaging needs to consider dexterity and decreased strength of older patients, but in turn, the packaging also needs to be child resistant.

A qualitative study focussing on the practical problems older patients experience noted patients trying to cut packaging with scissors or knives. The most challenging issue was removing tablets from a blister pack, resulting in broken or crumbled tablets [41].

To overcome this, caregivers will often arrange tablets in a daily pill dispenser. This conquers one challenge but causes another! When tablets are removed from their primary packaging, the tablet is exposed to oxygen, humidity and moisture. We've already discussed how coatings provide protection to the tablet, but for some very sensitive APIs, more expensive protective packaging such as Alu/Alu may be necessary.

4 Ideas for Future Consideration

No matter the patient's age, reminders to take medication will inevitably improve patient adherence. Numerous mobile phone apps now exist to remind patients to order and take their medications as prescribed. Some patients, particularly the elderly, may not own such devices or be reluctant to use such tools, but its use should be promoted as a positive way to improve adherence and ultimately improve the patient experience.

Invisible on-dose taggants such as Soteria RX® may be applied to the tablet during the film coating to minimize counterfeiting [71]. The technology is evolving to allow the use of a mobile phone app to scan the tablet, ensure it is authentic and set reminders to take the medication. This will allow the patient, the caregivers and pharmaceutical company to obtain information and insight around the use of the medicinal products.

5 Summary

From the regulatory review, it is apparent that existing guidance and legislation already require pharmaceutical drug products to be formulated to meet the needs of the patient, regardless of age and capability; however, in reality, not all drug products are suitable for their intended patient population, which has a negative effect on compliance and ultimately patient health. Strategies from regulatory bodies and specifically the EMA reflection paper [9] have been written to encourage the development of patient-centric medicines for the older adult population. The EMA Guideline uses words like "should" rather than "must" or "required" to allow some leeway to formulators in how they do this, but the expectation is a drug product that is ultimately suitable for the intended patient. The recommendation is that a greater understanding and analysis of the patient's needs and capabilities are incorporated into the drug development process to ensure that the resulting drug product is truly patient centric. The suitability of a drug product for the intended patient should be demonstrated through human factor trials that consider the patient, caregiver and location of administration.

The older adult population has specific requirements that need to be considered when formulating patient-centric medicines; these include difficulty with swallowing, physiological changes, handling difficulties, cognitive ability to understand instructions and identification of medicines among others. There are several ways that a formulation can be designed to address these needs such as the use of a film coating to aid swallowability, modified-release formulations to address pharmacokinetic changes, drug product size and packaging optimization, colour and shape for identification and clear usage instructions, for example. Many different types of dosage forms are available, and some might seem to be particularly suited to meeting the needs of the older adult population, but they still need to be assessed carefully in the context of patient capability and polypharmacy, e.g. liquid formulations may overcome swallowing difficulties but are likely to require higher levels of excipients for taste masking and preservation, which will need to be taken into account alongside the excipient burden from other medicines the patient is taking.

The use of modern technology does provide a useful tool to aid patients in remembering what to take and when, and there are limitations, as not every older adult patient will have a smartphone and ultimately even if the patient knows what to take and when, they still need to be able to get it out of the packaging and be able to swallow it as instructed. The considerations, design process and techniques discussed in this chapter should enable the development of medicines appropriate for the older adult population.

References

1. Califf RM. "Best Pharmaceuticals for Children Act (BPCA 2002,2007) and Pediatric Research Equity Act (PREA 2003/2007) FDA". 2016.
2. European Council, REGULATION (EC) No 1901/2006 OF THE EUROPEAN PARLIAMENT AND OF THE COUNCIL. 2006.
3. European Medicines Agency, "Human regulatory - medicines for older people". September 2019. [Online]. Available: https://www.ema.europa.eu/en/human-regulatory/research-development/medicines-older-people#geriatric-medicines-strategy-section.
4. European Medicines Agency, "Geriatric medicines strategy EMA/CHMP/137793/2011". 2011.
5. Tasi K-T, Chen J-H, Wen C-J, Kuo H-K, Lu I-S, Chiu L-S, Wu S-C, Chan D-C. Medication adherence among geriatric outpatients prescribed multiple medications. Am J Geriatric Pharmacotherapy. 10(1)
6. European Medicines Agency, "Reflection paper on the pharmaceutical development of medicines for use in the older population EMA/CHMP/QWP/292439/2017". 2020.
7. Stegemann S, Breitkreutz J, Gosch M. Swallowing dysfunction and dysphagia is an unrecognized challenge for oral drug therapy. Int J Pharm. 2015;1-2(430):197–206.
8. NHS, "Dysphagia (swallowing problems)". 2021. [Online]. Available: https://www.nhs.uk/conditions/swallowing-problems-dysphagia/causes/. Accessed Sep 2021.
9. Miller C. Dare to be different. Med Maker. 2015:36–8.
10. De Spiegeleer B, Wynendaele E, Bracke N, Veryser L, Taevernier L, Degroote A, Stalmans S. Regulatory development of geriatric medicines: to GIP or not to GIP? Aging Res Rev. 2016;27:23–6.
11. Cerreta F, Eichler H-G, Rasi G. Drug policy for an aging population - the European medicines Agency's geriatric medicines strategy. N Engl J Med. 2012;367(21):1972–4.
12. European Medicine Agency, "ICH guideline Q8 (R2) on pharmaceutical development. EMA/CH MP /I CH/ 16 7068/2004". 2014.
13. European Medicines Agency, "Geriatric medicines strategy report analysis of scientific guidelines. EMA/352591/2013". 2013.
14. van Riet-Nales DA, Sundberg K, de Boer A, Hirschlérova B. Developing patient centric medicines for older people: reflections from the draft EMA paper on the pharmaceutical development of medicines for use in the older population. Br J Clin Pharmacol. 2020;86(10):2008–13.
15. FDA, "Guidance for industry - guideline for the study of drugs likely to be used in the elderly". 1989.

16. FDA, “Preamble of the demographic rule (21 CFR 314.50(d)(5); 63 FR 6854)”. 1998.
17. FDA, “Guidance for industry and FDA staff collection of race and ethnicity data in clinical trials”. 2016.
18. FDA, “Fed register 2012 Feb 21;77(34):9948–9”. 2012.
19. FDA, “ICH guidance for industry E7 studies in support of special populations: geriatrics: questions and answers”. 2012.
20. Bernard MA, Clayton JA, Lauer MS. Inclusion across the lifespan: NIH policy for clinical research. JAMA. 2018;320(15):1535–6.
21. Helfand BK, Webb M, Gartaganis SL. The exclusion of older persons from vaccine and treatment trials for coronavirus disease 2019—missing the target. JAMA. 2020;180(11):1546–9.
22. Committee on improving the quality of cancer care, delivering high-quality cancer care: charting a new course for a system in crisis, Washington DC: National Academies Press (US), 2013, pp. 91–152.
23. Drumond N. Future perspectives for patient-centric pharmaceutical drug product design with regard to solid Oral dosage forms. J Pharm Innov. 2020;15:318–24.
24. Hughes CM. Medication non-adherence in the elderly: how big is the problem? Drugs Aging. 2004;21:793–811.
25. Chen J, Mullins D, Novak P, Thomas SB. Personalized strategies to activate and empower patients in health care and reduce health disparities. Health Educ Behav. 2016;43:25–34.
26. Stegemann S, Ternik RL, Onder G, Khan MA, van Riet-Nales DA. Defining patient centric pharmaceutical drug product design. AAPS J. 2016;18(5):1047–55. https://doi.org/10.1208/s12248-016-9938-6.
27. FDA, “Applying human factors and usability engineering to medical devices. Guidance Indust Food Drug Administrat Staff”. 2016.
28. FDA, “Safety considerations for product design to minimize medication errors guidance for industry”. 2016.
29. FDA, “Guidance for industry. Patient-reported outcome measures: use in medical product development to support labelling claims”. 2009.
30. European Medicines Agency, “Good practice guide on risk minimisation and prevention of medication errors. EMA/606103/2014”. 2015.
31. European Medicines Agency, “Medication-errors workshop Workshop Report EMA/144458/2013”. 2013.
32. “ISMP - Institute for safe medication practices”. [Online]. Available: https://www.ismp.org/resources. Accessed Sep 2021.
33. Chartered Institute of Ergonomics & Human Factors, “Actual v simulatged use in human factors testing of medical devices”. [Online]. Available: https://www.ergonomics.org.uk/Public/Resources/Articles/Actual_v_simulated_use_in_human_factors_testing_of_medical_devices.aspx. Accessed Sep 2021.
34. Hellier E, Tucker M, Kenny N, Rowntree A, Edworthy J. Merits of using color and shape differentiation to improve the speed and accuracy of drug strength identification on over-the-counter medicines by laypeople. J Patient Saf. 2010;6:158–64.
35. Shariff ZB, Dahmash DT, Kirby DJ, Missaghi S, Rajabi-Siahboomi A, Maidment ID. Does the formulation of Oral solid dosage forms affect acceptance and adherence in older patients? A mixed methods systematic review. J Am Med Dir Assoc.
36. European Medicines Agency, “ICH Q8 (R2) Pharmaceutical development”. 2009.
37. Heppner HJ, Sieber CC, Esslinger AS, Togner J. Drug administration and dosage forms in geriatric patients. Ther Umsch. 2006;6(63):419–22.
38. Breitkreutz J, Boos J, Breitkreutz J, Boos J. Paediatric and geriatric drug delivery. Expert Opin Drug Deliv. 2007;4(1):37–45. https://doi.org/10.1517/17425247.4.1.37.
39. British Medical Journal, “237+million medication errors made every year in England”. 11 06 2020. [Online]. Available: https://www.bmj.com/company/newsroom/237-million-medication-errors-made-every-year-in-england/. Accessed Sep 2021.

40. Daanish M. Preference and perception of geriatrics twoards oral solid dosage form characteristics. In: UK PharmSci; 2014.
41. Notenboom K, Beers E, van Riet-Nales DA, Egberts TCG, Leufkens HG, Jansen PA, Bouvy ML. Practical problems with medication use that older people experience: a qualitative study. J Am Geriatr Soc. 2014;12(62):2339–44.
42. FDA, "Size, shape, and other physical attribues of generic tablets and capsules: guidance for industry". 2015.
43. Schiele JT, Quinzler R, Klimm H-D, Pruszydlo MG, Haefeli WE. Difficulties swallowing solid oral dosage forms in a general practice population: prevalence, causes, and relationship to dosage forms. Eur J Clin Pharmacol. 2013;4(69):937–48.
44. Stegemann S, Riedl R, Sourij H. Identification of different shapes, colors and sizes of standard oral dosage forms in diabetes type 2 patients—a pilot study. Int J Pharm. 2017;517:112–8.
45. Shariff Z, Kirby D, Missaghi S, Rajabi-Siahboomi A, Maidment I. Patient-centric medicine design: key characteristics of Oral solid dosage forms that improve adherence and acceptance in older people. Pharmaeutics. 2020;10(12)
46. Hoogendam YY, van der Lijn F, Vernooji MW, Hofman A, Niessen WJ, van der Lugt A, Ikran M, van der Geest JN. Older age relates to worsening of fine motor skills: a population-based study of middle-aged and elderly persons. Front Aging Neurosci. 2014;6(259)
47. Yoder S, Rajabi J, Miller C, Hansell J, Oza K. Physical appearance preferences for oral solid dosage forms. In: AAPS. San Diego; 2014.
48. Wilson C, O'Mahony B, Farrell T, Friend B, Taylor D. Modern tablet film coatings and influence on ease of swallowing. In: AAPS; 2003.
49. Arnet I, Hersberger KE. Misleading score-lines on tablets: facilitated intake or fractional dosing. Swiss Med Wkly. 2010;140(7–8):105–10.
50. Noviasky J, Lo V, Luft DD, Saseen J. Clinical inquiries. Which medications can be split without compromising efficacy and safety? J Fam Pract. 2006;55(8):707–8.
51. Hill SW, Warker AS, Karlage K, Myrdal PB. Analysis of drug content and weight uniformity for half-tablets of 6 commonly split medications. J Manag Care Pharm. 2009;3(15):253–61.
52. "University of Loyola, Maryland Study".
53. Wan X, Woods AT, Salgado-Montejo A, Velasco C, Spence C. Assessing the expectations associated with pharmaceutical pill colour and shape. Food Qual Prefer. 2015;45:171–82.
54. de Craen AJ, Roos PJ, de Vries AL, Kleijnen J. Effect of colour of drugs: systematic review of perceived effect of drugs and of their effectiveness. Br Med J. 1996;313:1624–6.
55. Spence C, Wan X, Woods A, Velasco C, Deng J, Yousseff J, Deroy O, Spence C, Wan X, Woods A, et al. On tasty colours and colourful tastes? Assessing, explaining, and utilizing crossmodal correspondences between colours and basic tastes. Flavour. 2015;4:23. https://doi.org/10.1186/s13411-015-0033-1.
56. Kesselheim AS, Misono AS, Shrank WH, Greene JA, Doherty M, Avorn J, Choudhry NK. Variations in pill appearance of antiepileptic drugs and the risk of nonadherence. JAMA. 2013;173(3):202–8.
57. Hofmanová JK, Mason J, Batchelor HK. Tribology provides an in vitro tool that correlated to in vivo sensory data on the mouthfeel of coated tablets. Int J Pharm. 2021;597
58. Hofmanová JK, Rajabi-Siahboomi A, Haque S, Mason J, Teckoe J, D. To, Batchelor HK. Developing methodology to evaluate the Oral sensory features of pharmaceutical tablet coatings. Int J Pharm. 2019;562:212–7.
59. Yamamoto S, Taniguchia H, Hayashia H, Horia K, Tsujimuraa T, Nakamuraa Y, Sato H, Inoue M. How do tablet properties influence swallowing behaviours? J Pharm Pharmacol. 2013;66:32–9.
60. Overgaard AB, Højsted J, Hansen R, Møller-Sonnergaard J, Christrup LL. Patients' evaluation of shape, size and colour of solid dosage forms. Pharm World Sci. 2001;23(5):185–8.
61. Colorcon Inc., "Impact of moisture barrier film coating on stability of amoxicillin/clavulanic acid tablets," Pharm Technol, pp. 44–47, April 2015.

62. Mangoni AJS. Age-related changes in pharmacokinetics and pharmacodynamics: basic principles and practical applications. Br J Clin Pharmacol. 2004;57:6–14.
63. Shi KUS. Age-related changes in pharmacokinetics. Curr Drug Metab. 2011;12(7):601–10.
64. Błeszyńska E, Wierucki L, Zdrojewski T, Renke M. Pharmacological interactions in the elderly. Medicina (Kaunas). 2020;56(7):320.
65. Commission of the European Communities. Characteristics and perspectives of the market for food supplements containing substances other than vitamins minerals. European commission staff working document. Belgium: Brussels; 2008.
66. Mill D, Dawson J, Lee Johnson J. Managing acute pain in patients who report lactose intolerance: the safety of an old excipient re-examined. Therapeut Adv Drug Saf. 2018:227–35.
67. Sultana J, Musazzi UM, Ingrasciotta Y, Giorgianni F, Ientile V, Fontana A, Minghetti P, Perrotta M, Santoro D, Savica V, Trifiro G. Medication is an additional source of phosphate intake in chronic kidney disease patients. NUMECD. 2015;25(10):959–67.
68. Roy J. An introduction to pharmaceutical sciences. Cambridge, UK: Woodhead Publishing Limited; 2011.
69. Nobili A, Marengoni A, Tettamanti M, Salerno F, Pasina L, Franchi C, Iorio A, Marcucci M, Corrao S, Licata G, Mannuccio Mannucci P. Association between clusters of diseases and polypharmacy in hospitalized elderly patients: results from REPOSI study. Eur J Int Med. 2011;22:597–602.
70. Kim PA. Polypharmacy and medication management in older adults. Nurs Clin N Am. 2017;52:457–68. https://doi.org/10.1016/j.cnur.2017.04.007.
71. Colorcon, "Product authentication". [Online]. Available: https://www.colorcon.com/solutions/product-authentication. Accessed Sep 2021.

Advanced Oral Sustained-Release Drug Delivery Systems for Older Patients

Kavil Patel and Fang Liu

Contents

Abstract Oral drug delivery remains the route of preference in the general population due to the convenience of administration. Older adults have unique and complex needs presented by an increase in morbidity in addition to a deterioration in swallowing capability and reduced capacity to adhere to complex dosing regimens. Conventional approaches to aid older adults and improve medication usability such as the prescribing of oral liquids or crushing of tablets can compromise the stability, efficacy and/or safety of medicines, calling for innovative approaches. Sustained-release dosage forms offer a potential opportunity for adherence

K. Patel · F. Liu (✉)
Department of Clinical, Pharmaceutical and Biological Sciences, School of Life and Medical Sciences, University of Hertfordshire, Hatfield, UK
e-mail: f.liu3@herts.ac.uk

M. Orlu, F. Liu (eds.), *Pharmaceutical Formulations for Older Patients*, AAPS Advances in the Pharmaceutical Sciences Series 51,
https://doi.org/10.1007/978-3-031-35811-1_6

improvement via dosing simplification, and there have been several recent advancements in this space to improve the accessibility and usability of medicines for older patients. The purpose of this chapter is to consider these advancements, outlining any potential benefits and drawbacks.

Keywords Oral · Sustained release · Novel · Long-lasting · Formulation

1 Introduction

The oral route is considered the most convenient for patients to take their medicines. For the older population, there are unique challenges in oral drug delivery, which also present opportunities for technology advancement. Ageing increases the prevalence of multiple chronic morbidities and the number of medicines required for treatment [1]. In 2018, the World Health Organization (WHO) stated in an article that as people age, they are more likely to experience several conditions at the same time categorised by what is commonly known as 'geriatric syndromes' [2]. This is further highlighted by data published by the Health and Social Care Information Centre in the UK who stated that in England, over 50% of those aged 65–74 and >70% of those aged 75 and over are being prescribed three or more medicines per day [3]. Data collected in the USA in 2011 mirrored this trend, where it was shown that approximately 36% of adults aged 62–85 took five or more prescribed medications, which was found to be an increase from 31% in 2001 amongst the same patient demographic [4].

In addition to the number of medicines being prescribed per patient, complex dosing regimens present a hurdle for all patient groups. Multi-dose per-day regimens can be required in order to maintain blood plasma drug concentrations and consequently therapeutic effects. Overall, there is strong evidence to show a correlation between the frequency of dosing schedules per day and hospital admissions with one source showing that hospitalisation between two patient groups differed significantly when dosing frequency was increased from once to twice daily [5, 6]. Various studies have shown that by increasing dosing frequency from once to twice daily, adherence rates can fall by as much as 10% [6, 7]. In elderly patients, the use of oral bisphosphonates is extensive for the prevention of osteoporosis (bone thinning), and this class of medicines is available as daily, weekly and even monthly (e.g. ibandronic acid) dosing regimens, providing a useful case to evaluate the relationship between dosing frequency and adherence. Similar to the trends outlined above, an increase of 14% in adherence rate was observed when reducing dosing frequency from daily to weekly [8, 9]. It was also noted that discontinuation of treatment was likely to occur earlier for formulations requiring daily dosing. No adherence rates were reported for monthly treatments; however, 71% of patients in the study showed preference for monthly dosing compared to 29% who preferred weekly treatment [8, 9].

Table 1 Formulation factors affecting oral medication adherence in older adults [10]

Formulation factor	Comment
Dimensions	
Size	Small sizes affect handling of tablets; large sizes can cause difficulty in swallowing especially in older patients with dysphagia
Shape	The shape of tablets and capsules affect the ease of swallowing
Break marks	Break marks enable dose flexibility; however, dividing tablets unevenly leads to inaccurate dosing
Physical properties	
Surface texture	Poor surface texture (e.g. 'chalkiness') contributes to swallowing difficulties in people with dysphagia
Mouthfeel	Coating and particle size can affect mouthfeel of a dosage form
Coating	The presence and nature of coating impact on the smoothness, stickiness, slipperiness and palatability of tablets
Taste	Taste significantly affects patient acceptability especially when tablets were crushed. Taste masking is needed for substances with inherent bitter taste
Appearance	
Visual distinction	Inability to distinguish different medicines and strengths can cause discomfort and clinical deterioration
Colour	Colour of a medicine may affect patient acceptability

The design of oral formulations has an impact on the experience of patients taking medicines and consequently patient acceptability and adherence. A systematic review carried out by Shariff et al. (2020) highlighted that three formulation-dependent themes significantly influence patient adherence in the older adult population, being dimensions, palatability and appearance. Table 1 shows formulation factors that relate to these themes [10].

2 Oral Sustained Release for Older Patients

Oral formulations that provide longer-lasting therapeutic effects than conventional immediate-release dosage forms are ideal to address the issues raised above, i.e. to reduce dosing frequency and alleviate pill burden for older patients. The concept of a 'long-lasting' formulation stems from the first commercial sustained-release formulation named Spansules® being introduced in 1952 by SmithKline Beecham [11]. This formulation comprised pellets coated with a water-soluble wax at various thicknesses in order to provide dextroamphetamine release for up to 12 h and twice-daily dosing. Since then, more advanced technologies have been introduced; however, the majority of sustained-release formulations still exist as monolithic matrices or coated tablets [11]. The main disadvantage of these systems is their large size, causing difficulty in swallowing for older patients, limiting patient acceptability.

Swallowability becomes a primary concern when considering formulations with prolonged therapeutic effects for older adults due to an increased prevalence of dysphagia. Two epidemiological studies based in the USA observed referrals into a tertiary acute care hospital and recorded the prevalence of dysphagia across various patient groups [12, 13]. The first study included annual observations between 2000 and 2007, and the second study was carried out as a continuation study in 2014. The first study showed that 70% of dysphagia referrals to hospital involved patients older than 60 years of age with more than 42% of these being aged 80 and over. Furthermore, it was shown that between 2000 and 2007, referrals in every age group increased year on year, showing an increasing prevalence of dysphagia in the general population. The continuation study showed a further 63% increase in dysphagia referrals for patients aged 60 and over in 2014 [12, 13].

To overcome swallowing difficulties for conventional immediate-release formulations, where switching to a liquid dosage form is not possible, it is common practice to crush tablets or open capsules to sprinkle the contents onto food for consumption. In 2002, a study was carried out to assess the common practices used in a nursing home environment to overcome the challenges posed by dysphagia with regard to medicine administration [14]. It was found that the practice of tablet crushing or capsule opening was occurring in over 80% of the nursing homes included in the study by more than 60% of the practicing nurses. It should also be noted that 58% of the nurse participants stated that this practice was recommended by prescribers with a reason of cost saving in comparison to liquid formulations being highlighted [14]. The widespread use of dosage form manipulation is exemplified by the existence of guidance such as the NEWT guidelines in England, which are widely used in specialist care settings such as hospitals or care facilities [15–17]. These guidelines make recommendations to carry out the manipulations as safe as possible, taking into account specific formulation requirements and intended mode of drug release; however, modifying the dosage form of a medicine can still be problematic. Table 2 lists examples of studies that reported issues related to dosage form alternation to assist administration.

When splitting or crushing immediate-release tablets, a study found that dosage uniformity can be questioned, with 24% of 180 split tablets containing drug content which fell outside of the accepted limits set by the United States Pharmacopeia (USP) [18]. Samples included warfarin tablets, which are often prescribed in the elderly and have a narrow therapeutic window with significant risks linked to under- or over-dosing. In addition to tablet crushing and capsule opening, formulations are often mixed with food and/or thickened fluid to facilitate swallowing. The implications of this practice are not well understood and can be drug and formulation design dependent. Using simplified in vitro testing, it was found that the co-administration of immediate-release tablets and foodstuffs or thickened liquids led to a retardation or delay in drug release, likely impacting in vivo bioavailability [19].

The implications of dosage form manipulation are magnified for prolonged-release formulations due to the increased drug content comprised within each dosing unit. The consequences of dose dumping (loss of drug release control

Table 2 Examples of dosage form manipulation and related consequences

Multi-Centre observational study on administration errors	Medication administration errors occurred in 21% dysphagic patients (excluding timing errors)	[24]
Small scale study on tablet splitting	The study concluded that using several splitting methods, large dosing variations were observed	[25]
Observational study of medicine administration by two nurses to a patient with dysphagia	Medicines were often compounded together irrespective of design, and this led to potential dosing errors and unlicensed use of medicines	[26]
Case study of medicine administration to dysphagic patients	In one case, sustained-release oxycodone was crushed, leading to excessive sedation and respiratory depression. In another, delayed-release omeprazole pellets were crushed, which led to drug degradation in the stomach and prolonged stay in hospital due to lack of improvement	[22]
Site study	Frequency of tablet crushing or capsule opening was 34%, with 17% of altered medicines having the potential, because of the alteration, to cause increased toxicity, reduced efficacy, unpalatability and safety or stability concerns	[27]
Patient death reports due to tablet crushing	Sustained-release cardiovascular drugs caused death in two reported cases due to crushing	[20, 21]

mechanism) can be fatal, and there is explicit advice to not carry out any dosing manipulation for modified-release products [20, 21]. Although this is widely noted, as evident by the 'Do Not Crush' list in the NEWT guidelines, there have been instances where administration errors and adverse events have occurred due to the patients having swallowing difficulties and modified-release formulations crushed [22, 23].

3 Conventional Sustained-Release Formulations Suitable for Older Patients

Sustained-release formulations that can facilitate ease of swallowing are more suitable for older patients especially those with dysphagia. An example of these is tablets that can be split without compromising drug release supported by relevant bioequivalence data [28]. Another commonly used approach is multi-particulate-based systems that can be administered as sprinkles onto foodstuff [28]. The risks involved using sustained-release 'sprinkles' are the possibility of chewing with consequence of dose dumping or potential of drug/excipient-food interactions. In addition, a level of dexterity is required to carry out tablet splitting or capsule opening safely, which is not always present in the older populations. Other sustained-release formulations with improved swallowability include mini-tablets, orally disintegrating tablets and chewable tablets.

3.1 Mini-Tablets

It was observed that swallowing becomes less acceptable by patients with dysphagia as tablet size increases [29]. Therefore, mini-tablets are an appealing option for oral sustained-release drug delivery. Mini-tablets are defined as tablets with a diameter of less than 3–4 mm and can be used as a single delivery unit or as part of a multiple unit delivery system where they share benefits with multi-particulate systems with regard to decreasing the risk of dose dumping and limiting local gastric irritation during drug release (Fig. 1) [30]. They are flexible in dose adjustment and use established tablet manufacturing methods, relatively simple in comparison to pellets and/or granules [31].

Mini-tablets usually comprise the active pharmaceutical ingredient (API) with a matrix former such as hydroxypropyl methylcellulose (HPMC) and can be prepared using conventional tablet manufacturing methods such as direct compression, wet granulation, dry granulation as well as other single-step methods such as hot melt extrusion. However, compared to conventional sized tablets (usually ≥5 mm), the surface area-to-volume ratio is much larger for mini-tablets, increasing the difficulty of controlling drug release via both erosion and diffusion mechanisms.

Prolonging drug release using soluble matrix-forming agents such as HPMC is a challenge as they can promote matrix erosion and create water-filled pores with increased drug permeability, especially for water-soluble compounds. Therefore, it is often the case that insoluble matrix-forming agents or coatings are used. Mehta et al. reported 2 mm extruded sustained-release pellets comprising a poorly soluble drug and matrix-forming methacrylate polymers, Eudragit® S and L, and demonstrated over 12 h drug release control affected by porosity of the matrix [33].

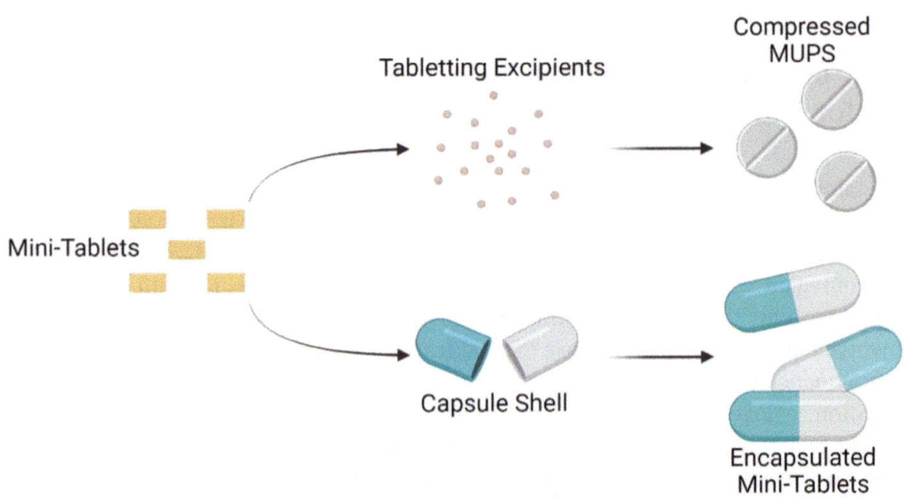

Fig. 1 Mini-tablet delivery options as (**a**) multi-unit pellet system (MUPS) or (**b**) capsule. (Reproduced with permission from [32]. Created with Biorender.com)

In another study, ethylcellulose (EC) was used as the matrix-forming polymer to produce 2 mm sustained-release ibuprofen mini-tablets manufactured via hot-melt extrusion [34]. Lopes et al. used EC in the matrix in comparison with HPMC and showed that EC was able to extend drug release time from approximately 8 h (50% HPMC) to more than 12 h (40% EC) for 2.5 mm diameter mini-tablets [32]. An alternative to including water-insoluble components into the mini-tablet matrix is to apply a semi-permeable coating layer to the surface of the mini-tablets in to order to further regulate drug release. Matrix mini-tablets comprising theophylline and HMPC were coated with a combination of Surelease® (aqueous EC dispersion) and Opadry® (HMPC-based) in order to prolong drug release up to 12 h [35]. This technique was also applied to provide 24 h sustained release for montelukast sodium [36]. Mini-tablets are thought to be preferred substrates for applying coating in comparison to pellets or granules as they tend to have smoother surface and more regular shape, which simplifies the coating process [37].

Sustained-release tolterodine mini-tablets are commercially available as an anti-muscarinic drug used in the treatment of urinary incontinence [38]. Mini-tablets are coated with EC and have a diameter less than 4 mm. Two or four mini-tablets, depending on the dose required, are encapsulated in a hard gelatin capsule shell for administration, and the patients are instructed to swallow the capsule whole, which does not take the advantage of ease of swallowing of the mini-tablets [39]. Another commercial example of sustained-release mini-tablets is Orfiril long® developed by Desitin Arzneimittel GmbH as 2 mm sodium valproate mini-tablets for anti-epileptic treatments. Mini-tablets are coated using a combination of EC and Eudragit® RS. Sodium valproate dosing can range from 150–1000 mg, and the mini-tablets are either encapsulated for the delivery of doses up to 300 mg, or single-dose sachets are available for doses upward of 500 mg [40].

In an open questionnaire study, a few areas of concern were highlighted with regard to the use of mini-tablets in older patients [41]. Mini-tablets were considered difficult to see for the visually impaired, and this can be widened to a general difficulty in handling, especially for those with neurodegenerative conditions or impaired motor function, both common in the older population. Further concerns were raised that mixing the mini-tablets with food to improve taste or facilitate swallowing was not acceptable by patients due to fear of inaccurate dosing during consumption.

3.2 Orally Disintegrating Tablets

The use of orally disintegrating tablets (ODTs) benefits older patients with dysphagia by removing the need to swallow intact tablets. ODTs comprise superdisintegrants, which cause the tablet to disintegrate rapidly in the mouth in the presence of saliva without the need of taking water, convenient for patients' use [42].

Multiple-unit pellet systems (MUPS) are commonly used in producing sustained-release ODTs by incorporating matrix or coated pellets in the tablet matrix comprising diluents, sweeteners and disintegrants. For example, spray-dried granules of ketoprofen comprising Eudragit® RS 30D, starch and polyethylene glycol were directly compressed into ODTs after mixing with lactose, mannitol and cross-linked polyvinyl pyrrolidone, which provided 24 h ketoprofen release following disintegration in 30 s [43]. A US patent described the production of a fixed-dose combination of desmopressin and tolterodine by combining immediate-release pellets and sustained-release sugar spheres coated with Eudragit® NE 30 D or Surelease® [44]. The pellets were then mixed with soluble excipients such as polysaccharide, mannitol and various polymers dissolved in water. The mixture was dosed directly into blister packs, which were subsequently subjected to lyophilisation in order to produce the final ODT.

Recent literature demonstrated the possibility of producing sustained-release matrix ODTs, without the need for incorporating MUPS, by using Soluplus® as a release modifier in combination with a superdisintegrant, croscarmellose sodium, using water-soluble naproxen as an example. When the concentration of Soluplus® was increased to up to 50% (w/w of the drug), sustained drug release was achieved for up to 2 h whilst maintaining tablet disintegration time within 30 s [45]. Soluplus® is a graft copolymer with amphiphilic properties conventionally used as a solubility enhancer for poorly soluble drugs; however, it was used in this scenario as a matrix-forming agent, which slowed down drug release following gelation. The potential of extending drug release for longer period using this approach needs to be further investigated to be clinically meaningful. Chitosan-based nanoparticles formed by covalent bond between chitosan and functional cross-linking agents such as polyethylene glycol, dicarboxylic acid or tripolyphosphate have been used to prepare sustained-release ODTs, with the advantage of biodegradability of chitosan. Nanoparticles were prepared using ionotropic gelation comprising promethazine hydrochloride, chitosan and tripolyphosphate, which were subsequently coated with one of polyethylene glycol, polyvinylpyrrolidone or polyethylene co-acrylic acid in order to provide sustained drug release for 24 h [46]. Nanoparticles were subsequently embedded inside an ODT matrix comprising l-substituted hydroxypropyl cellulose, which was shown to disintegrate within 40 s.

Drug-ion exchange resin complex has been used in producing sustained-release ODTs with commercial success, an example being drugs used in the treatment of attention-deficit hyperactivity disorder (ADHD), which, although is largely prevalent in the younger population, provides a technological insight into methods of manufacture sustained-release ODTs. Aytu Biopharma's Adzenys XR-ODT (amphetamine) and Contempla XR-ODT® (methylphenidate) are examples of commercial products that were developed using the technology, whereby the active pharmaceutical ingredients (APIs) are adsorbed onto resins via weak ionic interactions. The potential of applying drug-ion exchange resin complex as a sustained-release delivery tool is dependent on the rates of adsorption or 'affinity' of the drug to the resin and simultaneously the rates of desorption of the drug from the resin. The complexity of this relationship between drug and resin limits the number of combinations which are viable for commercial development. Furthermore, it is often the case that additional coating is required for ion exchange resins to provide suitable release profiles; for example, an additional ethylcellulose coating was used in Contempla XR-ODT® in order to provide sufficient sustained release [47].

As a relatively new type of dosage form, older patients might not be familiar with ODTs and reported difficulties in understanding handling instructions and removing the tablets from the blister strips [48]. Similarly, a lack of distinguishability between ODTs and conventional tablets was a concern for older patients and led to non-compliance to treatments [48]. Another challenge for the application of ODTs in older patients is that there is a high incidence of xerostomia (dry mouth) in this demographic group. Xerostomia occurs when saliva production is reduced, and this can be related to medicines such as antimuscarinics, which are known to commonly produce this effect but can also be a symptom of other conditions such as type II diabetes [49]. Hyposalivation as a result of xerostomia creates a problem for ODT administration, as tablet disintegration will be impeded, leading to an increase in the probability of chewing and a loss in formulation functionality [50].

In the presence of moisture, ODTs can disintegrate or be weakened. Special packaging, such as Alu Alu blister pack, and desiccants are often used to protect the tablets from moisture in the environment. The need for special packaging and storage requirements deems the formulation unsuitable for using in multi-dosing systems (MDS), which are often used by older patients who require multiple medications. It is also known that splitting ODTs is contraindicated due to fragility of the tablet, limiting dose flexibility to meet specific patient needs.

3.3 Chewable Tablets

Chewable tablets have been used in paediatrics for immediate-release drug delivery and can be prepared either through compression or moulding/extrusion of molten polymer-drug combinations dependent on whether the tablet is designed to be hard or soft, respectively. The main components include a gum or insoluble polymer, which would entrap drug particles (can be coated if used for sustained release) [51–55].

Sustained-release chewable tablets have been developed for the treatment of ADHD, which, although is more common in children, has increasing prevalence in older adults [54]. An example is Quillichew ER, manufactured by Tris Pharma [56, 57]. The formulation comprises coated methylphenidate-ionic resin complex compressed into tablets designed to be chewed before swallowing. The tablet maintains its extended-release properties after chewing and is effective for 2–8 h after dosing.

The drawbacks relating to chewable sustained-release formulations for older adults centre on the patient's ability to use the formulation as intended. The first of the four phases of the swallowing process is known as the oral preparatory phase where the bolus is formed. This is the only phase that requires voluntary use of several nerves such as the trigeminal and vagus nerves to control the masticatory muscles and tongue [58, 59]. Deficiencies in these nerve networks and lack of chewing strength are known to be of increasing prevalence as age increases, and this impairs the voluntary control required to carry out the oral preparatory phase of swallowing [60]. The degradation in chewing ability and issues such as the incompatibility of denture and chewable tablets highlighted by a patient questionnaire bring to question the suitability of using chewable tablets for sustained-release drug delivery in older adults [41].

4 Advanced Sustained-Release Technologies for Older Patients

Technology advancement in oral sustained-release formulations has offered new solutions to meet special patients' needs, particularly in providing ease of swallowing suitable for children and older adults with dysphagia. These formulations include oral sustained-release liquids, sustained-release microparticles and ultra-long-lasting formulations.

4.1 Oral Sustained-Release Liquids

Oral liquids are commonly used as alternatives to solid dosage forms for patients who cannot swallow tablets and capsules, such as children, older adults and patients with dysphagia. Conventional liquid formulations provide immediate drug release as solutions or suspensions. New technologies have emerged investigating oral liquids with extended drug release, facilitating ease of swallowing and reducing dosing frequency. Three approaches are introduced here for producing oral sustained-release liquid formulations using solid lipid nanoparticles, in situ gelling and emulsion-based systems.

4.1.1 Solid Lipid Nanoparticles

Lipid nanoparticles are solid matrix systems in the nano-scale and are categorised as either solid lipid nanoparticles (SLN) or nanostructured lipid carriers (NLC) [61]. SLNs are considered as the first generation of the two approaches and are described as colloidal dispersions based on biodegradable solid lipids. There are several advantages of SLNs, which include the ability to improve drug stability, increased bioavailability of poorly soluble drugs, no requirement for organic solvent during manufacturing and ease of scaling up for production at low cost. Although known for topical and parenteral applications, for older adults, SLNs and NLCs can be delivered orally as aqueous dispersions, tablets, capsules or pellets for controlled release. With regard to production methods, there are several options.

Hot homogenisation can be applied where the API is dissolved in melted lipid, and this mixture is added to an aqueous phase comprising surfactants and co-surfactants, which is heated to the same temperature as the molten lipid. Once the two phases are combined, they are homogenised and then allowed to cool at room temperature causing lipid recrystallisation. This approach is not impossible for heat-sensitive drugs as the duration of heating is short; however, if API stability remains a concern, the following approach can be applied.

Cold homogenisation is an alternative approach where the API is still dissolved in molten lipid; however, once combined, this mixture is cooled immediately. The cooled melt is grounded to achieve particles in the micron size range before being dispersed in cold surfactant solution and homogenised. The primary justification for cold homogenisation in place of hot homogenisation is to avoid API partitioning from the lipid into the aqueous phase, which occurs during heating.

Microemulsions are another form of SLN, and the preparation method is similar to hot homogenisation. Instead of homogenising the API-lipid melt with the hot aqueous phase, mild agitation is applied. The hot microemulsion is then dispersed in cold water (2–10 °C), which precipitates causing the formation of SLNs under mild agitation. The large volume of water needed to form SLNs is a drawback of this method as an additional step of ultrafiltration is required to increase the dispersion concentration. Emulsification-sonification is another preparation method similar to hot homogenisation. After homogenisation of the hot melt and aqueous phase, the mixture is ultrasonicated in order to produce a nanoemulsion, which is then cooled at room temperature to produce SLNs. During sonication, metal contamination from the probe is known to occur.

Some methods involve the use of organic solvent such as solvent emulsification-evaporation (SEE), solvent diffusion (SD) and solvent injection (SI) methods. For each method, lipid is dissolved in a solvent and then emulsified in an aqueous phase comprising surfactants. During emulsification, the solvent evaporates, leading to lipid precipitation and SLN formation. Similar to microemulsions, a filtration step is required to concentrate the dispersion due to the large volumes of diluent used. SEE and SI methods utilise water-immiscible solvents with stirring or injection used to mix the two phases, respectively, whereas water-miscible solvents and stirring are used during the SD method [62–64].

Disadvantages of SLNs include the high water content (70–95%) of the dispersion, which reduces dosing efficiency, limited drug loading capacity and issues regarding long-term stability including the tendency for drug expulsion during storage due to lipid crystallisation and rearrangement. This can also lead to dissolution changes, polymorphic transitions, particle growth and dispersion gelation over time. These stability issues are largely overcome by NLCs, which use a combination of spatially incompatible lipids to maintain the spacing between the fatty acid chains of glycerides and crystal imperfections. This larger internal spacing also allows for higher drug loading capacity and controlled release with a less pronounced burst release, which is apparent for SLNs. For both SLNs and NLCs,

Table 3 Literature examples of sustained-release solid lipid particle formulations

API	Sub-type	Method of manufacture	Maximum drug release (%, h)	References
Hyhdrochlorathiazide	SLN	Hot homogenisation and sonication	75, 6	[65]
Aciclovir	SLN	Emulsification-sonication	80, 35	[66]
Ibuprofen	SLN	Emulsification-evaporation	100, 10	[67]
Melatonin	SLN	Microemulsion	Not given	[68]
ODA-FITC (investigative API)	SLN	Solvent diffusion	Not given	[69]
Rifampicin, isoniazid and pyrazinamide	SLN	Emulsion solvent diffusion	20, 72	[70]
Puerarin	NLC	Solvent evaporation	Not given	[71]

Table 4 Drugs and polymers investigated for in situ gelling

API	Polymer(s)	References
Cefpodoxime proxetil	Pectin and gellan	[77]
Montelukast sodium	Pluronic F127 and xyloglucan	[78]
Nateglinide	Calcium-free alginate and alginate/chitosan	[79]
Cimetidine	Gellan gum, enzyme-degraded xyloglucan or sodium alginate	[80]
Paracetamol	Xyloglucan (1.5%w/v) and alginate (0.5%w/v)	[74]
Paracetamol	Alginate and methylcellulose	[72]
Ranitidine	Sodium alginate and calcium carbonate	[81]
Amoxicillin	Alginate with hydroxypropyl methylcellulose	[82]
Pregabalin		[83]
Nizatidine		[84]
Ofloxacin		[85]
Losartan potassium		[86]

Adapted from [76]

the burst release profile is difficult to negate, and so applications for sustained-release formulations may be inappropriate although there are some literature examples as shown below (Table 3).

4.1.2 In Situ Gelling

In situ gelling formulations comprise an aqueous solution of a combination of polymers to be administered or taken as an oral liquid, which then gels in specific conditions in the gastrointestinal tract. These products often involve the use of ion and thermo-responsive polysaccharides including gellan, alginate, pectin, methylcellulose and xyloglucan [72–75]. The gelation is triggered in response to changes in the environmental pH, temperature and/or ionic composition once the liquid has reached the stomach or the small intestine. Table 4 shows the list of polymers investigated as in situ gelling systems.

Xyloglucan forms thermoreversible gels in situ and can be combined with other thermo-responsive polymers, such as Pluronics, or ion-responsive polymers, e.g. alginate, to prepare in situ gelling formulations. The availability of xyloglucan was considered to be problematic for commercial development, and methylcellulose was preferred as an alternative in combination with either pectin or alginate. When methylcellulose was combined with pectin, a source of divalent ions and D-sorbitol were required in order to provide the required hardness and ensure thermo-responsive in situ gelation, respectively [87]. The replacement of pectin with alginate led to a simplification in the formulation as divalent ions were not required to provide satisfactory gelation, although D-sorbitol was still required. SABER™ and ORADUR™ are technologies in phase III clinical trials which utilise in situ gelling to provide sustained delivery of bupivacaine and methylphenidate, respectively, for pain management awaiting marketing approval [87]. The nature of this technology means that formulations will be sensitive to storage conditions, possibly making them unsuitable in warmer countries. Furthermore, the use of ion-responsive polymers will only be suitable for non-reactive APIs to avoid premature gelation.

4.1.3 Emulsion-Based Systems

Emulsion-based systems have been shown to be a viable option for the development of sustained-release formulations in the liquid format. More specifically, multiple emulsions, which are described as ‘emulsions of emulsions’, are being investigated and developed [88, 89]. These multiple emulsion systems are complex as they involve the sequential combination of immiscible phases with the aid of solvents, stabilisers and surfactants (Fig. 2). Several examples have been shown in literature to demonstrate viability for this platform; drug entrapment did not appear challenging to achieve, and emulsions with small particle sizes and high uniformity were obtained [88, 90–95]. The long-term stability has been a concern for these systems. Other limitations include the requirement for large amounts of chemicals and energy at the preparation stages of manufacturing as well as the difficulty in removing solvent-based components from the formulation post-production.

4.2 Sustained-Release Microparticles

Microparticles can potentially provide a suitable platform for sustained-release drug delivery to older adults with swallowing difficulties. They are defined as particles with a diameter less than 1 mm; however, particles smaller than 250 μm are

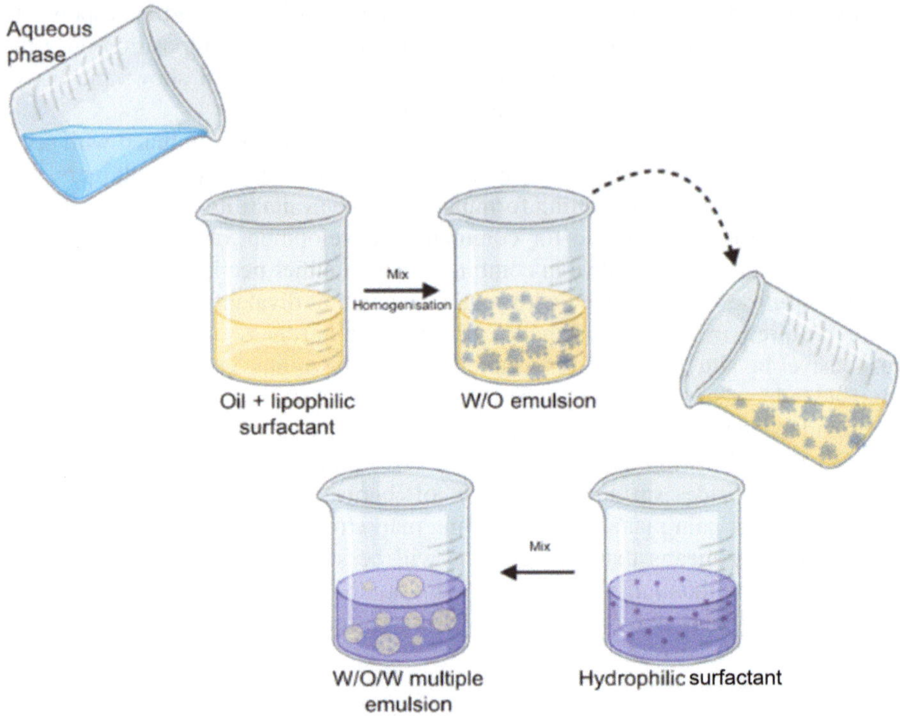

Fig. 2 Schematic for steps involved in emulsion-based formulation preparation. (Reproduced with permission from [88]. Created with Biorender.com)

preferred as they are more likely to provide an appealing mouthfeel with minimal grittiness, favourable in ensuring patient acceptability and compliance [96]. Producing microparticles of this size and simultaneously maintaining sustained-release properties are technically challenging and limit the number of commercially available products on the market. However, several manufacturing techniques have been investigated to overcome this challenge.

4.2.1 Spray Drying

Spray drying is a technique where solutions of polymer and drug are atomised at high temperature, leading to the formation of solid matrix particles. Several literature examples show that spray drying can be used to produce sustained-release microparticles of the requisite size for oral drug delivery [97–100]. Spray-dried microparticles (< 200 μm in diameter) of mirabegron were produced using sodium lauryl sulphate, forming lauryl sulphate-drug complex, which showed pH-independent drug release for up to 12 h. The particles were dispersed in water to provide an oral suspension suitable for administration to older adults with dysphagia

[101]. Similar research has shown the development of spray-dried micro-/nanoparticles for sustained release using polymers such as ethyl cellulose, hydroxypropyl methylcellulose phthalate and gelatin for reconstitution into oral suspensions [102, 103]. The disadvantage of using spray-drying technique is the high losses and low yield due to the vast range of process outcomes, which is undesirable as they can lead to the formation of particles with differing internal structure and morphology.

4.2.2 Ion Exchange Resins

As previously mentioned, ion exchange resins are used in producing sustained-release microparticles, which can be formulated into ODTs or oral suspensions for administration to older adults with dysphagia. Ion exchange resins reduce the rate of drug release by forming ionic bonds with drug molecules, which are then exchanged with ions present in the dissolution media (i.e. the gastrointestinal fluid). This technology relies on the affinity of the resin to the drug in order to retard drug release, which can be affected by the ionic strength of the dissolution media. If the affinity of the API molecule to the resin is low, the API will not form strong enough ionic bonds with the resin and will be released at a fast rate; if the ionic strength of the dissolution media is high, ions in the media will displace the API rapidly and again, fast drug release will be achieved [104].

In addition to the aforementioned ODTs, ion exchange resins have been applied in commercial products as sustained-release oral suspensions including MST Continus (morphine sulphate) by Napp Pharmaceuticals Ltd. and Quillivant XR (methylphenidate hydrochloride) by NextWave Pharmaceuticals Inc., both of which are available as dry powders for reconstitution [87]. In these examples, ion exchange resin-drug matrices are coated with an acrylate-based polymer to produce sustained-release particles with a particle size of 100–250 μm. To achieve a zero-order drug release profile, resin-drug complexes without coating are used in combination with the coated matrices. Delsym® (dextromethorphan cough syrup) and Phentuss® (codeine and chlorpheniramine syrup) are other marketed formulations of ion exchange resin sustained-release drug delivery systems. In these formulations, polyethylene glycol-treated resinates are coated with hydrophobic polymers to control drug release. This technology is known as Pennkinetic™ system and was originally patented by Pennwalt Corporation [105].

Resin purification is usually necessary for product development, which adds cost to the manufacturing process. The reliance on additional polymer coatings to control drug release makes the use of ion exchange resins somewhat redundant [106–108]. Another limitation of the ion exchange resin technology with regard to development into oral liquids for patients with dysphagia is that the vehicles that can be used must be non-ionic in order to prohibit leakage of drug into the vehicle.

4.2.3 Coacervation

Coacervation can be effectively used for the preparation of stable microspheres and is the process whereby two phases are combined and subsequently separated using an environmental influence such as temperature or pH [109, 110]. A classic example of this technique includes the combination of calcium and alginate, whereby cross links are formed between the calcium ions and the acidic subunits of alginate. This leads to the formation of what is known as an 'egg-box' structure [111]. These structures are then dried in order to produce the final microparticles; however, this is where difficulties can be encountered. As the particles are dried, they become porous, which makes providing a sustained-release profile challenging using this technology.

4.2.4 Spray Congealing

Another method used to produce sustained-release microparticles is spray congealing. This involves atomising a molten preparation of drug mixed with control release polymers or waxes; after cooling, droplets solidify to form particles of drug-polymer/wax matrix. Particles between 150 and 250 μm can be obtained using various materials such as carnauba wax or polyethylene glycol (PEG 4000) [112, 113]. To apply the technology, novel design of the apparatus is required to be able to nebulise highly viscous systems, which would allow for good particle manufacturing with a particle size less than 300 μm.

4.2.5 Coating

Fluidised bed coating of microparticles has been shown to be effective in developing sustained-release formulations, which can then be formulated into oral suspensions. Wurster fluidised bed coating is widely used in producing coated multi-particulates for taste masking and modified drug release. The technique is established and easily scalable for commercial scale production compared to the previously mentioned technologies. However, coating particles smaller than 200 μm encounter particle agglomeration and aggregation due to the large surface area, rendering high surface cohesion. Advancement has been made utilising periodical addition of dry powder glidants during the coating process to improve coating efficiency for particles sized 150 μm and smaller (Fig. 3) [114]. The powdered glidant forms a surface coating surrounding the partially dried particles, modifying surface properties of the particles including decreasing surface energy and preventing the rubber-rubber interactions between coated particles. The coated microparticles provided up to 20 h drug release for metoprolol succinate once reconstituted into an oral suspension. The approach was applied to develop gliclazide sustained-release particles, which were incorporated into oral jellies for anti-diabetic treatments for older patients with dysphagia (Fig. 4) [115].

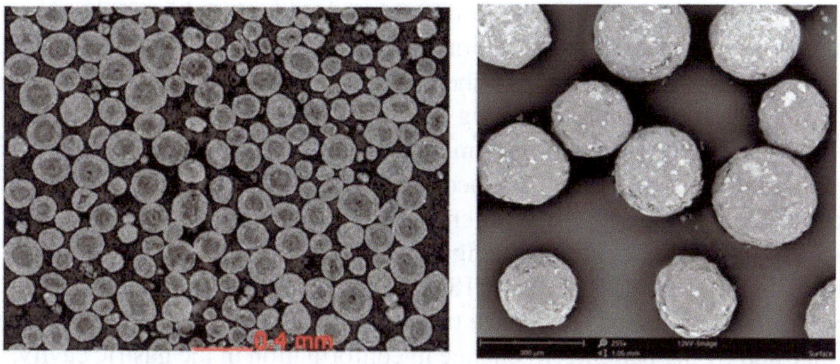

Fig. 3 Structure of metoprolol succinate-loaded and Eudragit® NM 30 D-coated microparticles: virtual cross sections using computed tomography (left) and SEM (right). (Reproduced with permission from [116], distributed under the terms of the Creative Commons Attribution 4.0 International License)

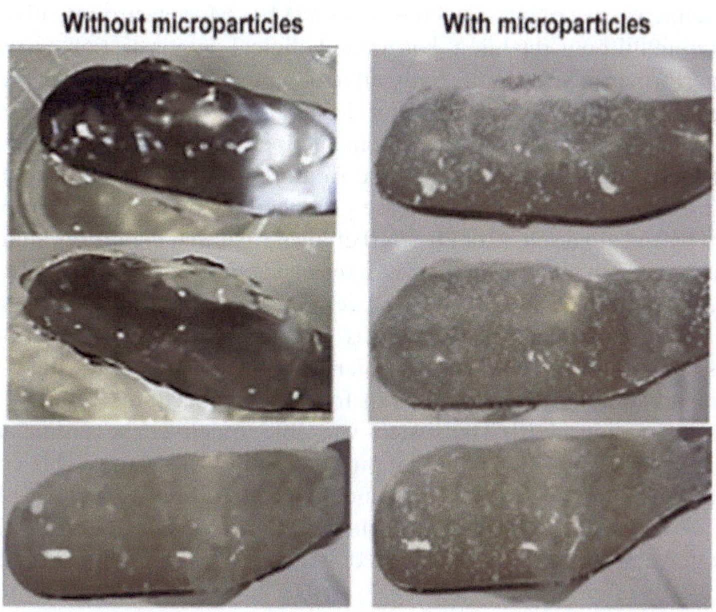

Fig. 4 Images of jellies with and without coated microparticles for older patients with dysphagia. (Reproduced with permission from [115])

4.3 *Ultra-Long-Lasting Oral Formulations*

Active substances with long half-lives due to slow rate of elimination provide opportunities of weekly or monthly dosing; examples include methotrexate and alendronic acid, which are to be taken weekly and ibandronic acid available for monthly administration. There is no definition for the duration of the therapeutic effect for a 'long-lasting' formulation; however, there is evidence to show that monthly dosing provided the best adherence rates [117]. For the majority of active substances which do not exhibit long half-lives, whilst achieving monthly dosing might be impractical for oral administration, prolonged-release formulations accomplishing weekly dosing have been investigated.

Gastro-retentive systems have been used as the main approach to obtain long-lasting oral administration, including formulations that float, sediment, expand, unfold or adhere to gastric mucosa [118]. In principle, formulations that are designed to have a prolonged gastric residence time should (i) be of a suitable size and shape to be ingested by the patient; (ii) change in conformation in the gastric cavity, which hinders passage through the pylorus; (iii) be able to carry large drug loads; (iv) deliver prolonged release for weeks or months with low or no dose dumping risk; (v) maintain API stability in gastric conditions; (vi) predictably break down for gastric emptying without risk of damage to the stomach and subsequent intestinal components; and (vii) have safety mechanisms in place to enable dissociation of the macrostructure in case of intestinal obstruction [119]. Most gastro-retentive systems have shown multi-hour retention, but limited clinical demonstration exists beyond 8 h, and performance was significantly affected in the fasted state [120, 121].

Few gastro-retentive sustained-release formulations are currently marketed, and there are examples that are in the development pipeline. One technological platform that has been used commercially is called Acuform®, which are expanding tablets able to provide 10 h sustained release. Another technology, Accordion Pill®, developed by IntecPharma, is in Phase III clinical trials and comprises a multi-layer folding tablet, which allows for up to 12 h residence time in the stomach [120, 122]. A design example of a modular gastric residence vehicle, in which a thermoset material Poly(ε-caprolactone) (PCL) was selected to form a rigid drug release matrix, is shown in the literature. An enteric polymer was used as linkers for the rigid elements to control safety and exit from the stomach, as well as to prevent intestinal obstruction [119]. The system was able to deliver the antimalarial drug ivermectin over a 10–14-day period in pigs during in vivo testing. This concept of utilising an elastomeric 'core' connected to drug-loaded rigid but detachable 'arms' has been explored in the treatment of human immunodeficiency virus (HIV) and demonstrated once-weekly delivery of oral anti-retrovirals (Fig. 5) [123]. This

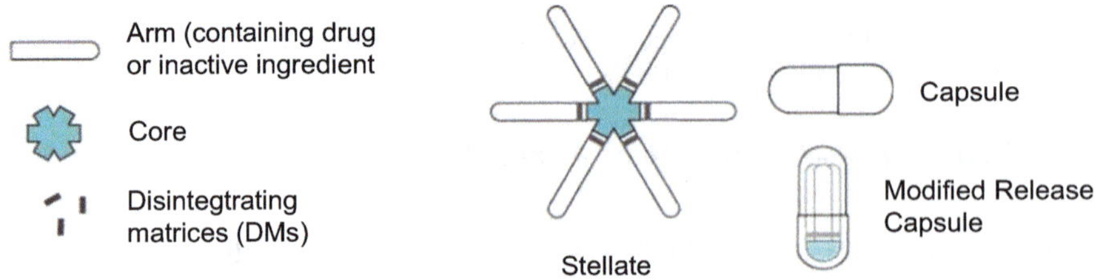

Fig. 5 Core and arm concept for ultra-long-lasting oral drug delivery. (Reproduced with permission from [122])

disease area is of increasing importance for older adults as patients with the condition are now living longer into older age. Investigators of this technology used thermoplastic polyurethane (Elastollan®) as the structural polymer and a combination of poly(sebacic anhydride), poly(adipic anhydride) and poly(ethylene glycol) to adjust drug release. The core and arms were prepared by a micro compactor extruder and injection moulding. After moulding, pockets were formed in the arms, which were then filled with drug and polymer mixtures. The core and arms were connected using enteric polymer (Eudragit® L100–55) with an adhesive plasticizer butyl methacrylate-methyl methacrylate copolymer (Plastoid® B) [119, 123, 124].

Lyndra Therapeutics used a similar stellate design of the core and arm approach with an additional sustained-release coating applied through a dipping method [122]. The coatings investigated include EC, Eudragit® RS, cellulose acetate and PCL. All coatings were shown to be successful with regard to achieving drug release over 7 days; however, PCL was found to be the most resistant to alcohol-induced dose dumping (in 40% ethanol) and therefore favoured for development. Prolonged release of memantine, a drug used in the treatment of Alzheimer's disease, was achieved using the design. General risks of dose dumping were deemed insignificant due to the use of drug particles as a crystalline suspension and the insoluble nature of the matrix material. The impact of food on memantine absorption and formulation residence time was observed, and apart from a slight reduction in drug release over the first 8 h in the fed state, there was no difference between fasted and fed states after the first 24 h of administration [122].

A drawback for the ultra-long-lasting systems is that the minimum capsule shell size required to deliver the formulation would be size 00 or approximately 2.3 cm long. Depending on the patient's swallowing capability, this may offset the benefits of reduced dosing frequency. Furthermore, translation of animal in vivo studies to therapeutic outcomes in humans is not always direct [125, 126]. Using high-resolution pill tracking, the gastric emptying forces in dogs were found to be three to four times higher than in humans, and it was suggested that dog models might underestimate the gastric residence in humans [127]. Pig models have been used to provide closer representation of the gastric environmental conditions in humans; however, there is contradicting evidence as to the effectiveness and success of this approach [119, 124]. In addition, it remains a challenge to achieve gastric residence time beyond 24 h, and the application of gastro-retentive systems is limited to APIs that can display stability in the gastric environment for a prolonged period of time and exhibit acceptable absorption at this site.

5 Concluding Remarks

Oral sustained-release drug delivery systems reduce pill burden and dosing frequency, as such improving acceptability and adherence in older adults. The high prevalence of dysphagia in the older population challenges the administration of conventional sustained-release dosage forms that are often large size monolithic. Oral sustained-release multi-particulates ('sprinkles'), mini-tablets, ODTs and chewable tablets have been used in commercial products to improve swallowability for older adults. Recent advancements have seen the development of oral sustained-release liquids such as in situ gelling systems and multiple emulsions as promising solutions. Microparticles were shown to be a versatile approach to provide sustained release and to be incorporated into a range of patient-friendly dosage forms such as oral liquid suspensions, ODTs and jellies. Various technical methods can be used to produce sustained-release microparticles by spray drying, coacervation, ionic resin complexation, spray congealing and fluidised bed coating. The design of sophisticated gastro-retentive systems has instigated an era of ultra-long-lasting oral drug delivery for ultimate dose frequency reduction to once weekly and even once monthly. The movement is still at its infancy with many barriers to be overcome including manufacturing complexity, large size, cost, safety, in situ stability and duration of retention time; however, the apparent benefit to patient acceptability and adherence will undoubtedly drive innovation in the field.

References

1. Sönnichsen A, Trampisch US, Rieckert A, Piccoliori G, Vögele A, Flamm M, et al. Polypharmacy in chronic diseases–reduction of inappropriate medication and adverse drug events in older populations by electronic decision support (PRIMA-eDS): study protocol for a randomized controlled trial. Trials. 2016;17(1):1–9. https://doi.org/10.1186/S13063-016-1177-8.
2. World Health Organisation. Ageing and health. 2021. https://www.who.int/news-room/fact-sheets/detail/ageing-and-health. Accessed 13 Aug 2021.
3. Briefing: Health and Care of Older People in England 2019. 2019.
4. Qato DM, Wilder J, Schumm LP, Gillet V, Alexander GC. Changes in prescription and over-the-counter medication and dietary supplement use among older adults in the United States, 2005 vs 2011. JAMA Intern Med. 2016;176(4):473–82. https://doi.org/10.1001/JAMAINTERNMED.2015.8581.
5. Toh MR, Teo V, Kwan YH, Raaj S, Tan SYD, Tan JZY. Association between number of doses per day, number of medications and patient's non-compliance, and frequency of readmissions in a multi-ethnic Asian population. Prev Med Rep. 2014;1:43–7. https://doi.org/10.1016/J.PMEDR.2014.10.001.
6. Laliberté F, Bookhart BK, Nelson WW, Lefebvre P, Schein JR, Rondeau-Leclaire J, et al. Impact of once-daily versus twice-daily dosing frequency on adherence to chronic medications among patients with venous thromboembolism. Patient: Patient-Center Outcom Res. 2013;6(3):213–24. https://doi.org/10.1007/S40271-013-0020-5.
7. Dezii CM, Kawabata H, Tran M. Effects of once-daily and twice-daily dosing on adherence with prescribed glipizide oral therapy for type 2 diabetes. South Med J. 2002;95(1):68–72.
8. Reginster JY, Rabenda V, Neuprez A. Adherence, patient preference and dosing frequency: understanding the relationship. Bone. 2006;38(4):S2–6. https://doi.org/10.1016/J.BONE.2006.01.150.
9. Reid DM. Once-monthly dosing: an effective step forward. Bone. 2006;38(4):S18–22. https://doi.org/10.1016/J.BONE.2006.01.153.

10. Shariff ZB, Dahmash DT, Kirby DJ, Missaghi S, Rajabi-Siahboomi A, Maidment ID. Does the formulation of oral solid dosage forms affect acceptance and adherence in older patients? A mixed methods systematic review. J Am Med Dir Assoc. 2020;21(8):1015–1023.e8. https://doi.org/10.1016/J.JAMDA.2020.01.108.
11. Reza Rezaie H, Esnaashary M. Aref arjmand A, Öchsner A. Hist Drug Deliv Syst. 2018:1–8. https://doi.org/10.1007/978-981-10-0503-9_1.
12. Leder SB, Suiter DM. An epidemiologic study on aging and dysphagia in the acute care hospitalized population: 2000–2007. Gerontology. 2009;55(6):714–8. https://doi.org/10.1159/000235824.
13. Leder SB, Suiter DM, Agogo GO, Cooney LM. An epidemiologic study on ageing and dysphagia in the acute care geriatric-hospitalized population: a replication and continuation study. Dysphagia. 2016;31:619–25. https://doi.org/10.1007/s00455-016-9714-x.
14. Wright D. Medication administration in nursing homes. Nurs Stand. 2002;16(42):8.
15. Smyth J. The NEWT guidelines for administration of medication to patients with enteral feeding tubes or swallowing difficulties. 2nd ed. Wrexham: Betsi Cadwaladr University Local Health Boad (East); 2010.
16. Wright D, Chapman N. Medication management of adults with swallowing difficulties Identifying patients with swallowing difficulties. 2013;50. www.eGuidelines.co.uk
17. Fair R, Proctor B. Guidelines for the administration of drugs through enteral feeding tubes. 2004.
18. Hill S, Varker AS, Karlage K, Myrdal PB. Analysis of drug content and weight uniformity for half-tablets of 6 commonly split medications. J Manag Care Pharm. 2015;15(3):253–61. https://doi.org/10.18553/jmcp.2009.15.3.253.
19. Manrique YJ, Lee DJ, Islam F, Nissen LM, Cichero JAY, Stokes JR, et al. Crushed tablets: does the administration of food vehicles and Thickened Fluids to aid medication swallowing alter drug release? J Pharm Pharm Sci. 2014;17(2):207–19.
20. Stubbs J, Haw C, Dickens G. Dose form modification – a common but potentially hazardous practice. A literature review and study of medication administration to older psychiatric inpatients. Int Psychogeriatr. 2008;20(3):616–27. https://doi.org/10.1017/S1041610207006047.
21. Schier JG, Howland MA, Hoffman RS, Nelson LS. Fatality from administration of labetalol and crushed extended-release nifedipine. Ann Pharm. 2003;37(10):1420–3.
22. Cornish P. "Avoid the crush": hazards of medication administration in patients with dysphagia or a feeding tube. CMAJ. 2005;172(7):871–2. https://doi.org/10.1503/CMAJ.050176.
23. Nissen LM. Australian don't rush to crush handbook. Aust Prescr. 2012;35(5):147.
24. Kelly J, Wright D, Wood J. Medicine administration errors in patients with dysphagia in secondary care: a multi-centre observational study. J Adv Nurs. 2011;67(12):2615–27. https://doi.org/10.1111/J.1365-2648.2011.05700.X.
25. Verrue C, Mehuys E, Boussery K, Remon J, Petrovic M. Tablet-splitting: a common yet not so innocent practice. J Adv Nurs. 2011;67(1):26–32. https://doi.org/10.1111/J.1365-2648.2010.05477.X.
26. Kelly J, Eggleton A, Wright D. An analysis of two incidents of medicine administration to a patient with dysphagia. J Clin Nurs. 2011;20(1–2):146–55. https://doi.org/10.1111/J.1365-2702.2010.03457.X.
27. Paradiso LM, Roughead EE, Gilbert AL, Cosh D, Nation RL, Barnes L, et al. Crushing or altering medications: what's happening in residential aged-care facilities? Australas J Ageing. 2002;21(3):123–7. https://doi.org/10.1111/J.1741-6612.2002.TB00432.X.
28. Stockis A, De Bruyn S, Deroubaix X, Jeanbaptiste B, Lebacq E Jr, Nollevaux F, et al. Pharmacokinetic profile of a new controlled-release isosorbide-5-mononitrate 60 mg scored tablet (Monoket multitab). Eur J Pharm Biopharm. 2002;53(1):49–56. https://doi.org/10.1016/S0939-6411(01)00210-7.

29. Vallet T, Michelon H, Orlu M, Jani Y, Leglise P, Laribe-Caget S, et al. Acceptability in the older population: the importance of an appropriate tablet size. Pharmaceutics. 2020;12(8):746. https://doi.org/10.3390/PHARMACEUTICS12080746.
30. Lennartz P, Mielck JB. Minitabletting: improving the compactability of paracetamol powder mixtures. Int J Pharm. 1998;173(1–2):75–85. https://doi.org/10.1016/S0378-5173(98)00206-3.
31. Schmidt C, Kleinebudde P. Influence of the granulation step on pellets prepared by extrusion/Spheronization. Chem Pharm Bull. 1999;47(3):405–12. https://doi.org/10.1248/CPB.47.405.
32. Lopes CM, Lobo JMS, Pinto JF, Costa P. Compressed mini-tablets as a biphasic delivery system. Int J Pharm. 2006;323(1–2):93–100. https://doi.org/10.1016/J.IJPHARM.2006.05.063.
33. Mehta KA, Kislalioglu MS, Phuapradit W, Malick AW, Shah NH. Effect of formulation and process variables on porosity parameters and release rates from a multi unit erosion matrix of a poorly soluble drug. J Control Release. 2000;63(1–2):201–11. https://doi.org/10.1016/S0168-3659(99)00193-5.
34. de Brabander C, Vervaet C, Remon JP. Development and evaluation of sustained release mini-matrices prepared via hot melt extrusion. J Control Release. 2003;89(2):235–47. https://doi.org/10.1016/S0168-3659(03)00075-0.
35. Mohamed FAA, Roberts M, Seton L, Ford JL, Levina M, Rajabi-Siahboomi AR. Film-coated matrix mini-tablets for the extended release of a water-soluble drug. Drug Dev Ind Pharm. 2015;41(4):623–30. https://doi.org/10.3109/03639045.2014.891128.
36. Hadi MA, Raghavendra Rao NG, Hadi A. A novel approach to sustained montelukast sodium release: differentially coated mini-tablets in HPMC capsules. Int J Pharm Biom Sci. 2011;2(2):90–7.
37. Turk M, Šibanc R, Dreu R, Frankiewicz M, Sznitowska M. Assessment of mini-tablets coating uniformity as a function of fluid bed coater inlet conditions. Pharmaceutics. 2021;13(5):746.
38. van Kerrebroeck P, Kreder K, Jonas U, Zinner N, Wein A. Tolterodine once-daily: superior efficacy and tolerability in the treatment of the overactive bladder. Urology. 2001;57(3):414–21. https://doi.org/10.1016/S0090-4295(00)01113-4.
39. Detrusitol XL 4mg – Summary of Product Characteristics (SmPC) – (emc). 2021. https://www.medicines.org.uk/emc/product/1535/smpc#POSOLOGY. Accessed 10 Sep 2021.
40. Stefan H, Hermann S. Sustained release valproate in the treatment of epilepsy-experience with sustained release minitablets (Orfiril ® long) in a once-daily evening dosage. Neuronews. 2006;6(15):1–18.
41. Liu F, Ghaffur A, Bains J, Hamdy S. Acceptability of oral solid medicines in older adults with and without dysphagia: a nested pilot validation questionnaire based observational study. Int J Pharm. 2016;512(2):374–81. https://doi.org/10.1016/J.IJPHARM.2016.03.007.
42. Battu SK, Repka MA, Majumdar S, Madhusudan RY. Formulation and evaluation of rapidly disintegrating fenoverine tablets: effect of superdisintegrants. Drug Dev Ind Pharm. 2007;33(11):1225–32. https://doi.org/10.1080/03639040701377888.
43. Wei Q, Yang F, Luan L. Preparation and in vitro/in vivo evaluation of a ketoprofen orally disintegrating/sustained release tablet. Drug Dev Ind Pharm. 2013;39(6):928–34. https://doi.org/10.3109/03639045.2012.664146.
44. Ahuja V, Verma R, Barabde UV, Hagsten A, Wannerberger K, Boorugu R, et al. Composition for immediate and extended release. 2013.
45. Hussain A, Misbah M, Abbas N, Irfan M, Arshad MS, Shamim R, et al. Design and in vitro characterization of orally disintegrating modified release tablets of naproxen sodium. Turk J Pharm Sci. 2020;17(5):486. https://doi.org/10.4274/TJPS.GALENOS.2019.24445.
46. Elwerfalli AM, Al-Kinani A, Alany RG, Elshaer A. Nano-engineering chitosan particles to sustain the release of promethazine from orodispersables. Carbohydr Polym. 2015;131:447–61. https://doi.org/10.1016/J.CARBPOL.2015.05.064.
47. Neos Therapeutics. Full prescribing information cotempla XR-ODT®. 2017. https://www.accessdata.fda.gov/drugsatfda_docs/label/2017/205489s000lbl.pdf. Accessed 13 Sep 2021.

48. Ogata I, Yamasaki K, Tsuruda A, Tsuzaki S, Ishimatsu T, Hirayama H, et al. Some problems for dosage form based on questionnaire surveying compliance in patients taking tamsulosin hydrochloride. J Pharm Soc Jap. 2008;128(2):291–7. https://doi.org/10.1248/YAKUSHI.128.291.
49. Dahlin C, Cohen AK. Physical aspects of care: Pain and gastrointestinal symptoms – Google Books. 9th ed. New York: Oxford University Press; 2015. https://books.google.co.uk/books?hl=en&lr=&id=Spz4BQAAQBAJ&oi=fnd&pg=PT77&dq=xerostomia+%22ODT%22+older+patients&ots=umgEUaSUBh&sig=aZlNLXACdmBkUnHiluuIt5ijNUs&redir_esc=y#v=onepage&q&f=false. Accessed 13 Sep 2021.
50. Szakonyi G, Zelkó R. Prediction of oral disintegration time of fast disintegrating tablets using texture analyzer and computational optimization. Int J Pharm. 2013;448(2):346–53. https://doi.org/10.1016/J.IJPHARM.2013.03.047.
51. Overholt S. Chewable soft capsule. United States Patent and Trademark Office; US6258380B1. 2000. https://patents.google.com/patent/US6258380B1/en. Accessed 6 June 2022.
52. Dille MJ, Hattrem MN, Draget KI. Soft, chewable gelatin-based pharmaceutical oral formulations: a technical approach. Pharm Dev Technol. 2017;23(5):504–11. https://doi.org/10.1080/10837450.2017.1332642.
53. Matulyte I, Marksa M, Bernatoniene J. Development of innovative chewable gel tablets containing nutmeg essential oil microcapsules and their physical properties evaluation. Pharmaceutics. 2021;13(6):873. https://doi.org/10.3390/PHARMACEUTICS13060873.
54. Michielsen M, Semeijn E, Comijs HC, van de Ven P, Beekman ATF, Deeg DJH, et al. Prevalence of attention-deficit hyperactivity disorder in older adults in The Netherlands. Br J Psychiatry. 2012;201(4):298–305. https://doi.org/10.1192/BJP.BP.111.101196.
55. El-Gazayerly ON, Rakkanka V, Ayres JW. Novel chewable sustained-release tablet containing verapamil hydrochloride. Pharm Dev Technol. 2004;9(2):181–8. https://doi.org/10.1081/PDT-120030248.
56. Tu YH, Perumal A, Kathala K. Methylphenidate extended release chewable tablet – Google Patents. US9295642B2. 2016. https://patents.google.com/patent/US9295642B2/en. Accessed 13 Sep 2021.
57. Chandrasekaran P, Kandasamy R. Solid oral flexible formulations for pediatric and geriatric patients: age-appropriate formulation platforms. Indian J Pharm Sci. 2018;80(1):14–25. https://doi.org/10.4172/PHARMACEUTICAL-SCIENCES.1000326.
58. Matsuo K, Palmer JB. Anatomy and physiology of feeding and swallowing: normal and abnormal. Phys Med Rehabil Clin N Am. 2008;19(4):691–707. https://doi.org/10.1016/J.PMR.2008.06.001.
59. Hughes CV, Baum BJ, Fox PC, Marmary Y, Yeh CK, Sonies BC. Oral-pharyngeal dysphagia: a common sequela of salivary gland dysfunction. Dysphagia. 1(4):173–7. https://doi.org/10.1007/BF02406913.
60. McCulloch TM, Jaffe D. Head and neck disorders affecting swallowing. GI motility. 2006; https://doi.org/10.1038/gimo36.
61. Das S, Chaudhury A. Recent advances in lipid nanoparticle formulations with solid matrix for oral drug delivery. AAPS PharmSciTech. 2011;12(1):62. https://doi.org/10.1208/S12249-010-9563-0.
62. Müller RH, Mäder K, Gohla S. Solid lipid nanoparticles (SLN) for controlled drug delivery – a review of the state of the art. Eur J Pharm Biopharm. 2000;50(1):161–77. https://doi.org/10.1016/S0939-6411(00)00087-4.
63. Müller RH, Radtke M, Wissing SA. Nanostructured lipid matrices for improved microencapsulation of drugs. Int J Pharm. 2002;242(1–2):121–8. https://doi.org/10.1016/S0378-5173(02)00180-1.
64. Saupe A, Gordon KC, Rades T. Structural investigations on nanoemulsions, solid lipid nanoparticles and nanostructured lipid carriers by cryo-field emission scanning electron microscopy and Raman spectroscopy. Int J Pharm. 2006;314(1):56–62. https://doi.org/10.1016/J.IJPHARM.2006.01.022.

65. Cirri M, Mennini N, Maestrelli F, Mura P, Ghelardini C, di Cesare ML. Development and in vivo evaluation of an innovative "hydrochlorothiazide-in cyclodextrins-in solid lipid nanoparticles" formulation with sustained release and enhanced oral bioavailability for potential hypertension treatment in pediatrics. Int J Pharm. 2017;521(1–2):73–83. https://doi.org/10.1016/J.IJPHARM.2017.02.022.
66. Ramakrishnan S, Abdur PB, Parthiban R, Prakash R. Design and evaluation of acyclovir-loaded solid lipid nanoparticles for sustained release. Drug Invent Today. 2020;14:108.
67. Omwoyo WN, Moloto MJ. Encapsulation of ibuprofen into solid lipid nanoparticles for controlled and sustained release using emulsification solvent evaporation technique. 2019;12. https://doi.org/10.22159/ajpcr.2019.v12i7.33652.
68. Priano L, Esposti D, Esposti R, Castagna G, de Medici C, Fraschini F, et al. Solid lipid nanoparticles incorporating melatonin as new model for sustained oral and transdermal delivery systems. J Nanosci Nanotechnol. 2007;7(10):3596–601. https://doi.org/10.1166/JNN.2007.809.
69. Yuan H, Chen J, Du YZ, Hu FQ, Zeng S, Zhao HL. Studies on oral absorption of stearic acid SLN by a novel fluorometric method. Colloids Surf B: Biointerfaces. 2007;58(2):157–64. https://doi.org/10.1016/J.COLSURFB.2007.03.002.
70. Pandey R, Sharma S, Khuller GK. Oral solid lipid nanoparticle-based antitubercular chemotherapy. Tuberculosis. 2005;85(5–6):415–20. https://doi.org/10.1016/J.TUBE.2005.08.009.
71. Ye Q, Li J, Li T, Ruan J, Wang H, Wang F, et al. Development and evaluation of puerarin-loaded controlled release nanostructured lipid carries by central composite design. Drug Dev Ind Pharm. 2021;47(1):113–25. https://doi.org/10.1080/03639045.2020.1862170.
72. Shimoyama T, Itoh K, Kobayashi M, Miyazaki S, D'Emanuele A, Attwood D. Oral liquid in situ gelling methylcellulose/alginate formulations for sustained drug delivery to dysphagic patients. Drug Dev Ind Pharm. 2012;38(8):952–60. https://doi.org/10.3109/03639045.2011.634809.
73. Itoh K, Hatakeyama T, Shimoyama T, Miyazaki S, D'Emanuele A, Attwood D. In situ gelling formulation based on methylcellulose/pectin system for oral-sustained drug delivery to dysphagic patients. Drug Dev Ind Pharm. 2011;37(7):790–7. https://doi.org/10.3109/03639045.2010.541465.
74. Itoh K, Tsuruya R, Shimoyama T, Watanabe H, Miyazaki S, D'Emanuele A, et al. In situ gelling xyloglucan/alginate liquid formulation for oral sustained drug delivery to dysphagic patients. Drug Dev Ind Pharm. 2010;36(4):449–55. https://doi.org/10.3109/03639040903244480.
75. Itoh K, Yahaba M, Takahashi A, Tsuruya R, Miyazaki S, Dairaku M, et al. In situ gelling xyloglucan/pectin formulations for oral sustained drug delivery. Int J Pharm. 2008;356(1–2):95–101. https://doi.org/10.1016/J.IJPHARM.2007.12.049.
76. Essa EA, Elebyary TT, Abdelquader MM, el Maghraby GM, Elkordy AA. Smart liquids for oral controlled drug release: an overview of alginate and non-alginate based systems. J Drug Deliv Sci Technol. 2021;61:102211. https://doi.org/10.1016/J.JDDST.2020.102211.
77. Patel A, Shah D, Modasiya M, Ghasadiya R, Patel D. Development and evaluation of cefpodoxime Proxetil gellan gum based in situ gel. Int J Res Pharm Biom Sci. 2012;1(2):179–90.
78. Mathews R. Development of a sustained release, liquid oral in situ gelling system of Montelukast sodium with improved bioavailability: equivalence testing using earth Mover's distance. Asian J Pharm. 2019;13(02):93. https://doi.org/10.22377/AJP.V13I02.3104.
79. Draget KI, Skjåk-Bræk G, Stokke BT. Similarities and differences between alginic acid gels and ionically crosslinked alginate gels. Food Hydrocoll. 2006;20(2–3):170–5. https://doi.org/10.1016/J.FOODHYD.2004.03.009.
80. Miyazaki S, Kawasaki N, Kubo W, Endo K, Attwood D. Comparison of in situ gelling formulations for the oral delivery of cimetidine. Int J Pharm. 2001;220(1–2):161–8. https://doi.org/10.1016/S0378-5173(01)00669-X.

81. Rohith Bhimagoni Keshavamurthy Sridhar Anegundha Srinatha G. Floating drug delivery of a locally acting H 2-antagonist: an approach using an in situ gelling liquid formulation. Acta Pharma. 2009;59(3):345–54. https://doi.org/10.2478/v10007-009-0021-z.
82. Patel DM, Patel DK, Patel CN. Formulation and evaluation of floating oral in situ gelling system of amoxicillin. Int Schol Res Netw Pharm. 2011; https://doi.org/10.5402/2011/276250.
83. Madan JR, Adokar BR, Dua K. Development and evaluation of in situ gel of pregabalin. Int J Pharm Investigat. 2015;5(4):226. https://doi.org/10.4103/2230-973X.167686.
84. Jadhav SL, Banerjee SK. Formulation and evaluation of floating in situ gel of Nizatidine. Int J Res Pharm Sci. 2013;4(2):250–5.
85. Chaniyara S, Modi D, Patel R, Patel J, Desai R, Chaudhary S. American journal of advanced drug delivery formulation & evaluation of floatable in-situ gel for stomach-specific drug delivery of Ofloxacin. Am J Adv Drug Deliv. 2013;1(3):285–99.
86. Kranthi Kumar K, Swathi M, Srinivas L, Shaik N. Formulation and evaluation of floating in situ gelling system of losartan potassium. Pharm Lett. 2015;7(1):98–112.
87. Trofimiuk M, Wasilewska K, Winnicka K. How to modify drug release in paediatric dosage forms? Novel technologies and modern approaches with regard to children's population. Int J Mol Sci. 2019;20(13):3200. https://doi.org/10.3390/IJMS20133200.
88. Garti N. Double emulsions – scope, limitations and new achievements. Colloids Surf A Physicochem Eng Asp. 1997;123–124:233–46. https://doi.org/10.1016/S0927-7757(96)03809-5.
89. Dhadde G, Raut ID, Gurunath DS, Hanmant MS, Indrayani RD, Manoj NM. An overview on multiple emulsions a review on microspheres: types, method of preparation, characterization and application view project assessment of in vitro anthelmintic activity of Prosopis Juliflora leaves view project an overview on multiple emulsions. Art Asian J Pharm Technol. 2021;11(2):2231–5713. https://doi.org/10.52711/2231-5713.2021.00026.
90. Morais JM, Santos ODH, Nunes JRL, Zanatta CF, Rocha-Filho PA. W/O/W multiple emulsions obtained by one-step emulsification method and evaluation of the involved variables. J Dispers Sci Technol. 2008;29(1):63–9. https://doi.org/10.1080/01932690701688391.
91. Mishra B, Sahoo BL, Mishra M, Shukla D, Kumar V. Design of a controlled release liquid formulation of lamotrigine. J Facul Pharm. 2011;19(2):126.
92. Saxena A, Mishra AK, Verma N, Bhattacharya SS, Ghosh A, Verma A, et al. Gelucire based in situ gelling emulsions: a potential carrier for sustained stomach specific delivery of gastric irritant drugs. Biomed Res Int. 2013; https://doi.org/10.1155/2013/436932.
93. Gaber DA, Alhawas HS, Alfadhel FA, Abdoun SA, Alsubaiyel AM, Alsawi RM. Mini-tablets versus nanoparticles for controlling the release of amoxicillin: in vitro/in vivo study. Drug Des Devel Ther. 2020;14:5405–18. https://doi.org/10.2147/DDDT.S285522.
94. Li M, Bi D, Yao L, Yi J, Fang W, Wu Y, et al. Optimization of preparation conditions and in vitro sustained-release evaluation of a novel nanoemulsion encapsulating unsaturated guluronate oligosaccharide. Carbohydr Polym. 2021;264:118047. https://doi.org/10.1016/J.CARBPOL.2021.118047.
95. Mizushima T, Nakayama RI, Imai M, Namiki N. Development of multi-cellular gel droplets using κ-carrageenan composited alginate for controlled release of hydrophobic components. J Polym Environ. 2021;29(9):2986–92. https://doi.org/10.1007/S10924-021-02092-8.
96. Lopez FL, Bowles A, Gul MO, Clapham D, Ernest TB, Tuleu C. Effect of formulation variables on oral grittiness and preferences of multiparticulate formulations in adult volunteers. Eur J Pharm Sci. 2016;92:156–62. https://doi.org/10.1016/J.EJPS.2016.07.006.
97. Desai KGH, Park HJ. Preparation of cross-linked chitosan microspheres by spray drying: effect of cross-linking agent on the properties of spray dried microspheres. J Microencapsul. 2008;22(4):377–95. https://doi.org/10.1080/02652040500100139.
98. Kristmundsdóttir T, Gudmundsson ÓS, Ingvarsdóttir K. Release of diltiazem from Eudragit microparticles prepared by spray-drying. Int J Pharm. 1996;137(2):159–65. https://doi.org/10.1016/0378-5173(96)04509-7.

99. Desai KGH, Park HJ. Preparation and characterization of drug-loaded chitosan–tripolyphosphate microspheres by spray drying. Drug Dev Res. 2005;64(2):114–28. https://doi.org/10.1002/DDR.10416.
100. Huang LX, Mujumdar AS. The effect of rotary disk atomizer RPM on particle size distribution in a semi-industrial spray dryer. Dry Technol. 2008;26(11):1319–25. https://doi.org/10.1080/07373930802330938.
101. Kasashima Y, Uchida S, Yoshihara K, Yasuji T, Sako K, Namiki N. Oral sustained-release suspension based on a lauryl sulfate salt/complex. Int J Pharm. 2016;515(1–2):677–83. https://doi.org/10.1016/J.IJPHARM.2016.10.032.
102. Emami J, Varshosaz J, Ahmadi F. Preparation and evaluation of a liquid sustained-release drug delivery system for theophylline using spray-drying technique. Res Pharm Sci. 2008;2(1):1–11.
103. Harsha S. Pharmaceutical suspension containing both immediate/sustained-release amoxicillin-loaded gelatin nanoparticles: preparation and in vitro characterization. Drug Des Devel Ther. 2013;7:1027. https://doi.org/10.2147/DDDT.S39956.
104. Sriwongjanya M, Bodmeier R. Effect of ion exchange resins on the drug release from matrix tablets. Eur J Pharm Biopharm. 1998;46(3):321–7. https://doi.org/10.1016/S0939-6411(98)00056-3.
105. Raghunathan Y. Prolonged release pharmaceutical preparations. 1979.
106. Adeyeye MC, Mwangi E, Katpally S, Fujioka K, Ichikawa H, Fukumori Y. Suspensions of prolonged-release diclofenac-Eudragit® and ion-exchange resin microcapsules: II. Improved dissolution stability. J Microencapsul. 2008;22(4):353–62. https://doi.org/10.1080/02652040500100865.
107. Ichikawa H, Fujioka K, Adeyeye MC, Fukumori Y. Use of ion-exchange resins to prepare 100 μm-sized microcapsules with prolonged drug-release by the Wurster process. Int J Pharm. 2001;216(1–2):67–76. https://doi.org/10.1016/S0378-5173(01)00573-7.
108. Kadam AU, Sakarkar DM, Kawtikwar PS. Development and evaluation of oral controlled release chlorpheniramine-ion exchange resinate suspension. Indian J Pharm Sci. 2008;70(4):531. https://doi.org/10.4103/0250-474X.44613.
109. Santos M, Bozza F, Thomazini M, Favaro-Trindade C. Microencapsulation of xylitol by double emulsion followed by complex coacervation. Food Chem. 2015;171:32–9. https://doi.org/10.1016/J.FOODCHEM.2014.08.093.
110. Chuang JJ, Huang YY, Lo SH, Hsu TF, Huang WY, Huang SL, et al. Effects of pH on the shape of alginate particles and its release behavior. Int J Polym Sci. 2017; https://doi.org/10.1155/2017/3902704.
111. Lengyel M, Kállai-Szabó N, Antal V, Laki AJ, Antal I. Microparticles, microspheres, and microcapsules for advanced drug delivery. Sci Pharm. 2019;87(3):20. https://doi.org/10.3390/SCIPHARM87030020.
112. Albertini B, Passerini N, Pattarino F, Rodriguez L. New spray congealing atomizer for the microencapsulation of highly concentrated solid and liquid substances. Eur J Pharm Biopharm. 2008;69(1):348–57. https://doi.org/10.1016/J.EJPB.2007.09.011.
113. Lo JB, Appel LE, Herbig SM, McCray SB, Thombre AG. Formulation design and pharmaceutical development of a novel controlled release form of azithromycin for single-dose therapy. Drug Dev Ind Pharm. 2009;35(12):1522–9. https://doi.org/10.3109/03639040903037223.
114. Mohylyuk V, Patel K, Scott N, Richardson C, Murnane D, Liu F. Wurster fluidised bed coating of microparticles: towards scalable production of oral sustained-release liquid medicines for patients with swallowing difficulties. AAPS PharmSciTech. 2019;21(1):1–10. https://doi.org/10.1208/S12249-019-1534-5.
115. Patel S, Scott N, Patel K, Mohylyuk V, McAuley WJ, Liu F. Easy to swallow "instant" jelly formulations for sustained release Gliclazide delivery. J Pharm Sci. 2020;109(8):2474–84. https://doi.org/10.1016/J.XPHS.2020.04.018.

116. Mohylyuk V, Patel K, Scott N, Richardson C, Murnane D, Liu F. Wurster fluidised bed coating of microparticles: towards scalable production of oral sustained-release liquid medicines for patients with swallowing difficulties. AAPS PharmSciTech. 2020;21(1) https://doi.org/10.1208/s12249-019-1534-5.
117. Rothen JP, Rutishauser J, Walter PN, Hersberger KE, Arnet I. Oral intermittent vitamin D substitution: influence of pharmaceutical form and dosage frequency on medication adherence: a randomized clinical trial. BMC Pharmacol Toxicol. 2020;21(1):1–8. https://doi.org/10.1186/S40360-020-00430-5.
118. Altreuter D, Kirtane AR, Grant T, Kruger C, Traverso G, Bellinger A. Changing the pill: developments towards the promise of an ultra long-acting gastroretentive dosage form HHS public access. Expert Opin Drug Deliv. 2018;15(12):1189–98. https://doi.org/10.1080/17425247.2018.1544615.
119. Bellinger AM, Jafari M, Grant TM, Zhang S, Slater HC, Wenger EA, et al. Oral, ultra-long-lasting drug delivery: application toward malaria elimination goals. Sci Transl Med. 2016;8(365) https://doi.org/10.1126/SCITRANSLMED.AAG2374.
120. Narayanan V. Gastro retentive drug delivery system of salbutamol sulphate: formulation and it's in vitro evaluation. 2008.
121. Prinderre P, Sauzet C, Fuxen C. Advances in gastro retentive drug-delivery systems. Expert Opin Drug Deliv. 2011;8(9):1189–203. https://doi.org/10.1517/17425247.2011.592828.
122. Kanasty R, Low S, Bhise N, Yang J, Peeke E, Schwarz M, et al. A pharmaceutical answer to nonadherence: once weekly oral memantine for Alzheimer's disease. J Control Release. 2019;303:34–41. https://doi.org/10.1016/J.JCONREL.2019.03.022.
123. Kirtane AR, Abouzid O, Minahan D, Bensel T, Hill AL, Selinger C, et al. Development of an oral once-weekly drug delivery system for HIV antiretroviral therapy. Nat Commun. 2018;9(1):1–12. https://doi.org/10.1038/s41467-017-02294-6.
124. Aoyagi N, Ogata H, Kaniwa N, Uchiyama M, Yasuda Y, Tanioka Y. Gastric emptying of tablets and granules in humans, dogs, pigs, and stomach-emptying-controlled rabbits. J Pharm Sci. 1992;81(12):1170–4. https://doi.org/10.1002/JPS.2600811208.
125. Cargill R, Caldwell LJ, Engle K, Fix JA, Porter PA, Gardner CR. Controlled gastric emptying. 1. Effects of physical properties on gastric residence times of nondisintegrating geometric shapes in beagle dogs. Pharm Res. 1988;5(8):533–6. https://doi.org/10.1023/A:1015981627525.
126. Cargill R, Engle K, Gardner CR, Porter P, Sparer RV, Fix JA. Controlled gastric emptying. II. In vitro erosion and gastric residence times of an erodible device in beagle dogs. Pharm Res. 1989;6(6):506–9. https://doi.org/10.1023/A:1015976709043.
127. Laulicht B, Tripathi A, Schlageter V, Kucera P, Mathiowitz E. Understanding gastric forces calculated from high-resolution pill tracking. Proc Natl Acad Sci. 2010;107(18):8201–6. https://doi.org/10.1073/PNAS.1002292107.

3D Printing: Advancements in the Development of Personalised Pharmaceuticals for Older Adults

Atheer Awad, Patricija Januskaite, Manal Alkahtani, Mine Orlu, and Abdul W. Basit

Contents

A. Awad (✉)
Department of Clinical, Pharmaceutical and Biological Sciences, University of Hertfordshire, Hatfield, UK

UCL School of Pharmacy, University College London, London, UK
e-mail: a.awad@herts.ac.uk

P. Januskaite · M. Alkahtani · M. Orlu · A. W. Basit
UCL School of Pharmacy, University College London, London, UK
e-mail: patricija.januskaite.18@ucl.ac.uk; manal.alkahtani.17@ucl.ac.uk; m.orlu@ucl.ac.uk; a.basit@ucl.ac.uk

M. Orlu, F. Liu (eds.), *Pharmaceutical Formulations for Older Patients*, AAPS Advances in the Pharmaceutical Sciences Series 51,
https://doi.org/10.1007/978-3-031-35811-1_7

Abstract Three-dimensional (3D) printing is a revolutionary additive manufacturing technology that constructs objects layer-by-layer. This versatile technology encompasses various processes and possesses unique properties. While originally not intended for this purpose, 3D printing has emerged as a promising approach for personalised pharmaceutical production, specifically Printlets™, which are customised dosage forms. The ageing population presents enduring challenges, and 3D printing offers innovative solutions tailored to this patient subgroup. These solutions encompass a wide range of dosage forms, including multi-drug formulations, orally disintegrating tablets, chewable medicines, multiparticulate systems, and formulations for visually or physically impaired patients. Moreover, besides solid oral dosage forms, 3D printing can also fabricate patient-specific devices for drug delivery via the auricular, topical, intravesical, or vaginal routes. This chapter provides an overview of different 3D printing technologies, highlights the various dosage forms suitable for older patients, and briefly discusses the integration of 3D printing with other advanced digital health technologies.

Keywords Three dimensional printing · Personalized medicines · Digitised healthcare and industry 4.0 · Rapid prototyping · 3D-printed drug products · Oral drug delivery systems

1 Introduction

It is estimated that the number of over 60-year-old individuals will increase from 900 million to 2 billion by 2050, based on the latest findings from the United Nations (UN) and World Health Organisation (WHO) [1]. As a result, the development of targeted medicines for older patients is an ever-growing worldwide necessity in the healthcare sector. Developing effective pharmaceutical products for the large elderly population presents several challenges, such as the complexity of the treatment regimen caused by comorbidities (i.e. polypharmacy), pharmacokinetic alterations (i.e. disturbances in drug elimination due to renal dysfunction), and noncompliance/nonadherence due to acceptability or accessibility concerns [2]. To overcome such challenges, frequent dose adjustments and changes in medicine/dosage form must be carried out, causing stress, discomfort, and lack of treatment effect to the patient. A growing body of evidence indicates that it is essential to move away from the common 'one-size-fits-all' approach and to tailor the treatment to the individual's needs, offering a more patient-centric approach [3, 4].

Three-dimensional (3D) printing, also known as additive manufacturing, is a novel approach to the personalisation of medicine, involving the layer-by-layer formation of a 3D object based on a computer-aided design (CAD) [5]. According to the American Society for Testing and Materials (ASTM), 3D printing is classified into seven distinct categories: binder jetting, vat photopolymerisation, powder bed fusion, material jetting, material extrusion, direct energy deposition and sheet lamination, which are further subdivided into multiple technologies (Fig. 1) [6]. Of the

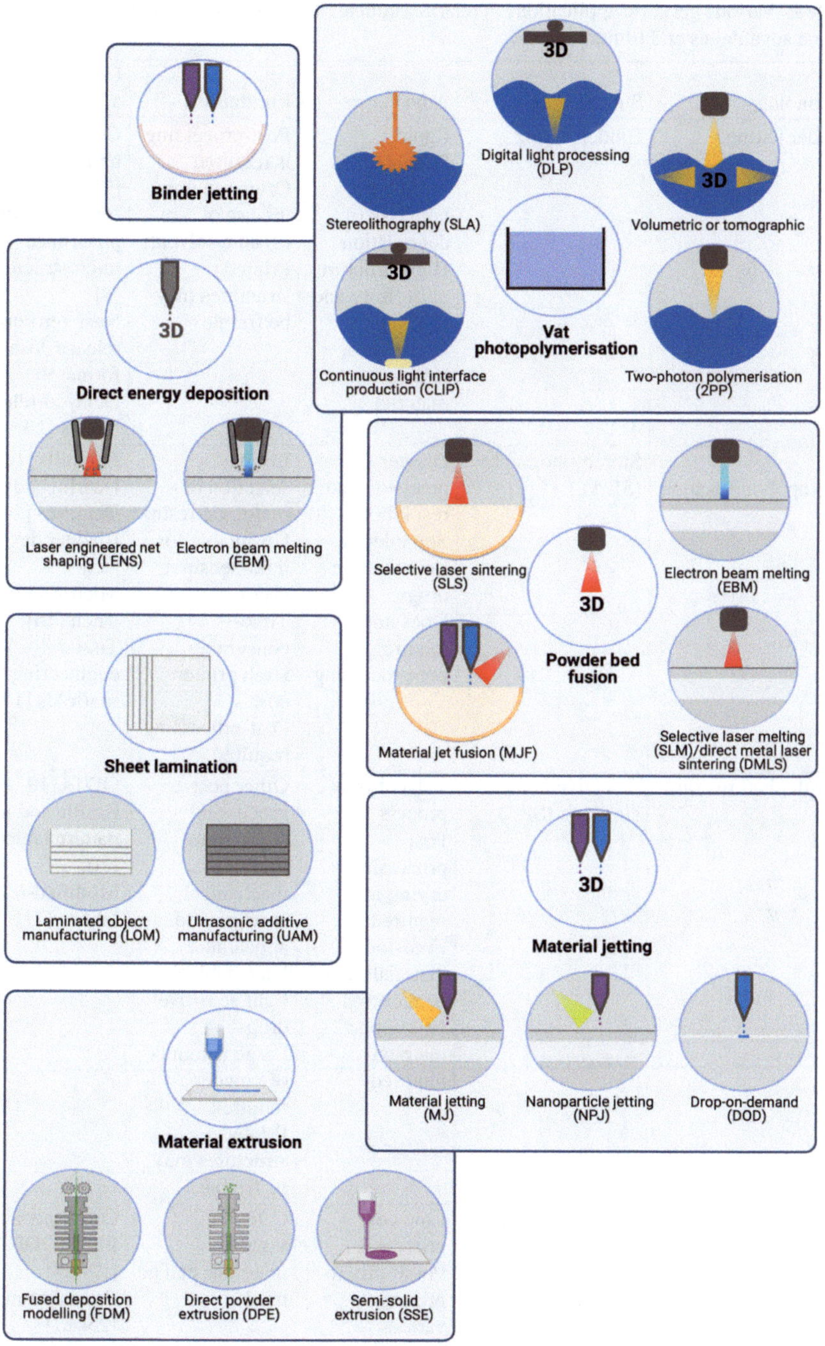

Fig. 1 Schematic representation of the seven 3D printing technologies and their subcategories based on the ASTM standards: binder jetting, vat photopolymerisation, powder bed fusion, material jetting, material extrusion, direct energy deposition and sheet lamination. Created using BioRender.com

Table 1 Various geriatric applications of pharmaceutical 3D printing technologies, alongside their unique advantages and limitations

Technology	Subcategory	Advantages	Limitations	Geriatric applications
Binder jetting	Binder jetting	Rapid Suitable for drugs with high risk of degradation High structure uniformity and porosity No need for a structure material	Post-processing is required Often requires the use of organic solvents Printed structures may be friable	Orodispersible tablets (ODTs) [7] Implants with predefined microstructures [8] Near zero-order-release dosage forms [9] Delayed-release tablets [10]
Vat photopolymerisation	Stereolithography (SLA)	Greater precision and resolution Suitable for thermolabile drugs Does not require pre-processing	Limited selection of non-toxic resins Unsuitable for light-sensitive drugs Time-consuming High printer cost Post-processing required	Polypills [11, 12] Hearing aids [13] Dentures [14] Bladder device [15] Microneedle patch [16] Tissue engineering scaffolds [17]
Powder bed fusion	Selective laser sintering (SLS)	Solvent free process Post-processing drying not required Feedstock materials can be recycled No support material required	Other post-processing required to improve mechanical strength and appearance Unsuitable for light-sensitive drugs Large amounts of powder required Printed structures may be friable	ODTs [18, 19] Braille and moon pattern tablets [20] Modified-release tablets [21]
Material jetting	Material jetting (MJ)	Low-cost process Possibility to print on various substrates High accuracy	Often only viscous materials can be used	Orodispersible films (ODFs) [22–24] Buccal films [25–27]

(continued)

Table 1 (continued)

Technology	Subcategory	Advantages	Limitations	Geriatric applications
Material extrusion	Fused deposition modelling (FDM)	Low cost Variety of materials available No post-processing required	Resolution relatively low Two-step process (time- and resource consuming) Unsuitable for thermolabile drugs	Polypills [28, 29] Chewable tablets [30] ODFs [31] Dentures [32] Catheters [33]
	Semi-solid extrusion (SSE)	Sterile process using disposable syringes Low-temperature printing	Low resolution Post-printing drying often required	ODFs [34, 35] Chewable tablets [36–38] Polypills [39, 40] Suppositories [41–43] Implantable patches [44, 45]
	Direct powder extrusion (DPE)	Single-step process Capable of printing at lower temperatures	Resolution relatively low Require support for complex structures Poor mixing of powders in nozzle	Taste-masking [46] Sustained-release tablets [47]

seven highlighted technology terms, five main ones are currently employed in the pharmaceutical field, and these include binder jetting, vat photopolymerisation, powder bed fusion, material jetting and material extrusion, each with its own unique advantages and limitations for geriatric applications (Table 1).

2 Overview of 3D Printing Technologies

2.1 Binder Jetting

Binder jetting is a 3D printing technology in which the main concept relies on the agglomeration of powder material using liquid binders in a layer-by-layer manner. The printing system consists of two adjacent platforms: a powder bed for powder delivery and a printing platform for object formation. The former rises, and a roller or blade pushes the powder toward the latter to form a thin and even layer of powder.

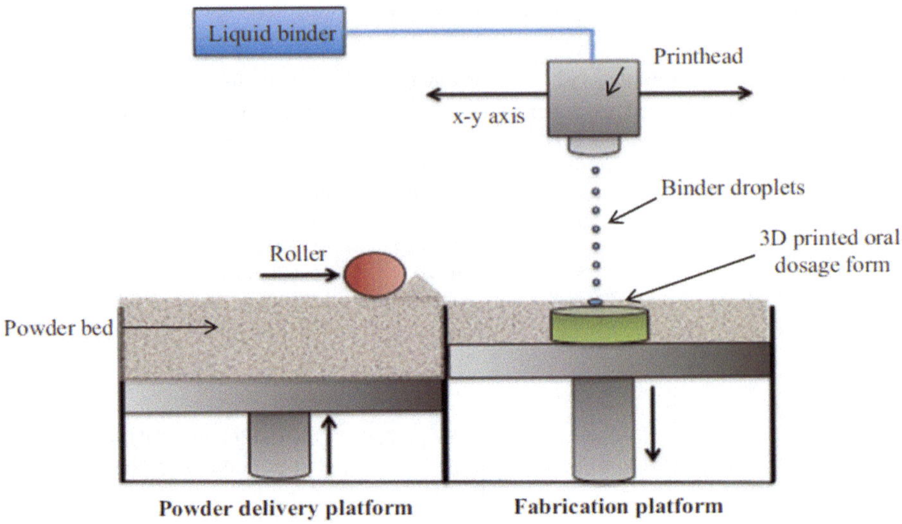

Fig. 2 Schematic illustration of the binder jetting 3D printing process. (Reprinted with permission from [49])

This is followed by the spraying of the liquid binder from the printer nozzle, which moves across the x–y axis, causing the powder particles to bind together in a predetermined pattern, and layer solidification takes place. The building platform then moves down on the z-axis, and a thin layer of powder is dispensed on the top of the existing one; the process repeats successively until the formation of the desired object is complete (Fig. 2). The active pharmaceutical ingredient (API) is either incorporated in the liquid binder solution or mixed with the excipients in powder form [48]. Both powder and liquid binder properties play a significant role in the quality of the product and the precision of the process; the liquid must be non-toxic and have a relatively fast evaporation rate. In case the API is mixed with the liquid binder, drug saturation must be considered to avoid its crystallisation and subsequent blockage of the nozzle. On the other hand, the powder characteristics such as particle size and shape will influence its flowability, wettability, and reactivity. For example, poor flowability may hinder the removal of excess powder [49].

This technology can form highly porous structures with high drug loadings, for example, orally disintegrating tablets (ODTs). This has led to the development of ZipDose®, a technology based on binder jetting by Aprecia Pharmaceuticals in 2015, resulting in the world's first United States Food and Drug Administration (FDA)-approved 3D printed tablet, Spritam® [50]. Spritam® only requires a small volume of saliva to dissolve in under 11 seconds and can be manufactured with a dose of up to 1000 mg [51]. Different types of formulations, such as near zero-order release, ODTs, and delayed-release tablets, have been fabricated by binder jetting 3D printing [7, 9, 10]. Older patients often exhibit swallowing or chewing issues, making ODT dosage forms a suitable and beneficial method of delivering medicines without the risk of choking [52]. Unlike other technologies that may require the application of heat or a light source, binder jetting is suitable for drugs that have a higher risk of degradation when processed with other techniques. The process produces structures with higher uniformity due to the accuracy of deposition of the liquid binder and is a rapid and cost-effective manufacturing technique, being the only 3D printing technology approved for mass manufacture of medicines [53]. Associated disadvantages with binder jetting include the requirement of a post-processing drying step to ensure the evaporation of the liquid binder. Mechanical strength and friability issues may also arise when formulating highly porous structures [54].

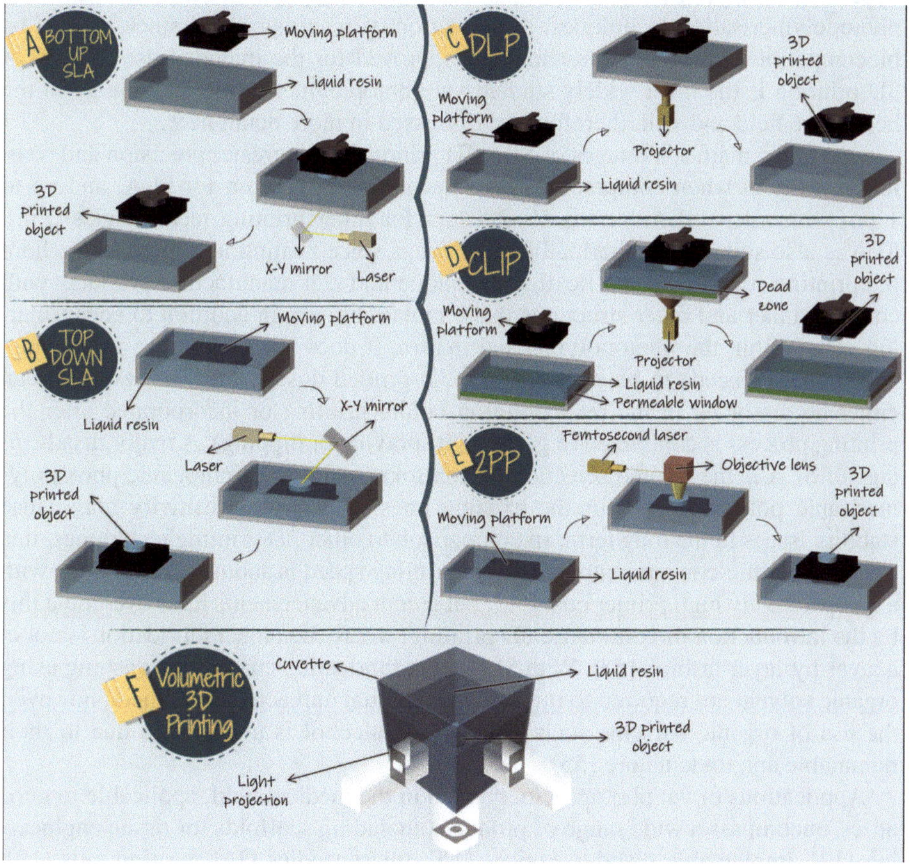

Fig. 3 Graphical illustrations of the vat photo polymerisation technologies, including (A and B) SLA, (C) DLP, (D) CLIP, (E) 2PP and (F) volumetric or tomographic 3D printing. (Reprinted with permission from [55])

2.2 Vat Photopolymerisation

Vat photopolymerisation is an umbrella term used to describe a variety of 3D printing techniques, which include stereolithography (SLA), digital light processing (DLP), continuous light interface production (CLIP), two-photon polymerisation (2PP) and tomographic (or volumetric) 3D printing (Fig. 3) [55, 56]. During the process, a vat containing liquid resin (mixture of API, monomers and photo initiator) is solidified by the irradiation of light to promote polymerisation in a layer-by-layer manner. The only exception is in the case of tomographic 3D printing, whereby the photosensitive liquid resin is simultaneously irradiated in three dimensions, bypassing the layer-by-layer method of production [57]. For all the above-stated vat photopolymerisation techniques, the selection of resin is critical since it must be biocompatible, biodegradable and FDA-approved for the intended use [55]. SLA 3D printing is the most widely studied vat photopolymerisation technology in the healthcare field and will, therefore, be discussed in more detail here.

One of the main advantages of SLA 3D printing is the greater precision and resolution it offers when compared to other technologies (20 μm for SLA, and up to 1 μm with micro-SLA, versus 50–200 μm for other printing technologies) [58]. SLA is also suitable for thermally labile drugs since heating is limited throughout the printing process. It is a flexible technique and can manufacture products with complex inner and outer structures that simulate tissues, in addition to encapsulating cells within the photopolymerised matrix. It does not require pre-processing, and if an API needs to be included in a 3D-printed dosage form or device, it can either be dissolved in the resin solution before printing or incorporated after the printing process is completed (e.g. through spraying or dipping). A major disadvantage of SLA is the limited selection of non-toxic and non-carcinogenic photopolymerisable polymers, and for the existing ones, their photosensitivity may cause stability issues in the long term. In comparison to other 3D printing techniques, it is a relatively time-consuming process (i.e. printing speed is about 1–3 cm/hour), with a comparatively high printer cost [59], but recent advancements have overcome this by the introduction of volumetric 3D printing, where 3D object fabrication is not in a layer-by-layer fashion [57]. With SLA, post-fabrication curing and washing using organic solvent are required to remove any residual unreacted monomers; however, the use of organic solvents such as isopropyl alcohol is undesirable due to their flammable and toxic nature [55].

Applications of vat photopolymerisation in the medical field, applicable to geriatrics, encompass a wide range of products, including scaffolds for tissue engineering [17], implantable bladder devices [15], microneedles [16], hearing aids [13], personalised medical devices [60, 61], and dental prostheses [14]. However, the field of oral drug delivery has not extensively explored this technology due to the previously discussed limitations. Nevertheless, there have been promising developments, such as oral modified-release 3D-printed tablets (Printlets™) containing a combination of 4-aminosalicylic acid (4-ASA) and paracetamol, which demonstrated good resolution and no drug degradation compared with a similar composition developed using FDM [11]. Aadditionally, a multi-layered polypill containing six different drugs was successfully printed using a modified commercial SLA printer, showcasing the potential for using this technology in the development of personalised medicines for older adults, for whom polypharmacy is a major cause of medication non-adherence [12].

2.3 *Powder Bed Fusion*

Powder bed fusion is a 3D printing technology that involves the selective aggregation of powder particles using a centred power source, such as a laser or electron beam [62]. Various subcategories of this technology exist, including selective laser sintering (SLS), selective laser melting (SLM)/direct metal laser sintering (DMLS), electron beam melting (EBM), and multi-jet fusion (MJF) (Fig. 4), with SLS 3D printing currently being the most applied to pharmaceuticals.

SLS is a type of powder bed fusion technique where the starting material is a powder mixture consisting of thermoplastic polymers, API, colourant and other excipients. During the printing process, the build platform is raised, and

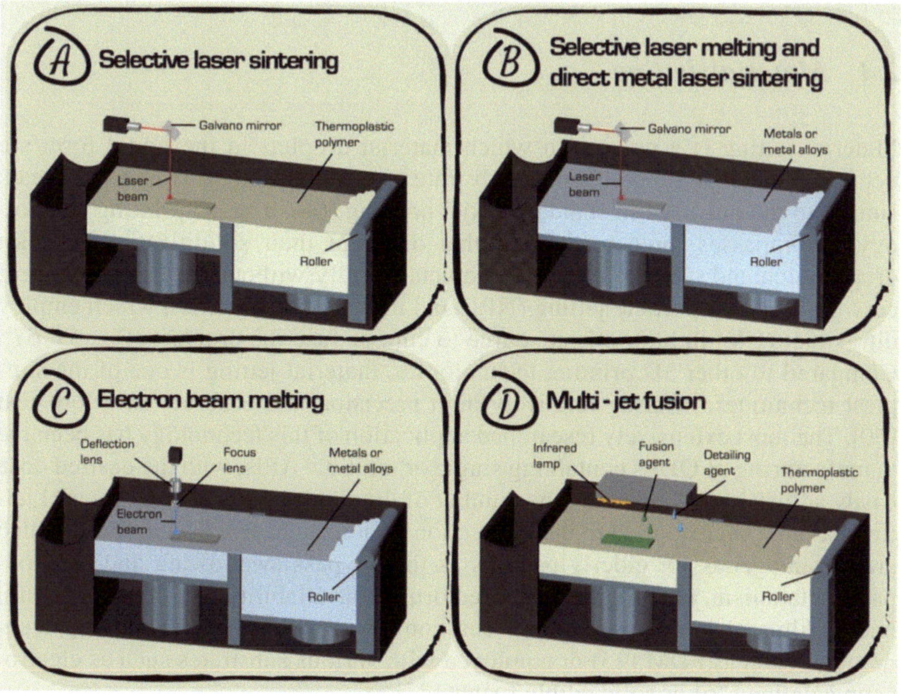

Fig. 4 Graphical illustration of the (**a**) SLS, (**b**) SLM and DMLS, (**c**) EBM and (**d**) MJF 3D printing technologies. (Reprinted with permission from [62])

a new layer of powder is deposited and flattened by a roller. The heating system and laser beam inside the printer are activated, causing the surfaces of the powder particles to fuse together according to the designed geometry, a process known as sintering. This layer-by-layer approach is repeated until the desired object is fully printed. Finally, excess powder is removed by brushing or exposing the object to compressed air [18]. Different types of laser beams are utilised, including fibre, diode, and most commonly carbon dioxide (CO_2) lasers. Important process parameters such as the laser scanning speed, powder bed temperature, and powder flow of the formulation mixture directly influence the porosity, printability, resolution and structure of the final print [63, 64].

SLS is a solvent-free process that does not require a post-fabrication drying stage like binder jetting, and the feedstock materials are recyclable, reducing waste and saving money. It can produce highly porous structures, which are essential for ODTs produced for patients with swallowing and chewing difficulties [18, 19]. SLS is also capable of producing intricate structures with high resolution, such as gyroid lattices for modified drug release profiles [21]. Unique dosage forms, such as mini-Printlets can be manufactured using SLS, and Printlets with braille or moon patterns for identification and added independence for visually impaired elderly patients [20, 65].

2.4 *Material Jetting*

Material jetting is a process in which material droplets in the liquid form are deposited onto a build platform or substrate (e.g., edible paper) from a printhead, similar to the one used in standard inkjet printing [66]. The 3D printing technology encompasses three techniques that differ in their solidification process: drop-on-demand (DOD), in which droplets solidify without requiring post-processing, and nanoparticle jetting (NPJ) and material jetting (MJ), which employ ultraviolet (UV) light or a heat source to cure or fuse the droplets (Fig. 5) [67]. Compared to other 3D printing technologies, material jetting is one of the simplest techniques, offering relatively high precision and accuracy at a low cost [50]. The most extensively researched application of this technology has been the manufacturing of ODFs containing single or multiple APIs, with the desired dose easily adjustable by changing the number of droplet layers [22–24]. Material jetting has also been applied to the fabrication of mucoadhesive buccal films, which are advantageous for elderly patients as they bypass swallowing and the first-pass metabolism, resulting in improved drug bioavailability [68]. For successful buccal films, material jetting has been combined with other 3D printing technologies such as FDM [25] or employed with various substrates such as electrospun gelatine [26] or solid edible foams [27].

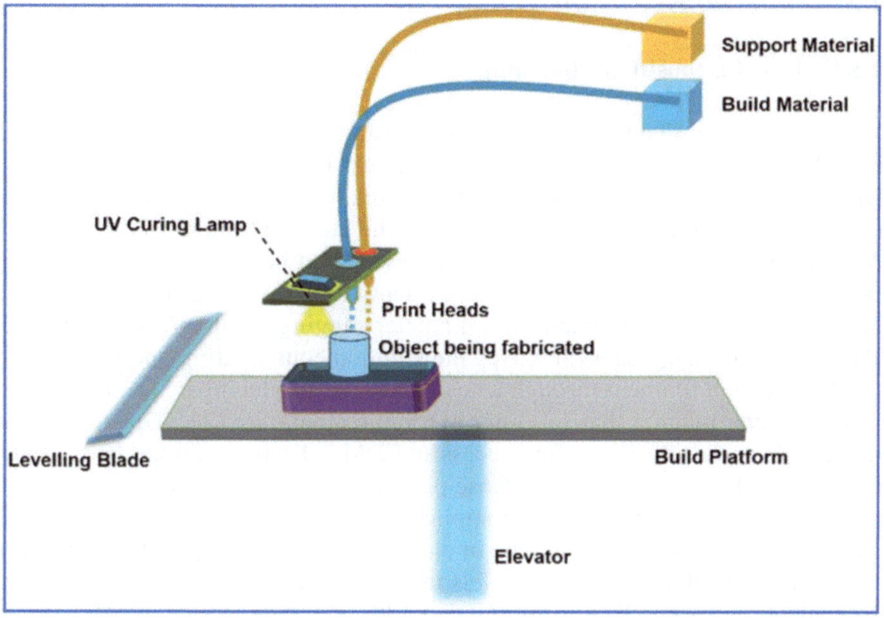

Fig. 5 Schematic representation of the MJ fabrication process, whereby an object is fabricated on a solid support. (Reprinted with permission from [66])

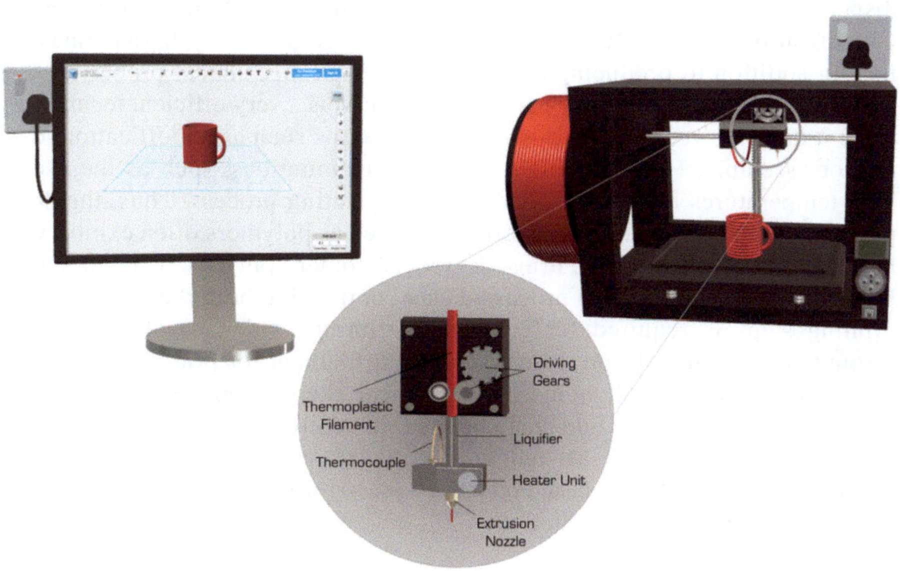

Fig. 6 Graphical illustration of the FDM process. (Reprinted with permission from [76])

2.5 *Material Extrusion*

2.5.1 Fused Deposition Modelling

The most well-known form of material extrusion 3D printing is fused deposition modelling (FDM), also known as fused filament fabrication (FFF) (Fig. 6) [69]. Due to its simplicity and low cost, FDM is one of the commonly used printing technologies in the pharmaceutical field. The starting material is a polymeric API-loaded filament, produced via a process called hot melt extrusion (HME), in which a screw-based system driven by a motor mixes and heats the raw ingredients and extrudes a filament in a diameter ranging from 1.75 to 3 mm [70]. The FDM printing process involves the simultaneous melting and extrusion of the drug-loaded filament through a heated nozzle, producing a 3D structure in a layer-by-layer manner that solidifies at room temperature [71]. Due to the wide range of materials available for successful FDM 3D printing, it has been used to fabricate tablets with different release profiles, geometries, and sizes, as well as polypills for elderly patients on complicated medication regimens [28, 29, 72–75]. To address swallowing issues in elderly patients, FDM has been shown to successfully produce chewable tablets in different shapes, flavours and colours, as well as orodispersible films (ODFs) [30, 31]. The use of FDM has also extended to the production of medical devices applicable to geriatric patients, such as catheters [33] and dentures [32].

FDM has several advantages, including the preparation of solid amorphous dispersions, resulting in increased water solubility of Biopharmaceutics Classification System (BCS) class II drugs (i.e., low solubility, high permeability). In addition to producing structures with high mechanical properties that withstand handling by elderly patients [77], FDM is a very efficient technology since it does not require any post-processing steps such as solidification, drying, or assembly. However, FDM has several limitations, such as the use of high temperatures in both the extrusion and printing process. Thus, the drug/polymer combination must be carefully selected as polymers often exhibit very high melting points at which drugs degrade. Another drawback associated with FDM is that it is a two-step process; hence, optimising both the HME and 3D printing steps is required for successful printing, which can be time- and resource-consuming. To produce a successful filament for printing, properties of the feedstock material, including mechanical, rheological, and thermal properties, must be optimised and characterised. In addition, the numerous printing process parameters such as temperature, layer height, printing speed, and extrusion rate are selected based on a trial-and-error principle or, ideally, using reliable computer-based predictions [78].

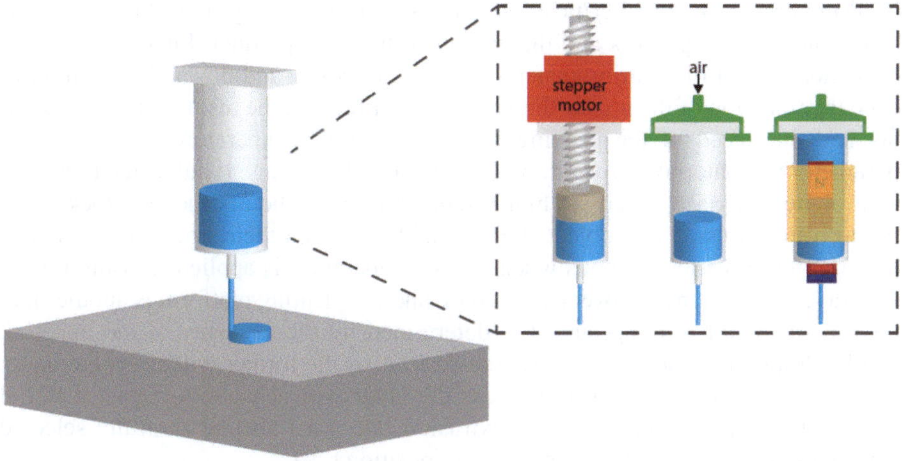

Fig. 7 Graphical illustration of the SSE process. (Reprinted with permission from [79])

2.5.2 Semi-Solid Extrusion

Semi-solid extrusion (SSE) is one of the most versatile material extrusion 3D printing technologies; its nomenclature comes from the type of feedstock used, which is a gel/paste in a semi-solid or semi-molten state. In SSE 3D printing, the feedstock material is extruded in a layer-by-layer manner, whereby layer solidification supports subsequent material deposition until a 3D object is formed. The printing formulation is filled in a disposable syringe, followed by material extrusion using one of three extrusion mechanisms: pneumatic, mechanical, or solenoid (Fig. 7) [79]. The last step involves object solidification, which includes cooling or drying to ensure solvent evaporation.

There are several advantages of SSE, including its ability to formulate a wide range of dosage forms such as polypills [39, 40], ODFs [34, 35], chewable Printlets [36, 38], rectal suppositories [41–43], and implantable patches [44, 45]. The printing process is a single step, fast, simple, and sterile; hence, compliance with Good Manufacturing Practice (GMP) is ensured, making this the first 3D printing technology to be used in a clinical study to manufacture and administer chewable dosage forms to paediatric patients in a hospital setting [37]. During the process, low temperature is consumed, making it suitable for printing thermolabile drugs, proteins, and live cells. On the other hand, SSE has some limitations; the printing resolution is relatively low whenever a wider nozzle diameter is used, which is unavoidable with highly viscous feedstock. Another common issue with SSE is the variability of the post-printing drying time between formulations depending on their viscosities. Drying time is important to ensure solidification of the formulation and evaporation of any included solvent. Upon exposure to moisture or higher temperatures, loss of structural integrity has been reported [79].

There are a number of critical parameters that must be optimised to ensure the smoothness of the process and the quality of the final product. First, the physical characteristics of the starting material, such as viscosity and flowability, will influence the selection of the nozzle diameter and, ultimately, the resolution of the product and the speed of the printing process. Printing temperature is an essential parameter for material extrusion; whenever used, the syringe is subjected to heat to control the material viscosity. A higher temperature will liquify the feedstock, leading to over-extrusion and objects being unable to retain their structure, whereas a material might be less viscous when a low temperature is applied, causing nozzle blockage. The distance between the nozzle and the printing platform is another factor for successful printing, and it is determined by the viscosity of the material. Neither longer nor shorter distances are desired, as the former will affect the material adherence to the platform, whereas the latter will negatively impact material flow. Both nozzle travel speed and extrusion rate must also be carefully selected since they influence the overall material deposition [79].

2.5.3 Direct Powder Extrusion

Direct powder extrusion (DPE) is a novel material extrusion technology that produces Printlets in a single-step process. A powder blend containing the API, polymer and other excipients is directly extruded through a heated nozzle with a rotating screw that applies pressure, producing Printlets in a layer-by-layer manner (Fig. 8) [47]. Unlike FDM, DPE is a simple one-step process that does not require the use of

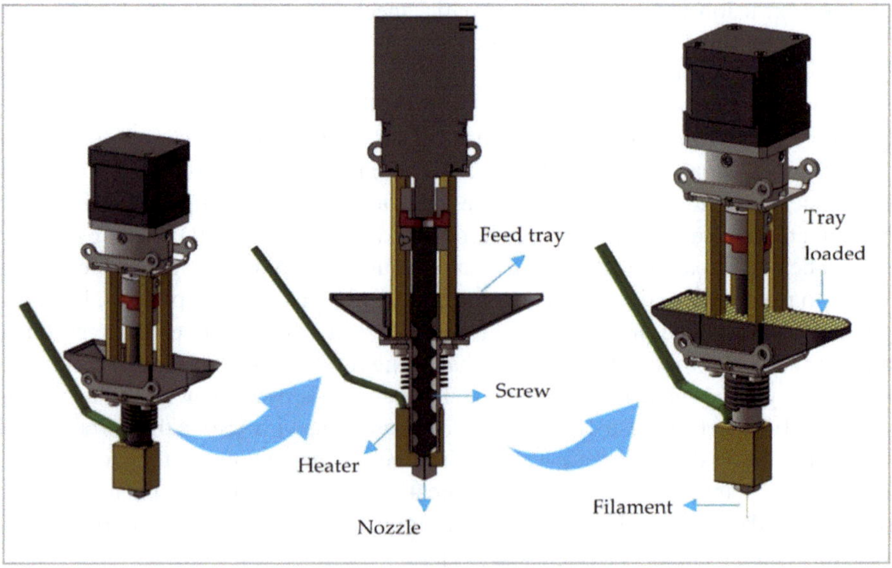

Fig. 8 Graphical illustration of the DPE process. (Reprinted with permission from [46])

HME, thus making it more suitable for manufacturing medicines on-site in a pharmacy or hospital setting, with comparable mechanical characteristics to FDM structures. DPE 3D printing fabricates structures that cannot be printed using traditional FDM, due to the latter being highly dependent on the physical and mechanical characteristics of filaments produced by HME. Despite being the improved and simpler version of FDM, DPE has similar drawbacks; in both, the printing resolution is relatively low, and when creating complex structures, both require support. Although heat is applied once in DPE compared to HME-FDM where it is used twice, there is no strong evidence indicating less drug degradation with DPE when high temperatures are used.

Nevertheless, researchers were able to use DPE at lower temperatures to formulate paracetamol-loaded Printlets that were formulated with low melting point excipients [80]. Drug content variability has been reported with DPE 3D printing due to the poor mixing since it is a single-screw printer mechanism; however, it has been proposed that increasing the number of screws could potentially overcome such drawback [81]. Several studies were able to successfully fabricate different formulations using DPE such as immediate-release, sustained-release and delayed-release dosage forms [47, 82]. Recently, this technology has been investigated in the fabrication and taste-masking of a bitter API, praziquantel, which requires constant dose adjustment in paediatric patients [46]. Such findings strongly suggest that DPE could be successfully utilised to produce personalised formulations for older adults where dosing adjustments are common due to variabilities in physical and pharmacokinetic properties.

3 Applications of 3D Printing in Personalised Medicines for Older Patients

3.1 Multi-Drug Dosage Forms

With the ageing population and high prevalence of chronic disorders, polypharmacy (i.e. the concurrent usage of five or more medications) is evidently becoming a persistent problem affecting patients' compliance to medications. The increase in the frequency of drug administration can significantly impact the compliance of a patient to treatment [83]. In particular, geriatric patients may face problems relating to the management of their treatment plan, causing them to skip doses or confuse medications. Consequently, medication nonadherence may have detrimental implications on patient health, leading to inefficacy of therapy and increased hospitalisations [84].

3D printing offers a simple method to prepare polypills or polyPrintlets, which are Printlets containing multiple APIs in a single dosage form. This helps in reducing the frequency of medication intake, reducing the pill burden on patients and facilitating their management of their treatment plan. Such a concept has been demonstrated using different 3D printing technologies, including FDM [85, 86], SSE

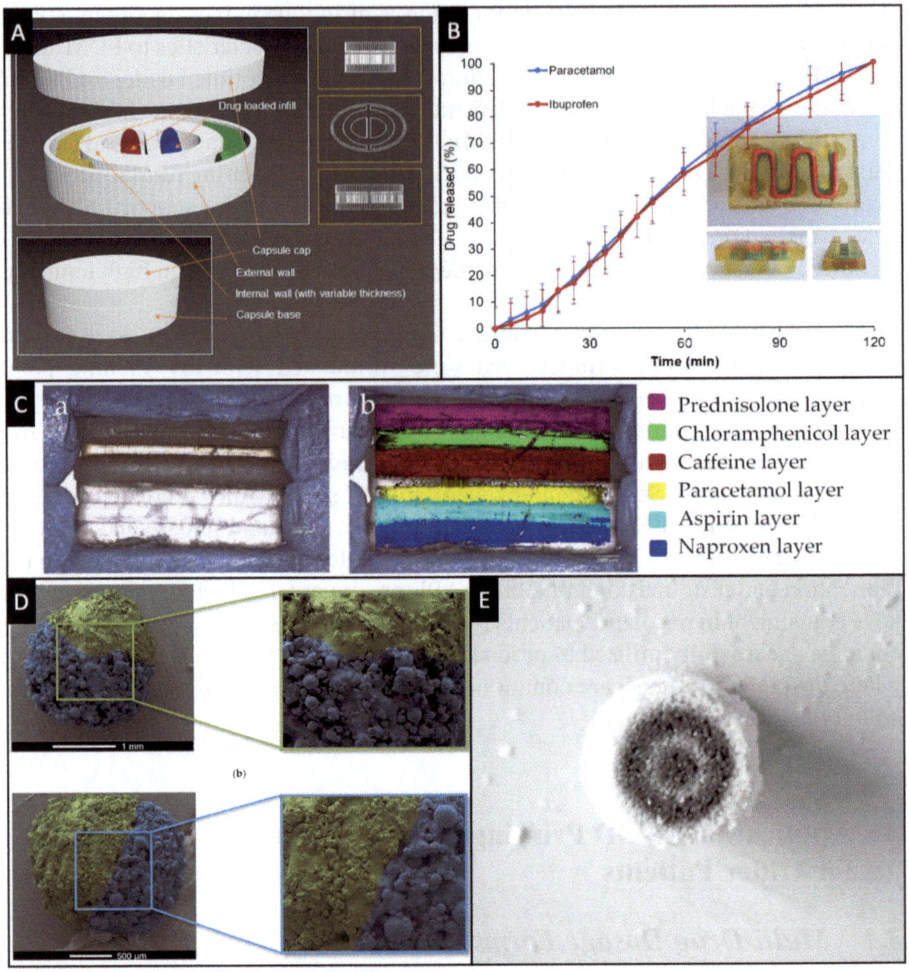

Fig. 9 Examples of multi-drug dosage forms and implants fabricated using 3D printing. (**a**) 3D design of FDM polyvinyl alcohol (PVA) capsules of concentric compartments design and varying internal wall thicknesses [94]. (**b**) Images and in vitro drug release profile from an SSE 3D-printed gelatine-based dosage form, containing (blue) paracetamol and (red) ibuprofen [95]. Drug release studies were performed in simulated intestinal fluids (pH 7.2) using a USP II dissolution apparatus (n = 3, 50 rpm, error bars represent standard deviation). (**c**) Visual imaging of an SLA 3D printed 6-layer polypill, using (**a**) optical light microscopy and (**b**) Raman mapping [12]. The images show the spatial separation of layers. (**d**) Scanning electron microscopy (SEM) images of SLS 3D-printed dual miniPrintlets (diameter, 2 mm) [65]. The green regions represent the ethyl cellulose matrices, whereas the blue regions represent the Kollicoat Instant Rplease (IR) matrices. (**e**) Image of binder jetted implants containing multi-layers of (white) isoniazid and (black) rifampicin [91]. (Images were reprinted with permission from their original sources)

[39, 87, 88], SLA [12, 89], SLS [65, 90], and binder jetting [91]. With the flexibility of the 3D printing process, polypills can be engineered to have the drugs mixed in a single layer [90] or spatially separated in discrete layers [89] or different compartments [92].

With FDM, a range of different polyPrintlets were fabricated, including core-shell caplets [85], multi-compartmental capsule shells [92, 93], and multi-layered Printlets [29]. A noteworthy example is a dual-compartmental dosage form containing the anti-tuberculosis drugs, rifampicin and isoniazid [93]. Although the two APIs interact with one another, the use of separate compartments ensures that the APIs are released at different timepoints, avoiding any loss of activity of the APIs. Similarly, FDM 3D printing was combined with hot-filling to create polypills containing four APIs (Fig. 9a) [94]. The separation of the drugs aimed at avoiding physical interactions during storage and enabling the control of the individual drug release profiles.

SSE polyPrintlets are mainly focused on multi-compartmental dosage forms. As an example, a polyPrintlet with five APIs (i.e. aspirin, hydrochlorothiazide, atenolol, pravastatin sodium and ramipril) was fabricated having compartments with distinctive release properties, three with immediate-release properties and three with extended-release profiles [39]. In another approach, a polyPrintlet containing three APIs (i.e. nifedipine, glipizide and captopril) was created based on different drug release mechanisms, namely, osmosis and diffusion [87]. A dietary supplement in the form of a core-shell polypill was also produced using SSE 3D printing [88]. The inner shell contained caffeine dispersed in a sustained release matrix, which was surrounded by an immediate-release outer shell containing vitamins B1, B3, and B6. Moving away from conventional tablets, the SSE technology was also used for the fabrication of chewable polyPrintlets in the shape of Lego™ bricks (Fig. 9b) [95]. The Printlets contained two model drugs, paracetamol and ibuprofen, which were embedded in a gelatine matrix. While the polyPrintlets were aimed at improving medication intake by paediatric patients, similar benefits, such as ease of swallowing and flexibility of drug dosing, can be foreseen in geriatric patients [96]. Apart from oral dosage forms, the SSE technology has also been used to create multi-drug suppositories aimed at the treatment of acute severe ulcerative colitis [42]. This involved the combination of a corticosteroid, budesonide, with a Janus-associated kinase (JAK) inhibitor, tofacitinib citrate, each having its individual mechanism of action.

Thus far, the polyPrintlet containing the highest number of APIs in a single dosage form was produced using SLA 3D printing. In particular, a six-layer polyPrintlet was produced containing paracetamol, caffeine, naproxen, chloramphenicol, prednisolone and aspirin (Fig. 9c) [12]. The multi-layered polyPrintlet was fabricated in various geometries, material compositions, and layer arrangements. This permitted more control over the drug release properties, enabling its tailoring to the individual needs of different patients. Although having shown favourable results, SLA 3D printing has revealed that some drugs may exhibit chemical interactions with other components of the resin mixture [89]. As an example, it has been shown that amlodipine interacts with the photoreactive monomer, poly(ethylene glycol) diacrylate. Thus, it is vital to select drug/excipient combinations that are compatible with one another.

Various multi-drug dosage forms were also created using SLS 3D printing. An example of such are films and cylindrical Printlets containing the two drugs amlodipine and lisinopril [90]. With the aid of a portable near-infrared (NIR) spectrometer, the dose content of both drugs was non-destructively analysed. The ability to integrate this at-line dose verification method even in the presence of more than one API served as a way to facilitate the integration of this novel concept into clinical practice. Another example involved the fabrication of miniPrintlets (i.e. 3D printed pellets) containing two drugs, paracetamol and ibuprofen (Fig. 9d) [65]. Despite the small size of the miniPrintlets (e.g. 1 or 2 mm in diameter), the drugs were contained in separate layers and had distinct release characteristics. In comparison to larger Printlets, the miniPrintlets can be sprinkled onto soft food to facilitate swallowing for older patients with dysphagia, although potential food-drug and food-excipient interactions should be considered.

Apart from dosage forms, 3D printing has also been applied for the creation of multi-drug implants. A noteworthy example is the use of the binder jetting 3D printing technology to fabricate bone tuberculosis implants containing two APIs: isoniazid and rifampicin (Fig. 9e) [91]. The cylindrical Printlets were composed of four concentric layers, where each layer contained one of the two drugs in the following arrangement (from outside to inside): isoniazid-rifampicin-isoniazid-rifampicin. This arrangement aimed at controlling the drug release such that the drugs are released in a consecutive manner starting from the outermost layer.

3.2 Overcoming Swallowing Difficulties

Due to its ease of intake and non-invasiveness, the oral route is the most preferable route of drug administration by patients and healthcare professionals [97]. However, age plays a vital role in the swallowing ability of individuals. In particular, dysphagia (i.e. swallowing difficulties) starts to manifest with ageing and advancement in disease state [98]. Thus, it is often challenging for geriatric patients to swallow conventional tablets and capsules, resulting in poor adherence to their therapeutic regimens [99]. In the long term, this leads to further health complications and increased morbidity and mortality.

3.2.1 Orally Disintegrating Formulations

As an alternative to tablets for swallowing, other solid dosage forms have been explored to facilitate drug intake by this patient subgroup. Orally disintegrating formulations (also known as orodispersible formulations) are an example of such. As these formulations disintegrate rapidly in the mouth, they are easier to swallow compared to conventional tablets and capsules. While ODTs can be produced using

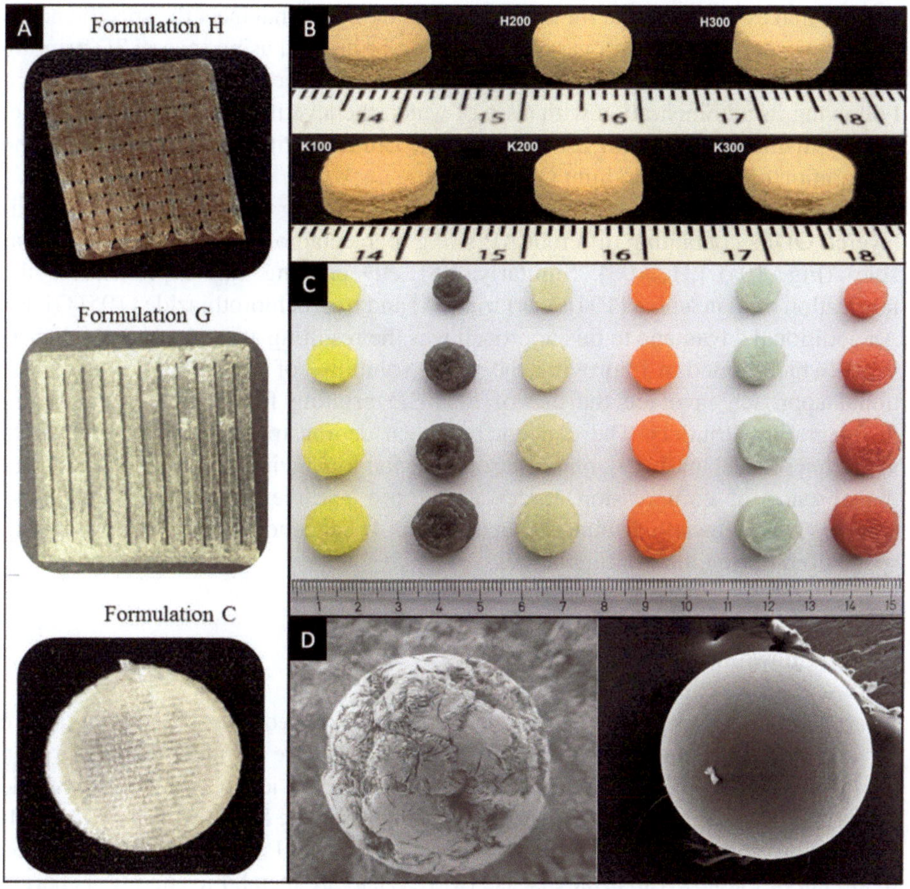

Fig. 10 Examples of different 3D-printed formulations that could aid in overcoming swallowing difficulties in elderly patients. (**a**) Images of FDM 3D-printed orodispersible films, including (from top to bottom) formulation H: a multi-layered, paracetamol-loaded film based on polyethylene oxide (MW 200,000, PEO 200 k) matrix, formulation G: a single-layered, paracetamol-loaded film based on PEO 200 k matrix, and formulation C: a single-layered, ibuprofen-loaded film based on a PEO 200 k and polyvinyl alcohol matrix [103]. (**b**) Images of SLS 3D printing orally disintegrating Printlets based on (top) hydroxypropyl methylcellulose (HPMC) and (bottom) Kollidon VA64 matrices, fabricated using laser speeds of (from left to right), 100, 200 and 300 mm/s [19]. (**c**) Images of SSE chewable Printlets containing (from top to bottom) 50, 100, 150 and 200 mg isoleucine, fabricated in different flavours/colours, including (from left to right) lemon-yellow, coconut-black, banana-light green, orange-orange, raspberry-light blue and strawberry-red [37]. (**d**) SEM images of ibuprofen-loaded SPHRINTS (3D printed microspheres) based on (left) polycaprolactone and (right) poly(lactic-co-glycolic) acid matrices [107]. Scales shown in cm. (Images were reprinted with permissions from their original sources)

traditional pharmaceutical production methods (e.g. direct compression, freeze-drying and moulding), most of these approaches have proven to be complex, costly and with high limitations [100]. In this regard, the flexibility of the 3D printing process makes it well suited for the fabrication of ODTs that meet the specific needs of geriatric patients. Indeed, various ODFs can be created using several 3D printing technologies. The most prominent example is the binder jetted Spritam® product. Following its administration with a sip of water, the levetiracetam-containing tablet swiftly disintegrates within 2 to 27 s [101]. Binder jetting was also used for the preparation of warfarin sodium ODTs [102].

Another strategy involved the use of FDM 3D printing to create single- or multi-layered ODFs containing different APIs (e.g. paracetamol, ibuprofen and aripiprazole) (Fig. 10a) [31, 103]. Similarly, SSE 3D printing was exploited for the fabrication of olanzapine [104], warfarin [35] and hydrochlorothiazide [105] ODFs. An additional advantage to this approach was the resulting amorphous solid dispersions, which aimed at improving the water solubility of the APIs. An unconventional approach involved the use of SLS 3D printing for preparation of orally disintegrating Printlets. The concept has been demonstrated using several APIs, including paracetamol [19], ondansetron [18] and lopinavir [106] (Fig. 10b). Due to the absence of compression forces and loose particle interconnections, these Printlets have shown to disintegrate in as low as 4 s, depending on the laser scanning speed used during the printing process.

3.2.2 Chewable Printlets

Apart from ODTs, chewable formulations have also gained interest as an alternative for conventional monolithic dosage forms [96]. As these formulations are broken down into smaller fragments in the mouth, they are easier to swallow. Chewable 3D-printed formulations are typically fabricated using the SSE 3D printing technology. A noteworthy example is the chewable isoleucine Printlets produced for the treatment of maple syrup urine disease (MSUD), a rare metabolic condition characterised by the deficiency of branched-chain α-ketoacid dehydrogenase complex (BCKD) (Fig. 10c) [37]. As the levels of isoleucine need to be titrated depending on the patient's disease condition, 3D printing has shown to be more efficient at providing the target therapeutic doses of isoleucine. Additionally, these Printlets can be produced in various flavours and colours, depending on the individual preferences of patients [108]. This can be beneficial for improving patients' adherence to their treatment plans. However, the use of chewable formulations in geriatric patients is typically limited by their chewing abilities [109], which often deteriorate with ageing [110, 111].

3.2.3 Multiparticulate Printlets

Other dosage forms that have also shown favourable outcomes in geriatrics are multiparticulates. These are typically in the form of powders, beads, granules or pellets which can be dissolved in liquids or sprinkled onto/mixed with soft food. As such, they can be swallowed without the need for being chewed or broken down into smaller fragments. Another attractive property of multiparticulates is their flexible dosing, wherein a defined volume, weight or number of particles is measured to obtain the required drug dose, avoiding the need to manipulate an existing formulation (i.e. crushing or splitting). Despite their benefits, producing multiparticulates, especially those with controlled release properties, is a lengthy and time-consuming process [112].

3D printing offers an alternative way to create multiparticulate systems in single-step processes. This concept has been exemplified using two technologies: DOD material jetting and SLS 3D printing. Using the DOD technology, microspheres with sustained-release properties were fabricated (Fig. 10d) [107]. Termed SPHRINTs, the ibuprofen-loaded microspheres displayed high uniformity in size (i.e. average size 372 ± 19 μm and 486 ± 15 μm for poly(lactic-co-glycolic) acid and polycaprolactone microspheres, respectively) and shape, as well as high drug encapsulation efficiency (> 95%). Another approach involved the SLS 3D printed miniPrintlets [65]. These miniPrintlets can be programmed to have immediate- or sustained-release properties. The sustained-release properties are achieved by decreasing the laser scanning speed, resulting in an increase in the contact duration between the laser beam and powder particles and consequently permitting more energy transmittance and a higher degree of sintering. Because the resulting system is a matrix system, unlike coating systems, damaging the surface does not alter their release properties. Furthermore, as SLS is a single-step, solvent-free process, it is much simpler and quicker to fabricate multiparticulates compared to conventional production methods.

3.3 Meeting Special Physical Needs

3.3.1 Blindness and Visual Impairment

Blindness and visual impairment are persistent health conditions that affect millions of people around the globe [113]. Previous studies shed light on the correlation between ageing and blindness, with data showing that more than 80% of the blind population are aged 50 years and over [114]. These findings, along with the prevailing polypharmacy in geriatrics, make it challenging for this patient subgroup to manage their daily medication regimens, increasing their dependency on others [115]. Moreover, it is often hard for patients to identify and distinguish medicines, even with the help of optical or low-vision aid devices [115, 116]. Due to that, there is a deterioration in medication compliance in blind and visually impaired patients, resulting in attenuated therapeutic outcomes and increased hospitalisations.

While blind patients currently identify medications using Braille patterns on packaging, this method of recognition is futile and infeasible once the medicines are removed from their containers. Indeed, a more practical way is to include the tactile patterns directly on the surface of the drug-loaded dosage forms. In this regard, 3D printing offers a quick and cost-effective way to achieve this. This concept has been

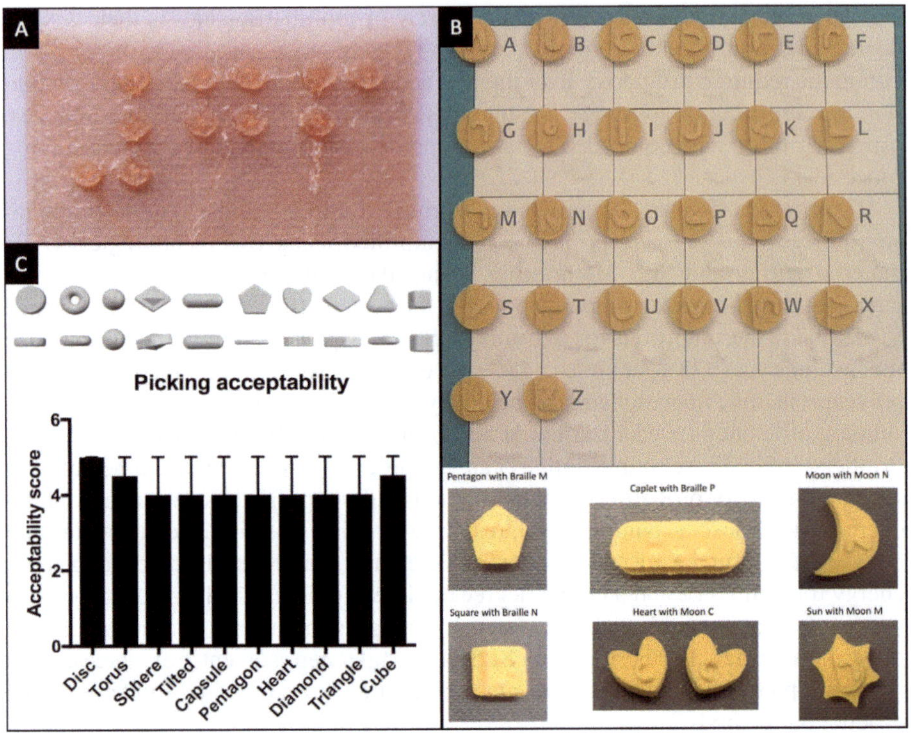

Fig. 11 (**a**) Image of an FDM-printed intraoral film with Braille dots on its surface [117]. (**b**) Images of (top) cylindrical Printlets containing the 26 Moon alphabets and (bottom) Printlets with different shapes having Braille or moon patterns on their surfaces [20]. (**c**) (top) 3D designs of Printlets in different shapes (from right to left): disc, torus, sphere, tilted diamond, capsule, pentagon, heart, diamond, triangle, and cube [118]. (bottom) Participant-reported outcome (PRO) scores for willingness and ability to pick Printlet samples (n = 50). (Images were reprinted with permissions from their original sources)

demonstrated using two 3D printing technologies: FDM and SLS. With FDM, ketoprofen-loaded intraoral films were fabricated with various Braille dot sizes and distribution on their surface (Fig. 11a) [117]. An in vivo haptic assessment involving visually impaired individuals has shown that the patterns were readable and remained stable during handling. Interestingly, when both the size and distance between the Braille dots were modified, the study participants reported confusion when attempting to read the patterns.

SLS Printlets for blind and visually impaired patients were also produced [20]. The cylindrical Printlets were designed to have Braille or Moon (i.e. a tactile writing system based on Latin Roman letters) patterns on their surface (Fig. 11b). Favourably, it has been shown that the presence of the patterns on the Printlets does not affect the mechanical or dissolution characteristics of the Printlets. However, due to the small surface, the Printlets could only fit one Braille or Moon alphabet at a time. Thus, Printlets with different shapes were also fabricated to provide extra information relating to the treatment indication or the dosing regimen. An additional advantage of these Printlets is their ability to disintegrate within ~5 s, avoiding the need to be taken with water and improving self-administration of medications in blind and visually impaired patients.

3.3.2 Picking and Handling

Picking and handling solid oral dosage forms is another problem impeding the adherence of geriatric patients to treatment [119, 120]. This is mainly associated with the loss of motor function due to arthritis or strokes, making it hard for patients to open medication containers or remove pills from containers [121]. To overcome such issues, FDM 3D printing has been suggested as novel way to design medicines with various shapes (i.e. disc, torus, sphere, tilted diamond, capsule, pentagon, heart, diamond, triangle, and cube) that have improved picking and handling

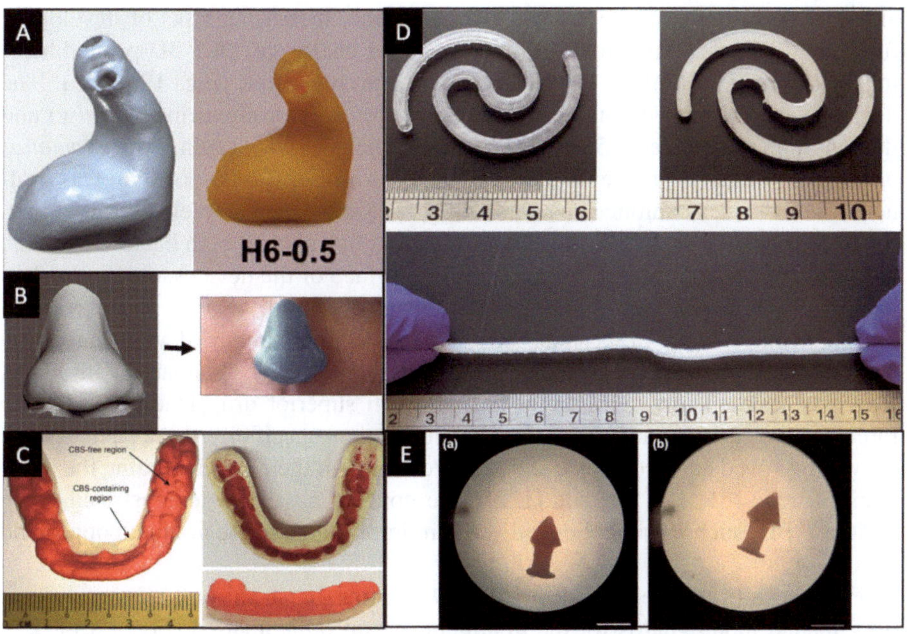

Fig. 12 Examples of 3D printed drug-laden medical devices and implants. (**a**) 3D design and image of a DLP 3D-printed hearing aid with ciprofloxacin-fluocinolone acetonide (6%–0.5%) [13]. (**b**) (left) 3D scan model of a nose and (right) image of an FDM 3D-printed wound dressing loaded with copper [122]. (**c**) Image of a 3D-printed mouthguard with (red) a drug-free and (white) dug-loaded bottom [123]. Scale shown in cm. (**d**) Image of the SLA 3D-printed hollow bladder device in its relaxed conformation, (top left) before and (top right) after filling with 10% lidocaine hydrochloride; and (bottom) the hollow 10% device under stretching. Scale shown in cm. (**e**) Light microscope images of the DLP 3D-printed punctal plugs containing (left) 10% dexamethasone and (right) 10% dexamethasone+17.4% PEG 400. Scale bar is equivalent to 1 mm [124]. (Images were reprinted with permissions from their original sources)

(Fig. 11c) [118]. The outcome of this hypothesis was evaluated through an open label trial in 50 adults using placebo Printlets. Overall, all the Printlets were easy to pick, with the disc shape having the highest score, followed by the torus and cube shapes.

3.4 Drug-Laden Devices

Aside from solid dosage forms, 3D printing could be utilised to fabricate drug-laden medical devices and implants. For such purposes, 3D models can be generated using various imaging techniques such as 3D scanning, computed tomography (CT) or magnetic resonance imaging (MRI). In doing so, the generated devices and implants are patient-specific and thus provide better fitting and superior activity. On this basis, a multitude of devices have been created, providing remarkable benefits for geriatric patients. As an example, loss of hearing is a common disability in geriatrics, requiring the use of hearing aids. However, with continuous usage, hearing aids are likely to affect the ear canal microbiome, increasing risks of developing fungal and bacterial infections. To overcome this challenge, DLP 3D-printed hearing aids with anti-biofilm properties have been suggested (Fig. 12a) [13]. The devices were fabricated containing a combination of ciprofloxacin (6% w/w) and fluocinolone acetonide (0.5% w/w), wherein their activity against *Pseudomonas aeruginosa* (Gram-negative) and *Staphylococcus aureus* (Gram-positive) was evaluated. The devices exhibited sustained drug release for over 2 weeks. Furthermore, the drug combination has been shown to be effective against both bacterial strains, where inhibition in biofilm formation on the surface of the devices and diminished bacterial growth in the surrounding medium was observed.

Within the same vein, bespoke nose- and ear-shaped topical devices have also been produced using SLA and FDM 3D printing (Fig. 12B). Compared with FDM, SLA has resulted in better-quality devices with superior drug loading and more rapid drug release characteristics. Due to their patient specificity, these devices have shown improved adhesion at the site of infection, with their uses including the treatment of acne [61] and wounds [122]. The concept has been demonstrated using various APIs, such as salicylic acid or antimicrobial metals (e.g. zinc, copper and silver).

Elsewhere in the field of dentistry, 3D printing has been utilised to fabricate drug-loaded mouthguards for the treatment of oral inflammation (Fig. 12c) [123]. The wearable devices were made using FDM and contained the model drug clobetasol propionate. A study involving six patients has shown that the mouthguards are well tolerated by patients, providing sustained drug release for the duration of wearing.

With regard to site-specific devices, a number of examples have been provided. For instance, indwelling lidocaine-loaded intravesical devices for the treatment of local diseases of the bladder have been produced (Fig. 12d) [15]. The devices were designed such that they are straight and rigid, easing their insertion and removal from the bladder using a urethral catheter that undergoes in situ shape change. The latter is mainly concerned with the retaining of the device in the bladder. Similar to bladder devices, vaginal rings have been produced for contraception [125]. The devices were created using the CLIP technology and were loaded with hormones and microbicides. For ocular drug delivery, DLP 3D-printed dexamethasone-loaded punctal plugs have been fabricated (Fig. 12e) [124]. Such devices are regarded as advantageous for the treatment of dry eye diseases, providing sustained drug release for up to 7 days.

4 Digital Healthcare and 3D Printing

3D printing of pharmaceuticals is a prolonged process relying on a trial-and-error approach for formulation optimisation, which can often be time- and resource-consuming and requires expert knowledge. To accelerate this, artificial intelligence (AI) in the form of machine learning (ML), a technology capable of pattern identification from complex datasets, has been recently explored [126]. Its accuracy depends mostly on the size of the dataset, with greater and more varied results leading to more accurate predictions, with data being derived from in-house experiments and literature-based data [127]. The development of ML techniques can be performed using programming languages, such as Python and R, or using third-party software, for example, Apple's Create ML, both of which are easily accessible to researchers today. Its application in different sectors has contributed to major innovations; in the pharmaceutical field, it has been used in different stages of drug development including drug discovery, small-molecule design and development, drug repurposing, target identification and validation, model design to predict pharmacokinetic properties and in clinical trials (Fig. 13) [128, 129].

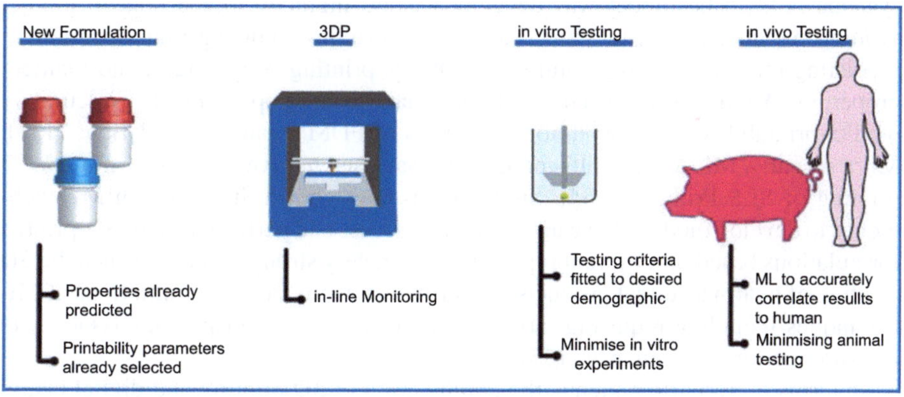

Fig. 13 Schematic illustration showing the stages where AI can be incorporated in the pharmaceutical 3D printing process. (Reprinted with permission from [129])

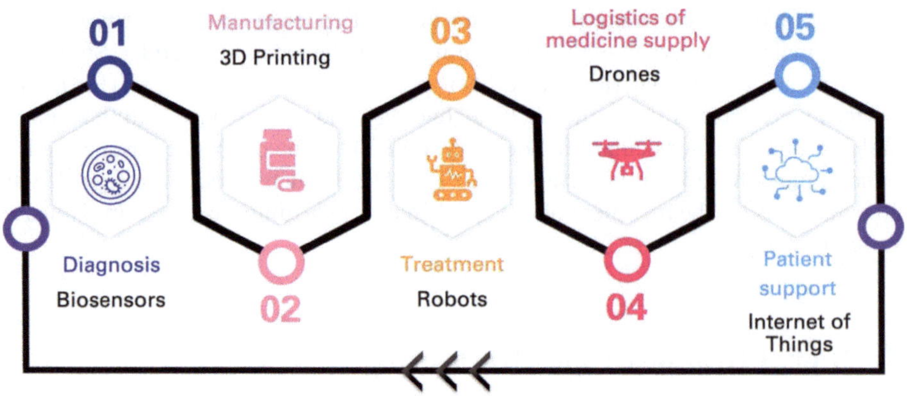

Fig. 14 The virtuous cycle of opportunities for digital health in pharmacy. (Reprinted with permission from [137])

Recently, researchers have shown an interest in utilising ML to accelerate the overall 3D printing process [130]. To produce a successful 3D printing formulation, properties of the feedstock material including mechanical, rheological and thermal must be optimised and characterised, alongside printing process parameters based on the 3D printing technology used. ML can accurately predict the optimal fabrication parameters and formulation in vitro and in vivo performances resulting in the minimisation of the consumption of resources, money, labour, and experimental animals. An example of this is the development of a novel software tool using AI ML techniques, namely, M3DISEEN, which can be used to guide the FDM 3D printing formulation development process [131]. The model was built based on the data of 614 drug-loaded formulations comprised of 145 different excipients, capable of predicting the printability, both printing and extrusion temperatures, and filament mechanical characteristics. Upon succession of M3DISEEN, the use of AI ML was extended to explore the performance of more than 900 3D-printed drug delivery systems [132]. The model was developed using literature-mined data to predict printing parameters and in vitro drug release profiles, showing high accuracy in predicting the process temperature, printability, printing temperature, and filament properties. Another study used rheological data to develop a model to accurately predict printability and dissolution properties of FDM printed formulations [133]. More recently, ML using multi-modal data was used to give an insight into the 3D printing of SLS Printlets [134]. Such studies have shown that utilising ML techniques to develop models that can predict the process and performance of 3D-printed formulations based on different types of data is a very strong tool to advance the 3D printing of pharmaceutical products. Nevertheless, there are some limitations of ML techniques including requiring large volume of datasets, transparency issues, and post-production maintenance [135].

In terms of geriatric patients, the combination of 3D printing and digital health technologies has the potential to further personalise the overall treatment regimen, from the formulation development of the 3D printed dosage forms to the monitoring of drug levels using biosensors [136]. Digital devices have the potential capability to diagnose, treat, and deliver treatment to geriatric patients in their home or care-home setting, allowing older adults to have more control over their health as well as helping caregivers make more informed choices on the treatment of the elderly, lessening the burden on healthcare facilities (Fig. 14) [137]. Recently, a study aimed at facilitating this concept by proposing the use of compact 3D printers that are operated using a smartphone device [138]. It was demonstrated that the novel system can be used for the on-demand printing of medications in different shapes, sizes, and dose strengths. To ensure patient safety, the printing could be remotely approved and monitored by a healthcare professional.

5 Conclusion

Geriatric patients often experience age-related health conditions and diseases, which are frequently accompanied by chronic illnesses. This highlights the importance of personalised medications that cater to the unique genetic makeup of each individual, moving away from the conventional 'one-size-fits-all' approach. 3D printing presents an innovative solution for producing small batches of medicines that are tailored to the specific needs of patients. Depending on the 3D printing technology utilised, a wide range of dosage forms and drug-laden devices can be designed and manufactured. Furthermore, advanced imaging technologies enable the production of patient-specific devices with distinct features. The creation of personalised medications allows for meeting the individual dosing requirements and special needs of geriatric patients. This significantly improves adherence to treatment plans and reduces side effects, ultimately leading to a decrease in hospitalisations. The positive outcomes can be further enhanced by integrating 3D printing with other digital health technologies, resulting in a novel digitised closed-loop healthcare model that offers improved efficacy. Although the 3D printing technology is still in its early stages, the current data show promising results. Only time will reveal whether this model will eventually be adopted on a larger scale.

References

1. Baraković S, et al. Quality of life framework for personalised ageing: a systematic review of ICT solutions. Int J Environ Res Public Health. 2020;17(8):2940.
2. Drenth-van Maanen AC, Wilting I, Jansen PAF. Prescribing medicines to older people—how to consider the impact of ageing on human organ and body functions. Br J Clin Pharmacol. 2020;86(10):1921–30.
3. Drumond N. Future perspectives for patient-centric pharmaceutical drug product design with regard to solid Oral dosage forms. J Pharm Innov. 2020;15(3):318–24.
4. Afsana, et al. 3D printing in personalized drug delivery. Curr Pharm Des. 2018;24(42):5062–71.

5. Awad A, et al. A review of state-of-the-art on enabling additive manufacturing processes for precision medicine. J Manuf Sci Eng. 2023;145(1):010802.
6. ASTM. Standard terminology for additive manufacturing technologies. F2792 – 12a. 2012.
7. Yu DG, et al. A novel fast disintegrating tablet fabricated by three-dimensional printing. Drug Dev Ind Pharm. 2009;35(12):1530–6.
8. Huang W, et al. Levofloxacin implants with predefined microstructure fabricated by three-dimensional printing technique. Int J Pharm. 2007;339(1):33–8.
9. Wang C-C, et al. Development of near zero-order release dosage forms using three-dimensional printing (3-DP™) technology. Drug Dev Ind Pharm. 2006;32(3):367–76.
10. Katstra WE, et al. Oral dosage forms fabricated by three dimensional printing™. J Control Release. 2000;66(1):1–9.
11. Wang J, et al. Stereolithographic (SLA) 3D printing of oral modified-release dosage forms. Int J Pharm. 2016;503(1–2):207–12.
12. Robles-Martinez P, et al. 3D printing of a multi-layered Polypill containing six drugs using a novel stereolithographic method. Pharmaceutics. 2019;11(6):274.
13. Vivero-Lopez M, et al. Anti-biofilm multi drug-loaded 3D printed hearing aids. Mater Sci Eng C. 2021;119:111606.
14. Hada T, et al. Effect of printing direction on the accuracy of 3D-printed dentures using Stereolithography technology. Materials (Basel, Switzerland). 2020;13(15):3405.
15. Xu X, et al. Stereolithography (SLA) 3D printing of a bladder device for intravesical drug delivery. Mater Sci Eng C Mater Biol Appl. 2021;120:111773.
16. Economidou SN, et al. 3D printed microneedle patches using stereolithography (SLA) for intradermal insulin delivery. Mater Sci Eng C. 2019;102:743–55.
17. Subbiah R, et al. 3D printing of microgel-loaded modular microcages as instructive scaffolds for tissue engineering. Adv Mater. 2020;32(36):2001736.
18. Allahham N, et al. Selective laser sintering 3D printing of orally disintegrating Printlets containing ondansetron. Pharmaceutics. 2020;12(2):110.
19. Fina F, et al. Fabricating 3D printed orally disintegrating printlets using selective laser sintering. Int J Pharm. 2018;541(1–2):101–7.
20. Awad A, et al. 3D printed tablets (Printlets) with braille and moon patterns for visually impaired patients. Pharmaceutics. 2020;12(2):172.
21. Fina F, et al. 3D printing of drug-loaded gyroid lattices using selective laser sintering. Int J Pharm. 2018;547(1–2):44–52.
22. Alomari M, et al. Printing T3 and T4 oral drug combinations as a novel strategy for hypothyroidism. Int J Pharm. 2018;549(1):363–9.
23. Thabet Y, Lunter D, Breitkreutz J. Continuous inkjet printing of enalapril maleate onto orodispersible film formulations. Int J Pharm. 2018;546(1–2):180–7.
24. Vakili H, et al. Application of a colorimetric technique in quality control for printed pediatric orodispersible drug delivery systems containing propranolol hydrochloride. Int J Pharm. 2016;511(1):606–18.
25. Eleftheriadis GK, et al. Inkjet printing of a thermolabile model drug onto FDM-printed substrates: formulation and evaluation. Drug Dev Ind Pharm. 2020;46(8):1253–64.
26. Palo M, et al. Development of Oromucosal dosage forms by combining electrospinning and inkjet printing. Mol Pharm. 2017;14(3):808–20.
27. Iftimi LD, et al. Edible solid foams as porous substrates for inkjet-printable pharmaceuticals. Eur J Pharm Biopharm. 2019;136:38–47.
28. Sadia M, et al. From 'fixed dose combinations' to 'a dynamic dose combiner': 3D printed bi-layer antihypertensive tablets. Eur J Pharm Sci. 2018;123:484–94.
29. Pereira BC, et al. 'Temporary Plasticiser': a novel solution to fabricate 3D printed patient-centred cardiovascular 'Polypill' architectures. Eur J Pharm Biopharm. 2019;135:94–103.
30. Tabriz AG, et al. Personalised tasted masked chewable 3D printed fruit-chews for Paediatric patients. Pharmaceutics. 2021;13(8):1301.

31. Jamróz W, et al. 3D printed orodispersible films with aripiprazole. Int J Pharm. 2017;533(2):413–20.
32. Lo Russo L, et al. Accuracy of trial complete dentures fabricated by using fused deposition modeling 3-dimensional printing: an in vitro study. J Prosthet Dent. 2021;129:908.
33. Mathew E, et al. Fused deposition modeling as an effective tool for anti-infective dialysis catheter fabrication. ACS Biomater Sci Eng. 2019;5(11):6300–10.
34. Oblom H, et al. Towards printed pediatric medicines in hospital pharmacies: comparison of 2D and 3D-printed Orodispersible warfarin films with conventional Oral powders in unit dose sachets. Pharmaceutics. 2019;11(7)
35. Sjöholm E, Sandler N. Additive manufacturing of personalized orodispersible warfarin films. Int J Pharm. 2019;564:117–23.
36. Karavasili C, et al. Pediatric-friendly chocolate-based dosage forms for the oral administration of both hydrophilic and lipophilic drugs fabricated with extrusion-based 3D printing. Eur J Pharm Sci. 2020;147:105291.
37. Goyanes A, et al. Automated therapy preparation of isoleucine formulations using 3D printing for the treatment of MSUD: first single-Centre, prospective, crossover study in patients. Int J Pharm. 2019;567:118497.
38. Herrada-Manchón H, et al. 3D printed gummies: personalized drug dosage in a safe and appealing way. Int J Pharm. 2020;587:119687.
39. Khaled SA, et al. 3D printing of five-in-one dose combination polypill with defined immediate and sustained release profiles. J Control Release. 2015;217:308–14.
40. Siyawamwaya M, et al. 3D printed, controlled release, tritherapeutic tablet matrix for advanced anti-HIV-1 drug delivery. Eur J Pharm Biopharm. 2019;138:99–110.
41. Seoane-Viaño I, et al. 3D printed tacrolimus rectal formulations ameliorate colitis in an experimental animal model of inflammatory bowel disease. Biomedicine. 2020;8(12)
42. Awad A, et al. 3D printed multi-drug-loaded suppositories for acute severe ulcerative colitis. International Journal of Pharmaceutics: X. 2023;5:100165.
43. Awad A, et al. 3D printed infliximab suppositories for rectal biologic delivery. International Journal of Pharmaceutics: X. 2023;5:100176.
44. Yi HG, et al. A 3D-printed local drug delivery patch for pancreatic cancer growth suppression. J Control Release. 2016;238:231–41.
45. Andriotis EG, et al. Development of bio-active patches based on pectin for the treatment of ulcers and wounds using 3D-bioprinting technology. Pharmaceutics. 2020;12(1):56.
46. Boniatti J, et al. Direct powder extrusion 3D printing of Praziquantel to overcome neglected disease formulation challenges in Paediatric populations. Pharmaceutics. 2021;13(8)
47. Goyanes A, et al. Direct powder extrusion 3D printing: fabrication of drug products using a novel single-step process. Int J Pharm. 2019;567:118471.
48. Sen K, et al. Pharmaceutical applications of powder-based binder jet 3D printing process – a review. Adv Drug Deliv Rev. 2021;177:113943.
49. Trenfield SJ, et al. Binder jet printing in pharmaceutical manufacturing. In: Basit AW, Gaisford S, editors. 3D printing of pharmaceuticals. Cham: Springer International Publishing; 2018. p. 41–54.
50. Vaz VM, Kumar L. 3D printing as a promising tool in personalized medicine. AAPS PharmSciTech. 2021;22(1):49.
51. Aprecia, Spritam, in https://www.aprecia.com/zipdose-platform/zipdose-technology.php. 2015.
52. Parkash V, et al. Fast disintegrating tablets: opportunity in drug delivery system. J Adv Pharm Technol Res. 2011;2(4):223–35.
53. Abaci A, et al. Additive manufacturing of Oral tablets: technologies, materials and printed tablets. Pharmaceutics. 2021;13(2):156.
54. Vithani K, et al. A proof of concept for 3D printing of solid lipid-based formulations of poorly water-soluble drugs to control formulation dispersion kinetics. Pharm Res. 2019;36(7):102.

55. Xu X, et al. Vat photopolymerization 3D printing for advanced drug delivery and medical device applications. J Control Release. 2021;329:743–57.
56. Rodríguez-Pombo L, et al. Volumetric 3D printing for rapid production of medicines. Addit Manuf. 2022;52:102673.
57. Loterie D, Delrot P, Moser C. High-resolution tomographic volumetric additive manufacturing. Nat Commun. 2020;11(1):852.
58. Ngo TD, et al. Additive manufacturing (3D printing): a review of materials, methods, applications and challenges. Compos Part B. 2018;143:172–96.
59. Alhnan MA, et al. Emergence of 3D printed dosage forms: opportunities and challenges. Pharm Res. 2016;33(8):1817–32.
60. Bloomquist CJ, et al. Controlling release from 3D printed medical devices using CLIP and drug-loaded liquid resins. J Control Release. 2018;278:9–23.
61. Goyanes A, et al. 3D scanning and 3D printing as innovative technologies for fabricating personalized topical drug delivery systems. J Control Release. 2016;234:41–8.
62. Awad A, et al. Advances in powder bed fusion 3D printing in drug delivery and healthcare. Adv Drug Deliv Rev. 2021;174:406–24.
63. Awad A, et al. 3D printing: principles and pharmaceutical applications of selective laser sintering. Int J Pharm. 2020;586:119594.
64. Thakkar R, et al. Impact of laser speed and drug particle size on selective laser sintering 3D printing of amorphous solid dispersions. Pharmaceutics. 2021;13(8)
65. Awad A, et al. 3D printed pellets (Miniprintlets): a novel, multi-drug, controlled release platform technology. Pharmaceutics. 2019;11(4):148.
66. Gülcan O, Günaydın K, Tamer A. The state of the art of material jetting-a critical review. Polymers. 2021;13(16):2829.
67. Seoane-Viaño I, et al. Translating 3D printed pharmaceuticals: from hype to real-world clinical applications. Adv Drug Deliv Rev. 2021;174:553–75.
68. Hanif M, Zaman M, Chaurasiya V. Polymers used in buccal film: a review. Designed Monomers and Polymers. 2015;18(2):105–11.
69. Awad A, Gaisford S, Basit AW. Fused deposition modelling: advances in engineering and medicine. In: Basit AW, Gaisford S, editors. 3D printing of pharmaceuticals. Cham: Springer International Publishing; 2018. p. 107–32.
70. Shaqour B, et al. Production of drug delivery systems using fused filament fabrication: a systematic review. Pharmaceutics. 2020;12(6):517.
71. Goyanes A, et al. PET/CT imaging of 3D printed devices in the gastrointestinal tract of rodents. Int J Pharm. 2018;536(1):158–64.
72. Goyanes A, et al. 3D printing of modified-release aminosalicylate (4-ASA and 5-ASA) tablets. Eur J Pharm Biopharm. 2015;89:157–62.
73. Xu, X., et al., 3D printed polyvinyl alcohol tablets with multiple release profiles. Sci Rep, 2019. 9(1): p. 12487ss.
74. Fastø MM, et al. Perceptions, preferences and acceptability of patient designed 3D printed medicine by polypharmacy patients: a pilot study. Int J Clin Pharm. 2019;41(5):1290–8.
75. Goyanes A, et al. Effect of geometry on drug release from 3D printed tablets. Int J Pharm. 2015;494(2):657–63.
76. Awad A, et al. 3D printed medicines: a new branch of digital healthcare. Int J Pharm. 2018;548(1):586–96.
77. Okwuosa TC, et al. Fabricating a Shell-Core delayed release tablet using dual FDM 3D printing for patient-Centred therapy. Pharm Res. 2017;34(2):427–37.
78. Mwema FM, Akinlabi ET. Basics of fused deposition modelling (FDM). Fused Deposition Modeling: Strategies for Quality Enhancement; 2020. p. 1–15.
79. Seoane-Viaño I, et al. Semi-solid extrusion 3D printing in drug delivery and biomedicine: personalised solutions for healthcare challenges. J Control Release. 2021;332:367–89.

80. Mendibil X, et al. Direct powder extrusion of paracetamol loaded mixtures for 3D printed pharmaceutics for personalized medicine via low temperature thermal processing. Pharmaceutics. 2021;13(6):907.
81. Fanous M, et al. Simplification of fused deposition modeling 3D-printing paradigm: feasibility of 1-step direct powder printing for immediate release dosage form production. Int J Pharm. 2020;578:119124.
82. Ong JJ, et al. 3D printed opioid medicines with alcohol-resistant and abuse-deterrent properties. Int J Pharm. 2020;579:119169.
83. Barat I, Andreasen F, Damsgaard E. Drug therapy in the elderly: what doctors believe and patients actually do. Br J Clin Pharmacol. 2001;51(6):615–22.
84. Malhotra S, et al. Drug related medical emergencies in the elderly: role of adverse drug reactions and non-compliance. Postgrad Med J. 2001;77(913):703–7.
85. Goyanes A, et al. 3D printing of medicines: engineering novel Oral devices with unique design and drug release characteristics. Mol Pharm. 2015;12(11):4077–84.
86. Ghanizadeh Tabriz A, et al. 3D printed bilayer tablet with dual controlled drug release for tuberculosis treatment. Int J Pharm. 2021;593:120147.
87. Khaled SA, et al. 3D printing of tablets containing multiple drugs with defined release profiles. Int J Pharm. 2015;494(2):643–50.
88. Goh WJ, et al. 3D printing of four-in-one oral polypill with multiple release profiles for personalized delivery of caffeine and vitamin B analogues. Int J Pharm. 2021;598:120360.
89. Xu X, et al. Stereolithography (SLA) 3D printing of an antihypertensive polyprintlet: case study of an unexpected photopolymer-drug reaction. Addit Manuf. 2020;33:101071.
90. Trenfield SJ, et al. Non-destructive dose verification of two drugs within 3D printed polyprintlets. Int J Pharm. 2020;577:119066.
91. Wu W, et al. A programmed release multi-drug implant fabricated by three-dimensional printing technology for bone tuberculosis therapy. Biomed Mater. 2009;4(6):065005.
92. Maroni A, et al. 3D printed multi-compartment capsular devices for two-pulse oral drug delivery. J Control Release. 2017;268(Supplement C):10–8.
93. Genina N, et al. Anti-tuberculosis drug combination for controlled oral delivery using 3D printed compartmental dosage forms: from drug product design to in vivo testing. J Control Release. 2017;268(Supplement C):40–8.
94. Pereira BC, et al. Additive manufacturing of a point-of-care "Polypill:" fabrication of concept capsules of complex geometry with bespoke release against cardiovascular disease. Adv Healthc Mater. 2020;9(13):2000236.
95. Rycerz K, et al. Embedded 3D printing of novel bespoke soft dosage form concept for pediatrics. Pharmaceutics. 2019;11(12)
96. Rodríguez-Pombo L, et al. Innovations in chewable formulations: the novelty and applications of 3D printing in drug product design. Pharmaceutics. 2022;14(8):1732.
97. Awad A, Trenfield SJ, Basit AW. Chapter 19 - Solid oral dosage forms. In: Adejare A, editor. Remington (Twenty-third Edition). Academic Press; 2021. p. 333–58.
98. Baijens LW, et al. European Society for Swallowing Disorders–European Union Geriatric Medicine Society white paper: oropharyngeal dysphagia as a geriatric syndrome. Clin Interv Aging. 2016;11:1403.
99. Stegemann S, Gosch M, Breitkreutz J. Swallowing dysfunction and dysphagia is an unrecognized challenge for oral drug therapy. Int J Pharm. 2012;430(1):197–206.
100. Nagar P, et al. Orally disintegrating tablets: formulation, preparation techniques and evaluation. J Appl Pharm Sci. 2011;1(4):35–45.
101. Aprecia Pharmaceutials. What is ZipDose technology? . 2016 [cited 2020 23/01/2020]; Available from: https://www.spritam.com/#/patient/zipdose-technology/what-is-zipdose-technology
102. Tian P, et al. Oral disintegrating patient-tailored tablets of warfarin sodium produced by 3D printing. Drug Dev Ind Pharm. 2018;44(12):1918–23.

103. Ehtezazi T, et al. The application of 3D printing in the formulation of multilayered fast dissolving Oral films. J Pharm Sci. 2018;107(4):1076–85.
104. Cho H-W, et al. Orodispersible polymer films with the poorly water-soluble drug, olanzapine: hot-melt pneumatic extrusion for single-process 3D printing. Pharmaceutics. 2020;12(8)
105. Eduardo D-T, Ana S-E, José B. A micro-extrusion 3D printing platform for fabrication of orodispersible printlets for pediatric use. Int J Pharm. 2021;605:120854.
106. Hamed R, et al. 3D-printing of lopinavir printlets by selective laser sintering and quantification of crystalline fraction by XRPD-chemometric models. Int J Pharm. 2021;592:120059.
107. Shpigel T, Uziel A, Lewitus DY. SPHRINT–printing drug delivery microspheres from polymeric melts. Eur J Pharm Biopharm. 2018;127:398–406.
108. Januskaite P, et al. I spy with my little eye: a Paediatric visual preferences survey of 3D printed tablets. Pharmaceutics. 2020;12(11):1100.
109. Liu F, et al. Patient-centered pharmaceutical design to improve acceptability of medicines: similarities and differences in Paediatric and geriatric populations. Drugs. 2014;74(16):1871–89.
110. Kida IA, et al. Chewing problems and dissatisfaction with chewing ability: a survey of older Tanzanians. Eur J Oral Sci. 2007;115(4):265–74.
111. Peltola P, Vehkalahti MM. Chewing ability of the long-term hospitalized elderly. Spec Care Dentist. 2005;25(5):260–4.
112. Ghebre-Sellassie I, Knoch A. Pelletization techniques. In: Swarbrick J, editor. Encyclopedia of pharmaceutical technology. New York: Informa Healthcare USA, Inc; 2007. p. 2651–63.
113. Pascolini D, Mariotti SP. Global estimates of visual impairment: 2010. Br J Ophthalmol. 2012;96(5):614.
114. Florence AT, Lee VH. Personalised medicines: more tailored drugs, more tailored delivery. Int J Pharm. 2011;415(1–2):29–33.
115. McCann RM, et al. Help needed in medication self-management for people with visual impairment: case–control study. Br J Gen Pract. 2012;62(601):e530–7.
116. Windham BG, et al. Impaired vision and the ability to take medications. J Am Geriatr Soc. 2005;53(7):1179–90.
117. Eleftheriadis GK, Fatouros DG. Haptic evaluation of 3D-printed braille-encoded intraoral films. Eur J Pharm Sci. 2021;157:105605.
118. Goyanes A, et al. Patient acceptability of 3D printed medicines. Int J Pharm. 2017;530(1):71–8.
119. Notenboom K, et al. Practical problems with medication use that older people experience: a qualitative study. J Am Geriatr Soc. 2014;62(12):2339–44.
120. Stegemann S, et al. Geriatric drug therapy: neglecting the inevitable majority. Ageing Res Rev. 2010;9(4):384–98.
121. Darnell JC, et al. Medication use by ambulatory elderly: an in-home survey. J Am Geriatr Soc. 1986;34(1):1–4.
122. Muwaffak Z, et al. Patient-specific 3D scanned and 3D printed antimicrobial polycaprolactone wound dressings. Int J Pharm. 2017;527(1):161–70.
123. Liang K, et al. 3D printing of a wearable personalized oral delivery device: a first-in-human study. Science. Advances. 2018;4(5):eaat2544.
124. Xu X, et al. 3D printed Punctal plugs for controlled ocular drug delivery. Pharmaceutics. 2021;13(9):1421.
125. Janusziewicz R, et al. Design and characterization of a novel series of geometrically complex intravaginal rings with digital light synthesis. Advanced Materials Technologies. 2020;5(8):2000261.
126. Baker DJ. Chapter 11 - artificial intelligence: the future landscape of genomic medical diagnosis: dataset. In: Lambert CG, Baker DJ, Patrinos GP, editors. Silico artificial intelligent clinical information, and machine learning systems, in human genome informatics. Academic Press; 2018. p. 223–67.
127. Cohen JP, et al. Problems in the deployment of machine-learned models in health care. CMAJ : Canadian Medical Association journal = journal de l'Association medicale canadienne. 2021;193(35):E1391–4.

128. Vamathevan J, et al. Applications of machine learning in drug discovery and development. Nat Rev Drug Discov. 2019;18(6):463–77.
129. Elbadawi M, et al. Harnessing artificial intelligence for the next generation of 3D printed medicines. Adv Drug Deliv Rev. 2021;175:113805.
130. O'Reilly CS, et al. Machine learning and machine vision accelerate 3D printed Orodispersible film development. Pharmaceutics. 2021;13(12):2187.
131. Elbadawi M, et al. M3DISEEN: a novel machine learning approach for predicting the 3D printability of medicines. Int J Pharm. 2020;590:119837.
132. Muñiz Castro B, et al. Machine learning predicts 3D printing performance of over 900 drug delivery systems. J Control Release. 2021;337:530–45.
133. Elbadawi M, et al. 3D printing tablets: predicting printability and drug dissolution from rheological data. Int J Pharm. 2020;590:119868.
134. Abdalla Y, et al. Machine learning using multi-modal data predicts the production of selective laser sintered 3D printed drug products. Int J Pharm. 2023;633:122628.
135. Elbadawi M, et al. Disrupting 3D printing of medicines with machine learning. Trends Pharmacol Sci. 2021;42:745.
136. Trenfield SJ, et al. Advancing pharmacy and healthcare with virtual digital technologies. Adv Drug Deliv Rev. 2022;182:114098.
137. Awad A, et al. Connected healthcare: improving patient care using digital health technologies. Adv Drug Deliv Rev. 2021;178:113958.
138. Xu X, et al. Smartphone-enabled 3D printing of medicines. Int J Pharm. 2021;609:121199.

The Ageing Microbiome, Pharmaceutical Considerations, and Therapeutic Opportu nities

Alessia Favaron, Laura E. McCoubrey, Moe Elbadawi, Abdul W. Basit, and Mine Orlu

Contents

Abstract Research in recent years has illuminated the importance of the microbiome for human wellbeing and healthy ageing. Numerous age-related conditions have now been linked to microbiome dysfunction, in which the symbiotic host-microbe relationship breaks down and disease ensues. Microbiome composition is being increasingly recognised as an indicator for healthy ageing. As microbiome composition changes over the life course, maintaining a healthy microbiome could be the key to promoting human health in later life. As such, the microbiome may represent a largely untapped therapeutic target for the prevention and treatment of age-related conditions. Novel interventions include prevention of medication-induced dysbiosis and development of new microbiome-targeted therapeutics. Due to the personalised nature of the microbiome, innovative solutions will likely harness precision medicine techniques to meet the ageing microbiome at an individual level. Such strategies will require considered formulation design,

A. Favaron · L. E. McCoubrey (✉) · M. Elbadawi · A. W. Basit · M. Orlu
UCL School of Pharmacy, University College London, London, UK
e-mail: a.favaron@ucl.ac.uk; laura.mccoubrey.18@ucl.ac.uk; m.elbadawi@ucl.ac.uk; a.basit@ucl.ac.uk; m.orlu@ucl.ac.uk

M. Orlu, F. Liu (eds.), *Pharmaceutical Formulations for Older Patients*, AAPS Advances in the Pharmaceutical Sciences Series 51,
https://doi.org/10.1007/978-3-031-35811-1_8

development, and prescribing. This chapter will explore the key pharmaceutical considerations for achieving microbiome health in later life.

Keywords Gut microbiome · Healthy ageing · Microbiome-targeted therapeutics · Drugs-induced dysbiosis · Colonic drug delivery

1 Unravelling the Intestinal Microbiota

During 4 billion years of evolution, prokaryotic microorganisms [1] have occupied habitats on earth with profoundly different physiochemical conditions and growth substrates. Some microbial communities have learned how to proliferate in harsh environments, with low or high pH, extreme temperatures, and salinity. It is therefore unsurprising that microbial colonisation also occurred in and on humans and animals, whose bodies offer an ideal temperature for bacterial growth. Bacteria are present all over the human body, skin, nose, vagina, etc., but the largest population of microbiota is found in the intestine with about 10^{13}–10^{14} microbes per millilitre [2]. This high density of microorganisms consists of bacteria, archaea, eukarya, and viruses, which collectively encode over 3 million genes, generating thousands of metabolites, whilst comparatively, the human genome contains 23,000 genes [3].

In the last decade, scientists strived to understand how our microbiota affects health and disease, but one of the main obstacles has been the ability to culture the bacteria. Only in the last years, new technologies have promoted the phylogenetic identification and quantification of the gut microbiome components. The most utilised technology today is the analysis of nucleic acids directly extracted from stools, specifically the amplification of the 16S ribosomal RNA gene (rRNA), which allows the identification of the specific microbial strains present in the sample [4] and, thus, the characterization of the gut microbiota profile. A healthy adult gut shows an overall predominancy of the anaerobic phyla *Bacillota*, *Bacteroidota*, and *Pseudomonadota*, which account for about 98% of the microorganisms. Other gut bacteria, in minor percentages, mainly belong to phyla *Actinomycetota*, *Verrucomicrobiota*, *Fusobacteriota*, or *Acidobacteriota* [5]. These phyla are present all along the gut, with temporal and spatial differences in the distribution of bacteria at the genus level [6]. Indeed, bacterial concentration has an ascending trend along the various segments of the gut in response to changes, i.e. substrate availability and physiochemical conditions. The small intestine is a zone of transition between the sparsely populated acidic stomach and the great variety of bacteria living in the colon. Most of the microbes in the small intestine are facultative anaerobes where families *Lactobacillaceae* and *Enterobacteriaceae* dominate [7]. They must withstand the bile acids secreted in the proximal end of the small intestine, which are bactericidal to certain species, and the rapid transit time of food, which could disturb the durable colonisation of this gut section. Only in the terminal end of the small intestine, in the ileum, the bacterial density is comparable to the levels observed in the large intestine [7]. The caecum and the colon have the highest density and diversity of microbial communities of the intestine, with 10^{12} bacteria per millilitre of intestinal content [5]. Besides longitudinal variations, microbial patterns are different between the gut lumen and the epithelium. The mucus layer, formed by mucin proteins, mostly produced from goblet cells, prevents harmful bacteria from penetrating the gut barrier. Only some bacterial species (*Clostridium*, *Lactobacillus*, *Enterococcus* specialized to adhere to the mucus) use mucus as a nutrient source, increasing their possibility of reaching epithelial cells. In contrast, most species found in faeces, belonging to *Bacteroides*, *Bifidobacterium*, *Streptococcus*, *Enterobacteriaceae*, *Enterococcus*, *Clostridium*, *Lactobacillus*, and *Ruminococcus* are the species located in the central luminal compartment of the colon [5].

1.1 Functions of the Gut Microbiota

The gut microbiota plays a central role in many hosts' physiological functions: besides secreting enzymes that increase the metabolic capacity of healthy individuals, the gut microbiota protects the host against pathogens and imparts substantial immunological functions. The major activity of the microbiota is the breakdown of dietary substrates, such as carbohydrates, proteins, plant metabolites, and xenobiotics that have not been digested in the small intestine. In particular, the fermentation of polysaccharides is the most researched metabolic activity in the large intestine because it provides rich sources of energy, namely, short-chain fatty acids (SCFAs) butyrate, propionate, and acetate [8]. Whilst members of the gram-negative *Bacteroidota* mostly produce acetate and propionate, the gram-positive *Bacillota* mostly produce butyrate [9]. The majority of the SCFAs produced in the colon permeate the gut wall to reach human tissues and implement significant beneficial processes for host metabolic and physiologic health. Roediger stated that around 70% of the energy needed by colonic epithelial cells is provided by butyrate, where it acts as a regulator of cell differentiation and growth and epithelial integrity [10]. Luminal bacteria convert some acetate into butyrate; however, most of the acetate produced reaches peripheral tissues where it promotes fat oxidation in adipose tissue, improves glucose homeostasis, and may even regulate blood pressure [11]. Propionate is mainly sequestrated in the liver where it may act as a gluconeogenic substrate, help the regulation of cholesterol synthesis, and even discourage the development of liver cancer [12].

Moreover, resident bacteria provide an essential antimicrobial protection to the host. The healthy gut mucosal immune system is constantly preventing the overgrowth of opportunistic pathogens and promoting tolerance to beneficial commensals. The gut's first antimicrobial mechanism is the presence of a thick mucus layer, characteristic of the large intestine, that keeps the luminal microorganisms away from epithelial cells. Furthermore, the gut microbiota is able to induce the production of local immunoglobulins, IgA, that can coat the microbiota to restrict the translocation of pathogenic strains from the intestinal lumen to the circulation; with this mechanism, a potential systemic immune response is prevented [13].

Considering that oral administration is the most common approach to drug delivery, due to its better patient compliance, safety, convenience, and flexibility, it is wise to consider the high chance of interactions between orally administered medications and intestinal microbiota [14]. A large and growing body of literature has now provided ample insights on the capability of gut microbiota to influence the bioavailability of oral drugs, either directly or indirectly [14–16]. Direct interactions refer to the biochemical transformation of the drug by the microbial enzymatic activity and possible bioaccumulation of drugs by bacteria [17]. Indirect interactions include microbiome effects on bile acid biotransformation (which can alter the pharmacokinetics of lipophilic drugs), effects on the reactivation of inactive drug metabolites, and effects on drug transporters, in addition to the influence that *Helicobacter pylori* infections or probiotics supplements can have on oral drug bioavailability [14, 15]. Despite the significant effects of gut microbiota on drug metabolism, few efforts have been spent in determining the role of the microbial enzymes in drug pharmacokinetics during drug developments, with possible severe consequences in oral drug clinical outcomes.

1.2 Establishment of the Intestinal Microflora from Early Life to Old Age

1.2.1 Gestation and Parturition

The microbial colonisation of the gut is the result of an elaborate interaction between microbiota – host – and external and internal factors. The host-related factors are associated with anatomical development of the gut, intestinal pH, bile acids, and peristalsis, in addition to drug therapy and microbial interactions. External factors include the mode of delivery at birth, microbiota composition of the mother, diet, medication, and the bacterial composition of individuals' environment. Thus, the gut microbiome co-evolves with its host, and its microbial composition varies dynamically throughout the life of an individual [18].

Pregnancy is the key biological process for the development of healthy progeny. It involves many physiological changes including hormonal fluctuations, immune system modulation, and weight gain, but the transformation of the microbiota composition has only recently been observed.

Traditionally, the foetal intestine and the amniotic fluid were hypothesized to be sterile until delivery, and there remains controversy around the matter [19]. Numerous pieces of evidence have suggested that the human placenta and some foetal tissues are colonised by a unique microbial niche of mostly beneficial bacteria, such as lactobacilli and bifidobacteria [20, 21]. The exposure of the foetus to these phyla could explain the early development of the immune system in infants [22]; recent findings observed an in vitro activation of memory T cells in foetal lymph node after microbial exposure, indicating the presence of live microbes in foetal organs, in particular from the second trimester of gestation [23].

With the rupture of the uterus membranes and the release of the amniotic fluids, bacteria ascend from the vaginal mucosa to the uterus, and microbial colonisation of the neonate fully commences. Newborns, born via vaginal delivery, pass through the birth canal for an efficient transfer of the mother's vaginal microbial contents. Consistently, these infants' enteric microbiota resemble the maternal vaginal microbiota: their gut is rich with lactose-digesting bacteria, such as *Lactobacillus* and *Prevotella* [24]. Neonatal microbial composition is appreciably different when delivery occurs by caesarean section: the enteric microbiota resembles the one of the maternal skin, which are sebum- and mucus-digesting bacteria, rather than lactose-digesting bacteria [25–27]. In addition, C-sections require more medical sterile interventions, often causing a delay in the start of breastfeeding, and a need for antibiotic prophylaxis to prevent infections.

1.2.2 Early Life and Infancy

The first year of life is pivotal to the development of gut microbiota. There are numerous exogenous factors that exert a major influence on the infant intestinal microecology: they can be roughly classified as dietary, environmental or therapeutic. The key factor influencing gut microbiota in early life is diet: babies fed breast milk have microbiota dominated by health-associated lactobacilli, bifidobacteria, and staphylococci. Human breast milk contains around 130 kinds of oligosaccharides and a great concentrations of secretory immunoglobulin A, which encourage beneficial species of microbes to proliferate and discourage colonisation by pathogenic bacteria [28]. Furthermore, the infant gut microbiota appears to be very susceptible to the compromising effects of antibiotics. Evidence of connection between antibiotics administration in the first years of life and the risk of developing diseases in childhood is growing with related cases of childhood asthma [29], obesity [30] and inflammatory bowel diseases [31] among other conditions. Exposure to antibiotics during both pregnancy and infancy clearly alters the enteric microbiota, in a way dependent on the substance used, its dose and regimen. In a general view, antibiotics increase the proliferation of bacterial species with intrinsic antibiotic resistance – *Enterococcus* spp., *Staphylococcus* spp., and members of *Enterobacteriaceae* – with a substantial drop in numbers of beneficial bifidobacteria. Since microbial communities are shaped by those we contact directly and by our surroundings, it is unsurprising that family members frequently share their intestinal bacteria. Indeed, it has been observed that children with older siblings host a positive bacterial richness and diversity of phyla *Bacillota* and *Bacteroidota* compared to children raised without contact with siblings [32]. Interestingly, early-life exposure to household furry pets also seems to increase the abundance of *Ruminococcus* and *Oscillospira* bacteria, which have been negatively correlated with the risk of obesity and allergies [33].

The development of the gut microbiota at early life is a chaotic process, varying widely between individuals. During early life, bacterial diversity is low, and colonisation from new incoming bacteria is generally successful since they do not have to compete for nutrients and space (a condition referred to as low colonisation resistance) [34]. This low resistance allows rapid establishment of a complex microbiota composition that assists the immune system and metabolic requirements of the developing child. In particular, one of the greatest changes that takes place between 6 and 18 months after birth is the change in the infant diet from breastfeeding to solid foods [35]. The anaerobic *Bifidobacterium*, abundant during breastfeeding, undergoes a slow and steady decline, and the introduction of solid food and plant oligosaccharides promotes expansion of the strictly anaerobic phyla *Bacteroidota*, *Bacillota*, and *Clostridia* [36]. This enriching process means that gut microbiota composition largely reflects that seen in adults by the age of 2–3 years [37].

1.2.3 Adulthood

The overall microbial profile of a healthy adult presents *Bacteroidota* and *Bacillota* as dominant phyla, contributing to approximately 90% of all microbiota, [6] along with a lower relative abundance of *Actinomycetota* and *Verrucomicrobiota* phyla. The *Bacillota* phylum includes more than 200 different genera such as *Clostridium*, *Lactobacillus*, *Bacillus*, and *Enterococcus*, whilst the *Bacteroidota* is composed of *Bacteroides* and *Prevotella*. The *Bifidobacterium* genus accounts for most of the *Actinomycetoma* phylum [38]. From birth to adulthood, the *Bacillota*/*Bacteroidota* ratio increases, and some evidence suggests that when this ratio is altered, a dysbiotic microbiome can occur, with a particular predisposition to obesity and metabolic disorders [6]. For example, a study of Ukrainian populations showed significant association between the *Bacillota*/*Bacteroidota* ratio and body mass index; overweight and obese patients had higher level of *Bacillota* and lower level of *Bacteroidota* compared to lean and normal-weight individuals [39].

1.2.4 Older People

The composition of the gut microbiota of young adults (30 years) has been shown to remain stable and is comparable to that of 65–70-year-old healthy individuals [40], underlining the fact that the loss of microbiome richness and diversity in the elderly is rather related to altered gut physiology, infections, medications, a weakened immune system, lifestyle, and dietary schedule modifications [41–45], rather than with chronologic age.

Compared to younger adults, a compromised ageing microbiota suffers a reduction in microbial diversity, with depletions of taxa and thus their functions and major inter-individual variation in microbiota composition. Notably, the carriage of health-associated bacteria (e.g. *Bifidobacterium*, *Lactobacillus*, and *Bacteroides*), responsible for maintaining immune tolerance in the gut, are found to be reduced in aged people, whereas the levels of opportunistic bacteria, such as *C. perfringens*, *C. difficile* [35], *E. coli*, and *Staphylococcus*, responsible for the development of age-associated disease, increase [46]. Vaiserman et al. underlined how these age-related shifts of elderly patient microbiomes can be very different in different populations [47], but it is interesting to report that there are specific species involved in cellular ageing, demonstrated also in animal models [47]. The mouse ageing microbiome, for example, has shown some differences in bacterial diversity compared to younger littermates with an increased presence of clostridia, decreased levels of lactobacilli and an altered proportion of *Firmicutes* and *Bacteroidetes* [48]. Mice with an aged microbiome have been observed to show depressive-like behaviour and impaired short-term and spatial memory [49]. These same shifts were analysed in the faecal microbiota of laboratory rats, from 3 weeks to 2 years of age (with year 2 corresponding to the old biologic stage in their life) with a clear abundance of *Bacillota*, such as *Clostridium* and *Ruminococcus* [50].

There are several factors that could be responsible for the gut microbial changes in older individuals. Principally, the deteriorating condition of the gut microbiota of the elderly has been largely associated with lower colonic SFCA production; for instance, decreased levels of butyrate may perturb gut barrier integrity, with a higher possibility for pathobionts to enter the intestinal mucosa [40] and a higher susceptibility to inflammation [35]. The age-related decline in SCFA proportions is one of the underlying mechanisms of low-grade chronic inflammation, known as 'inflammaging', which accompanies many gastrointestinal diseases (IBD, *Clostridioides difficile* infection, colitis) and non-gastrointestinal pathologies (neurodegenerative conditions, diabetes, cancer, frailty) in the elderly [47]. Indeed, it has been demonstrated that SCFAs are essential in preventing and reducing inflammatory processes through the inhibition of inflammatory mediators (such as interleukins, TNF-α, NO) and through hormone modulation [51]. The aberration of gut wall integrity and enhanced proinflammatory cytokines can also underlie the pathogenesis and progression of some metabolic diseases prevalent in older people, such as hepatic steatosis, insulin resistance, adipose tissue plasticity, and even secondary cardiovascular events [52]. Finally, some studies have observed that decline in SCFAs could also modulate neuro-immune activation, which could possibly explain the higher incidence of gastrointestinal perturbations among patients with severe neurodegenerative diseases [53].

Elderly people often struggle with malnutrition, and with the greater need for drugs/antibiotics, they may follow a poor diet, rich in fats and low in fibre intake. Whilst a diet rich in fats, like the Western Diet, favours the proliferation of a pro-inflammatory microbiota, which eventually promotes pathological conditions, a high-fibre diet promotes the reproduction of health-associated bacterial strains [51]. In addition, poor diet may be combined with polypharmacy administration that has its evident effects on the microbiome. An elegant study from Maier et al. showed that approximately 24% of drugs approved by the FDA inhibit the growth of at least one bacterial strain in vitro [54]. Among these drugs, proton-pump inhibitors are particularly dangerous because of their capacity to reduce chloride acid excretion in the stomach responsible for preventing unwanted bacterial overgrowths, like *E. coli* and *C. difficile* [55]. Furthermore, statins [56, 57], nonsteroidal anti-inflammatory drugs [58] and antipsychotics [59, 60] may have a huge impact on decreasing many beneficial microbial strains. Last, the administration of antibiotics is well known to significantly alter gut microbiota composition, even after short therapies [61, 62]. These major effects of drugs on the gut microbiota should not be ignored if we consider that polypharmacy is extensively common among aged patients. In nursing houses, more than half of the residents take five or more medications [63]. Additional evidence has found a correlation between a compromised microbiome and the gut-brain axis, and such obviated communication could play a role in diseases of the central nervous system, e.g. depression and anxiety [64].

1.2.5 Centenarians

Although it has been observed that gut microbiota remain stable from the third to the seventh decade of life, centenarians have been observed to undergo a profound adaptive remodelling of gut microbial composition [40]. We need to consider centenarians not as the most robust individuals but those who better adapt to biological challenges during life, experience healthy ageing, and in this sense are free for longer from the major age-related and life-threatening pathologies [65]. Biagi et al. were the first in 2010 to introduce centenarians, alongside young and elderly adults, in their comparative study on the gut microbiota composition. The researchers managed to study the extremely old population of a restricted geographic area in the north of Italy, with comparable lifestyle and dietary habits. The microbiota of these subjects has been coexisting with their host for over 100 years and showed a great transition in terms of composition and diversity [40]. What is exceptional is the unexpected increase in bacteria linked to beneficial immunological and metabolic health, such as *Akkermansia muciniphila*, *Bifidobacterium*, and *Christensenellaceae* [41]. It is suggested that *A. muciniphila* preserves gut barrier dysfunction, hampers inflammation and mitigates metabolic dysfunctions such as insulin resistance [66]. Both *Bifidobacterium* and *Christensenellaceae* may prevent inflammation, with the last also generating lactate and acetic acid [67]. These microbial proliferations might be involved in a new homeostasis within the ageing host, facilitating survival to extreme age [40]. On the other hand, the gut microbiota of centenarians is enriched in *Pseudomonadata*, frequently prevailing over mutualistic symbionts and responsible for inflammatory conditions; indeed, changes in the microbiota observed in these very old subjects correlated with higher levels of pro-inflammatory cytokines in their peripheral blood [67]. However, despite their inflammatory status, centenarians live an overall healthy and long life, probably aided by a positive effect of counterbalanced anti-inflammatory physiological events and/or microbial species with anti-inflammatory properties. In conclusion, the gut microbiota composition observed in centenarians could be the result of a complex microbial remodelling to favour the balance between inflammatory and anti-inflammatory processes, to attain longevity [40]. Because of the lack of longitudinal studies in human ageing, it is ambitious to state whether the gut composition of the centenarians was already a characteristic during their earlier life or a microbial pattern acquired only later during their last decades.

2 The Medicine-Microbiome Relationship

Interactions between the gut microbiome and medications are bidirectional: gut microbes can affect drug pharmacokinetics (PK), and medications can affect the functioning of gut microbes (Fig. 1). This medicine-microbiome axis should be an important feature in the pharmaceutical care of elderly patients; however, in reality, it is often not recognised or overlooked. The medicine-microbiome relationship is important because it can account for interindividual differences in PK and could be a preventable source of dysbiosis or opportunity for drug repurposing [68]. Evidence highlights that elderly patients are more likely to have PK profiles that are difficult to anticipate, sometimes leading to increased risk of adverse effects [69]. This can be exemplified by the common PK variability seen with drugs such as warfarin, digoxin, and psychotropics [70]. Moreover, advanced age increases susceptibility to conditions associated with gut dysbiosis, such as *C. difficile* infection, atherosclerosis,

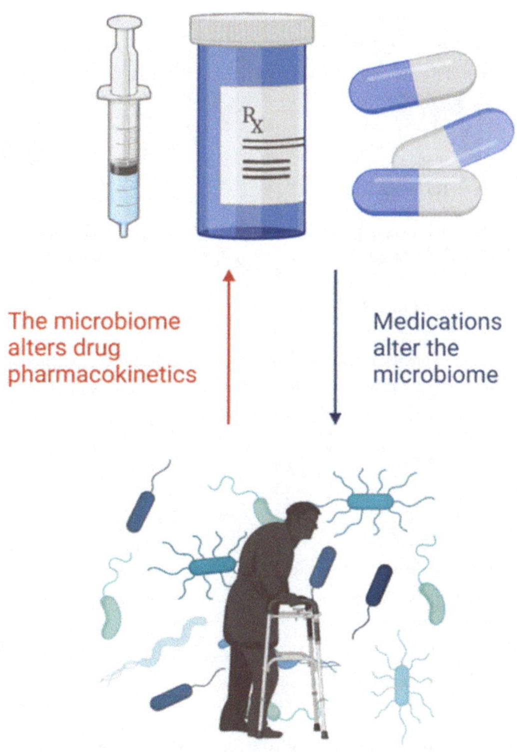

Fig. 1 The bidirectional interactions of the medicine-microbiome axis

and Parkinson's disease [71–73]. Through recognition of medicine-microbiome interactions, PK variability may become easier to predict, and patients' risk of dysbiosis-associated diseases could be reduced. Additionally, identification of beneficial drug effects on gut microbiome composition could lead to the development of novel or repurposed therapeutics for health ageing [74].

2.1 Bugs Vs. Drugs: Microbiome Effects on Pharmacokinetics

Patients' drug response can be significantly altered by the gut microbiome, in multifarious ways that may be surprising. Though the liver and kidneys are often exemplified as the main determinants of drugs' PK, gastrointestinal (GI) microbes also play an influential role. In fact, the metabolic power of the gut microbiome has been compared to that of the liver [75]. The gut microbiome can alter drugs' PK either directly, through chemical transformation by microbial enzymes, or indirectly, through effects on host physiology or functioning (Fig. 2). Until recently, the influence of the gut microbiome on PK has not been recognised or appreciated. In the late 2000s, only 30 drugs had been identified as being susceptible to chemical transformation by intestinal microbiota [76, 77]. In the following years, many more

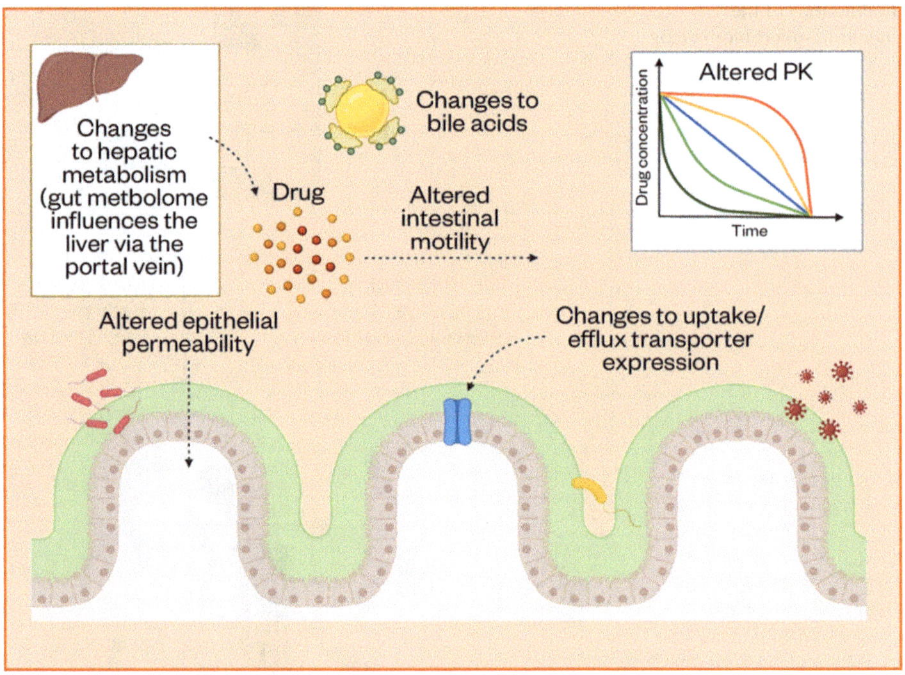

Fig. 2 Host factors modulated by the gut microbiome, which can lead to indirect microbiome effects on drug pharmacokinetics. (Image reproduced with permission from [88])

cases of microbial drug metabolism were uncovered, illuminating hundreds of further instances and encompassing many drugs used commonly in elderly patients [15, 78–85]. Whilst the scale of drug alteration by gut microbiota has begun to be elucidated, its importance has not yet been appreciated in pharmaceutical development or clinical practice. It is currently rare for the pharmaceutical industry to measure therapeutics' risk of microbial metabolism. When this is performed, it is usually solely for extended release formulations that are designed to release drugs in the distal GI tract, where microbiota density is highest [86]. For example, the research and development team within AstraZeneca proposed an in vitro method for identification of faecal microbiota drug degradation in 2014, specifically for colonic-release medicines [87]. There is also very little consideration for how microbiota may affect patients' drug responses within healthcare settings, despite differences in microbiome composition being a key driver of PK variability [68].

2.1.1 Enzymatic Microbial Metabolism

Microbiota inhabiting the GI tract produce enzymes with broad functional redundancy, capable of digesting many types of compounds for symbiotic processes including nutrient extraction, toxin inactivation, vitamin synthesis, and hormone regulation [89, 90]. These enzymes can also chemically transform the structure of drugs passing through the gut, leading to changes in compound activity and even toxicity. Over 180 drugs are currently known to be susceptible to structural alteration by microbiome-produced enzymes, with more examples being regularly published [91]. Common types of microbial enzymes responsible for these gut reactions include glucuronidases, transferases, sulfatases, and reductases [92]. Azoreductases were among the first microbial enzymes identified to directly interact with drugs. The azo-bonded prodrugs sulfasalazine, balsalazide, and olsalazine are activated to 5-aminosalicylic acid (5-ASA, also known as mesalazine and mesalamine) by bacterial azoreductases within the colon [84]. The liberated 5-ASA is then available at therapeutic concentrations for local treatment of ulcerative colitis [93]. Research suggests that age-associated changes in gut microbiome composition may affect the activation of such prodrugs [94]. A study by Taggart et al. in 1992 found that the elimination half-life of sulfasalazine is longer in elderly patients compared to younger adults [95]. This finding could highlight that bacterial conversion of sulfasalazine to 5-ASA occurs at a slower pace in people of advanced age.

In juxtaposition to prodrug activation, toxicity can also arise from direct microbial metabolism. The most infamous example of microbiome-mediated drug toxicity is the case of sorivudine in 1993 [96]. The antiviral drug was removed just 40 days after its Japanese market approval due to the associated deaths of 18 patients. It was later determined that the deceased patients had been administered sorivudine concurrently with an anti-cancer drug, 5-fluorouracil (5-FU). Sorivudine is hydrolysed by microbiota to its metabolite, bromovinyluracil, in the caecum and colon. Following this, bromovinyluracil is metabolised by the host to an inhibitor of dihydropyrimidine dehydrogenase, an enzyme necessary for the breakdown of

5-FU. As a result, patients experienced fatal accumulation of 5-FU, tragically suffering severe intestinal damage and pancytopenia [97]. Another example of microbiome-based toxicity is the intestinal damage associated with the antineoplastic irinotecan. In the liver, irinotecan is glucuronide-conjugated and subsequently excreted into the intestines in bile, for removal from the body via faeces. However, β-glucuronidases produced by colonic bacteria can deconjugate glucuronide-conjugated irinotecan, leading to dose-limiting diarrhoea [98, 99]. These examples of toxicity highlight the importance of identifying drug-microbiome interactions at an early stage of development. Frailty is commonly associated with old age, and frail patients will likely fare worse than younger, otherwise healthy individuals, following severe adverse drug events, such as irinotecan-associated diarrhoea. For this reason, the frail and elderly population should be especially considered when assessing the risk of new drugs' microbiome-mediated toxicity.

Many drugs susceptible to direct metabolism by the gut microbiome are indicated for conditions associated with advanced age (Table 1). For example, the cardiac glycoside digoxin, used to treat atrial fibrillation and heart failure, is inactivated by reductases produced by strains of the bacterium *Eggerthella lenta*. Formation of the metabolite, dihydrodigoxin, via lactone ring reduction is known to result in decreased drug bioavailability, possibly leading to higher-dose requirements in patients colonised by the bacterium [100, 101]. Another example of microbial drug inactivation is levodopa (L-dopa), an amino acid indicated for the treatment of Parkinson's disease. At its site of absorption, the jejunum, L-dopa is susceptible to conversion to dopamine by *Enterococcus faecalis* producing tyrosine decarboxylases. To effectively treat Parkinson's disease symptoms, it is essential that this L-dopa-to-dopamine conversion occurs within the brain; therefore, premature peripheral conversion by intestinal bacteria results in drug inactivation. In fact, intestinal abundance of tyrosine decarboxylase has been associated with increased L-dopa dose requirements. Following intestinal formation of dopamine, the compound is further inactivated to m-tyramine by dehydroxylases produced by *E. lenta* [102, 103].

Two recent high-throughput studies have greatly increased the number of drugs thought to undergo structural alteration by intestinal microbiota. The first, performed by Zimmerman et al., measured the in vitro degradation of 271 drugs in the presence of 76 isolated strains of gut bacteria [78]. The researchers found that 176 (two-thirds) of drugs were significantly transformed, decreasing by ≥20% in starting concentration, by at least one of the strains. Drugs with lactone, azo, urea, or nitro groups were observed to be most susceptible to biotransformation. Notable examples relevant to age-associated diseases include entacapone (containing a nitro group), nitrendipine (nitro group), and gliclazide (urea group). The second study, by Javdan et al., incubated 438 drugs in human faecal microbiota and found 13% to be significantly metabolised [15]. Affected drugs covered 28 therapeutic classes, including the antiepileptic clonazepam, the immunosuppressant mycophenolate mofetil, and the anti-cancer prodrug, capecitabine. The group also measured how variability in microbiome composition leads to inter-individual differences in microbial drug metabolism by studying reactions using faeces from 20 healthy donors. In some cases, drugs were unanimously degraded by all individuals'

Table 1 Examples of drugs susceptible to direct transformation by enzymes produced by gut microbiota. Reproduced with permission from [91]

Drug	Reaction	Causative agent	Experimental model	Effect
Brivudine	Cleaving of tetrahydrofuran ring	*Bacteroides thetaiotaomicron* encoding *bt4554* gene	Mice (sex unspecified)	Increased conversion to hepatotoxic metabolite, bromovinyluracil (BVU) in the caecum, resulting in higher BVU serum levels [104]
Dexamethasone	Desmolysis (side chain cleaving)	*Clostridium scindens*	Mice (both sexes)	Reduced drug concentration in the caecum and increased androgen metabolite concentration in the caecum and serum [78]
Digoxin	Lactone ring reduction	*Eggerthella lenta* producing cardiac glycoside reductase enzyme	Mice (male)	Formation of an inactive metabolite, dihydrodigoxin [100]. Reduction in digoxin bioavailability [101]
Diltiazem	Deacetylation	*Bacteroides thetaiotaomicron* encoding *bt4096* gene	Ex vivo human microbiota from faeces (64% male)	Differences in diltiazem metabolising capacity, correlating with *bt4096* homolog abundance [78]
Doxifluridine	Deglycosylation	*Escherichia coli* encoding *deoA* or *upd* genes	In vitro incubation with bacterial strains	Premature activation to 5-fluoruracil, potentially increasing risk of intestinal toxicity [79]
Hydrocortisone	Deacetylation (by unidentified enzyme) and subsequent ketone reduction by 20β-HSDH	*Bifidobacterium adolescentis* encoding the 20β-HSDH gene	Ex vivo human microbiota from faeces (sex unspecified)	Formation of 20β-dihydrocortisone [15]
Levodopa	Decarboxylation	Bacterial tyrosine decarboxylases	Humans (both sexes)	Peripheral conversion of levodopa to dopamine. Abundance of intestinal tyrosine decarboxylase explains increased oral levodopa dose requirements in Parkinson's disease patients [102]

(continued)

microbiota. One such drug was spironolactone, which underwent thioester hydrolysis to the active compound 7α-thiospironolactone. In juxtaposition, other drugs were completely stable in all donors' samples, including ketoconazole and

Table 1 (continued)

Drug	Reaction	Causative agent	Experimental model	Effect
Mycophenolate mofetil	Ester hydrolysis	Unknown	Ex vivo human microbiota from faeces (sex unspecified)	Formation of mycophenolic acid, a metabolite linked to GI toxicity. Metabolism shows inter-individual variability [15]
Progesterone	Likely reduction	Unknown	Ex vivo human microbiota from faeces (males)	Progesterone is degraded by faecal microbiota within 2 h. Potential metabolites include 5α and 5β-pregnanolone [85]
Sulfasalazine	Cleavage of azo bond	Bacterial azoreductases (widely produced across species)	Ex vivo human microbiota from faeces (sex unspecified)	Rapid metabolism of the prodrug sulfasalazine (within 120 min) to its active compound, 5-aminosalicylic acid [84]
Tacrolimus	C9 keto-reduction	*Faecalibacterium prausnitzii*	Humans (both sexes)	Production of metabolite, M1, with 15-fold lower immunosuppressant activity [105]. *F. prausnitzii* abundance positively correlates with oral tacrolimus dose requirements in adult kidney transplant patients [106]

ropinirole. Interestingly, several drugs were variably metabolised between individuals. Hydrocortisone is an example of variable metabolism, which was seen to undergo ketone reduction to the androgenic 20β-dihydrocortisone in a fraction of cases.

Whilst both of these studies have broadly expanded knowledge concerning drug-microbiome reactions, it is important to recognise their context and limitations. Neither study measured in vivo microbial metabolism in humans; thus, results cannot be assumed to totally outline microbial drug metabolism in patients. Zimmerman et al. incubated drugs with bacterial isolates, which does not reflect the multi-species communities of interacting microbiota found in the human gut [7]. Alone, individual bacterial isolates may behave differently than when in the presence of other microorganisms. In comparison, Javdan et al. used bacteria cultured from faecal samples, which better reflects the heterogenous populations found in the intestines. However, neither study accounted for other features of the intestinal environment, such as the presence of mucus, food, bile, and even disease, which can all affect microbial functioning [107–109]. Whilst neither investigation characterises the total nature of microbial drug metabolism in vivo, they do provide a platform from which future in vivo work can be based. Such work may begin in animals; however, due to differences between the human microbiome and that of other species, only observations performed in humans can fully validate drug-microbiome interactions [110]. Within humans, interactions should be considered across the spectrums of age, sex, ethnicity, and health status [111–114].

2.1.2 Indirect Microbiome Effects on Pharmacokinetics

Aside from direct biotransformation by enzymes, the gut microbiome can significantly affect drugs' PK via other means. Though researched less, microbiome modulation of bile acids, epithelial permeability, hepatic metabolism, gut motility, and intestinal transporter expression illustrate examples of characterised indirect effects (Fig. 2) [115–117]. Overarchingly, when these aspects of human physiology and functioning are altered, the bioavailability of orally administered drugs can change. Bile acids secreted into the intestines help support healthy microbiome composition, through inherent antibacterial properties, and are metabolised by resident bacteria [118]. Bile acids also facilitate the solubilisation of luminal lipids, including lipophilic drugs. If intestinal bile acid composition is altered, e.g. due to changes in bacterial metabolism, then the luminal solubility of drugs and subsequently their epithelial absorption may be affected [119]. One study has demonstrated that the absorption of ciclosporin, a lipophilic immunosuppressant, is affected by the presence of a secondary bile acid, ursodeoxycholate [120]. Further, work by Enright et al. has demonstrated microbial enzymes to affect bile salt solubilisation of nine oral drugs, including the narrow therapeutic index antiepileptic, phenytoin [121]. It is likely that the true number of lipophilic drugs affected by microbiome bile modulation is higher than nine. It is also recognised that hepatic synthesis of bile acids decreases with advancing age [94, 122]. If compounded by microbiome effects on bile acids, elderly populations may struggle to solubilise very lipophilic drugs.

Changes in the metabolic capacity of the liver can have substantial effects on drugs' plasma concentration profiles. The liver is responsible for the two-step metabolism of many drugs, often producing water-soluble metabolites that can be safely excreted from the body. In this way, the liver greatly contributes towards the detoxification and clearance of drugs. Ageing itself can lead to effects on hepatic function. For example, blood flow to the liver declines by approximately 35% between young adulthood (under 40 years) and older adults (over 65 years) [123]. Elderly adults also more commonly have comorbid diseases and/or frailty, which can similarly affect hepatic function. As such, age-related changes to the liver can be a major source of inter-individual variability in PK [124]. Amitriptyline, diltiazem, levodopa, morphine, and verapamil are all examples of high hepatic clearance drugs that have lower clearance in elderly people [125]. In addition, the microbiome can exert significant influence on the hepatic metabolism of drugs. The gut microbiome and liver have direct lines of bidirectional communication via the portal vein and bile duct. In this manner, changes to microbiome composition can directly affect liver functioning. For example, a study found over 4000 gene transcripts to be differentially expressed in the livers of mice with and without microbiomes [126]. Several of these transcripts were identified as playing roles in drug metabolism, including the cytochrome P450 (CYP450) enzyme Cyp3a11. Over half of all marketed drugs are known to be metabolised by the CYP3A subfamily [127]. Another study also examining mice identified that a cluster of 112 hepatic genes associated with drug metabolism are affected by gut microbiota [128]. The researchers observed that the half-life of a common anaesthetic, pentobarbital, was significantly longer in mice colonised with microbiota compared to those that were germ-free.

Even lifestyle factors can affect physiological response to drugs. Liu et al. have recently demonstrated that diet can influence the therapeutic efficacy of antibiotics [129]. The researchers measured the efficacy of multiple bactericidal antibiotics, including doxycycline, vancomycin, and ciprofloxacin, in mice infected with methicillin-resistant *Staphylococcus aureus* (MRSA) or *Escherichia coli*, two pathogens that can be deadly in elderly patients. They found that mice fed high-fat diets as opposed to normal diets responded less well to antibiotic treatment, demonstrated by higher pathogen burden and lower pathogen susceptibility to antibiotics. In response to high-fat diets, the metabolic reactions and microbial populations present within the gut microbiome shifted. Chiefly, synthesis of an organic acid, indole-3-acetic acid (IAA) [130], was significantly decreased in response to high-fat diets. IAA is produced by microbiota following ingestion of fibre and is thought to act synergistically with antibiotics by rendering pathogens more susceptible to bactericidal attack. Though yet to be confirmed in human models, it may be that high-fibre diets support the efficacy of antibiotics through the production of IAA. This finding is important for people of advanced age, as fibre intake in elderly populations is often much lower than recommended. One reason for this is that hard-textured fibrous foods can be difficult for elderly people with masticatory problems to ingest [131]. In general, sufficient fibre intake is beneficial for older individuals, as it supports healthy GI and microbiome functioning. With the emerging evidence that a fibre-rich diet may also promote the efficacy of antibiotics, health- and social care professionals have even more reason to support elderly patients to meet the recommended 30 g fibre intake per day [132].

2.2 Drugs Vs. Bugs: Drug-Induced Microbiome Remodelling

It is fairly well recognised that diet influences which microbial species inhabit the gut, and this is also true for medicines [133, 134]. Understandably, drugs with intended antimicrobial actions can deplete whole populations of intestinal species practically overnight [135]. Traditional antimicrobial agents are indiscriminate and thus target susceptible microbiota and pathogens by the same mechanisms [136].

The impact that antimicrobial drugs have on the gut microbiome are manifold, influenced both by patient-specific factors (e.g. age, sex, previous exposure to antimicrobials) and medicine-specific factors (e.g. drug, formulation, dose) [137–139]. Elderly patients are specifically susceptible to adverse outcomes of antimicrobial-induced microbiome perturbations. *C. difficile* is a life-threatening condition in which the composition of the gut microbiome becomes unbalanced (a state also known as dysbiosis), and pathogenic bacteria multiply, leading to severe diarrhoea and systemic illness [140]. Unfortunately, the vast majority of *C. difficile*-related deaths are in hospitalised, elderly populations [141]. Guidelines aimed at preventing *C. difficile* recommend reducing the use of systemic antibiotics, particularly those with broad-spectrum anti-bacterial activity, in order to reduce the widespread destruction of patients' microbiomes [142]. Whilst *C. difficile* infection is typically an acute outcome of antimicrobial therapy, effects of antimicrobials on the gut microbiome can be sustained. A study measuring long-standing effects of antibiotics on human gut microbiota found that bacterial diversity can be reduced for up to 4 years following prescription of macrolides and lincosamides and a year for beta-lactams and quinolones [143]. The consequences of these changes can present as alterations in digestion, antibiotic resistance, autoimmune activity, reduced vitamin or hormone production, and even altered response to other medicines [144–146]. For example, immune response following vaccination may be impaired following courses of antibiotics [147]. Additionally, antibiotic administration in rats has been shown to increase bioavailability of the antipsychotic olanzapine [148]. Further, patients co-administered with warfarin and high-activity antibiotics against *Bacteroides fragilis* have a significantly increased anticoagulant bleeding risk [149].

Aside from antimicrobials, drugs with human targets can also exert effects on the microbiome (Table 2). In some cases, effects on microbiome composition or functioning could support a drug's mechanism of action. For instance, methotrexate (commonly used in elderly patients with rheumatoid arthritis) has been recently found to alter gut bacterial composition, leading to decreased immune activation and improved medication response [150]. Statins, often prescribed to older patients to reduce plasma cholesterol, have been found to be protective against a gut microbiome enterotype associated with obesity and systemic inflammation [151]. Metformin's relationship with the gut microbiome has also received substantial attention in recent years. Both human and animal studies have demonstrated that oral metformin alters microbial species inhabiting the gut, supporting the drug's actions in improving host glucose tolerance [152]. There is also mounting interest in metformin's possible effects on ageing. It is thought that metformin promotes growth of symbiotic gut bacteria whilst reducing the expansion of pathogenic species, resulting in beneficial immunomodulation. In fact, the Targeting Aging with Metformin (TAME) trial is currently underway in the United States, to specifically assess metformin as a repurposed health-ageing therapeutic [153].

In a high-throughput study, Maier et al. measured the ability of 1197 marketed drugs to impair the in vitro growth of 40 gut bacterial isolates [54]. They found that 24% of human-targeted drugs significantly impaired at least one of the 40 strains,

Table 2 Effects of drugs on the gut microbiome and health

Drug(s)	Effects	Experimental model
Atypical antipsychotics (PO) (including clozapine, olanzapine, risperidone, quetiapine, asenapine, ziprasidone, lurasidone, aripiprazole, paliperidone, and iloperidone)	Decreased bacterial species diversity in females (potentially explaining why females are more prone to antipsychotic-induced weight gain). Both sexes showed increased abundance of *Lachnospiraceae* and decreased abundance of *Akkermansia* and *Sutterella*	Adult humans (both sexes) [59]
Benzylpenicillin in combination with gentamicin (IV)	Reduced bacterial richness, particularly decreased abundance of bifidobacteria for 2 years. Attenuation of weight and height gain in boys for first 6 years of life. Higher body mass index in both sexes	Human neonates in first 48 h of life (both sexes) [137]
Fluoxetine (PO)	Decreased abundance of *Turicibacter sanguinis*, leading to increased serum triglyceride levels and reduced white adipose tissue in females (but not males)	Mice (both sexes) [89]
Metformin (PO)	Treatment for 4 months altered abundance of 86 bacterial strains, mostly γ-proteobacteria (e.g. *Escherichia coli*) and *Bacillota*. Increased abundance of *Akkermansia muciniphila*. Altered bacterial gene expression and improved host glucose tolerance	Human adults (both sexes) and mice (male) [152]
Methotrexate (PO)	Decreased abundance of *Bacteroidota* and increased abundance of actinobacteria. Expression of 6409 bacterial genes altered. Reduced inflammatory potential of microbiota	GF female mice colonised with human microbiota (both sexes); bacterial isolates; humans (both sexes) [150]
Omeprazole (PO)	Treatment for 4 weeks altered bacterial taxa associated with *C. difficile* infection (*Enterococcaceae* and *Streptococcaceae*, *Clostridiales*) and GI bacterial overgrowth (increased *Micrococcaceae* and *Staphylococcaceae*)	Humans (both sexes) [154]
Paracetamol (PO)	Higher abundance of *Streptococcaceae*	Humans (both sexes) [155]
Statins (PO) (simvastatin 48%, 31% atorvastatin, 21% other statins)	Protective against the *Bacteroides* 2 (Bact2) enterotype, a gut microbiome configuration associated with systemic inflammation and obesity. This may be due to attenuated inflammation	Human adults (both sexes) [151]

GF germ-free, *IV* intravenous, *PO* oral administration. Reproduced with permission from [91]

pointing to the potential magnitude of drug-induced microbiome remodelling in clinical practice. Antipsychotics, antineoplastic, and calcium channel blockers were found to make up the therapeutic classes with the highest anti-microbiota activity.

Based on this dataset, another study has built a machine learning algorithm to predict adverse effects of drugs on gut bacteria [134]. The work tested 13 types of machine-learning techniques, including neural networks and ensemble methods, to produce a model that can predict the anti-bacterial actions of any small-molecule drug. In silico methods for predicting drug-microbiome interactions using artificial intelligence will likely become more frequent in approaching years, potentially facilitating real-time predictions in clinical settings [91].

It is also important to recognise that excipients likely exert effects on the microbiome. Though much less characterised, PEG 4000, chitosan, saccharin, glycerol monolaurate, carboxymethylcellulose, polysorbate 80, and methylparaben have all been demonstrated to affect gut microbiota in some way [156–162]. Traditionally, pharmaceutical excipients are considered inert; however, increasing evidence is highlighting their multifarious biological impacts [163–165]. Because the majority of medicines are formed of excipients, it is important that this dominant portion of formulations is not overlooked when it comes to medicine-microbiome interactions. Excipient-microbiome interactions may even present untapped therapeutic opportunities [166].

3 Microbiome Medicine: Targeting the Ageing Microbiome

Increasingly, the therapeutic potential of the gut microbiome is coming to light. The gut microbiome's intrinsic connection to human health is presenting growing numbers of novel microbially dependent drug targets. The last several years has witnessed an upsurge in pharmaceutical start-ups focused solely on discovering, designing, and developing new microbiome medicines. Because the gut microbiome is linked with many age-associated diseases, there is great potential for leveraging microbiome-based targets for the promotion of healthy ageing [167–172]. However, to achieve successful development of microbiome therapeutics, advanced pharmaceutical formulation will likely be required to deliver active compounds to the correct niche of the microbiome at the correct time.

3.1 Delivering Microbiome Therapeutics with Smart Formulation

3.1.1 Single-Trigger Approaches

Because the colon houses the highest density of microbiota, it is probable that novel microbiome therapeutics will need to be delivered specifically to the colon. Delivery of medicines to their intended site of action often improves efficacy and may allow administration of lower doses [82, 86]. That said, site-specific drug delivery to the colon can be challenging without considered formulation. In the past, colonic drug delivery has harnessed one of three strategies: pH-dependent release, time-dependent release, or enzyme-dependent release.

pH mechanisms rely on the pH changes that naturally occur from the proximal to distal gut. The pH of the stomach marks the most acidic region of the GI tract due to the secretion of gastric acid from parietal cells, at around pH 2.0–4.5, depending on feed status [173]. Within the small intestines, pH progressively increases in response to secretion of pancreatic fluid rich in bicarbonate ions, starting at around pH 5.0–7.0 in the duodenum and peaking at around pH 7.0–8.0 by the ileum [174]. However, from the colon, average pH drops from 6.4 ± 0.6 in the proximal colon to 7.0 ± 0.7 at distal sites, due to the liberation of SCFAs by microbial fermentation of dietary fibre [173, 175]. This pH shift between the ileum and colon has been frequently exploited for colonic drug delivery [176]. Strategies often use formulation coatings that resist degradation in the stomach and small intestine and subsequently dissolve at colonic pH, releasing drug cargo locally. Examples of pH-sensitive polymers used for coatings include the Eudragit® polymers based on methacrylic acid (Evonik, Germany) and cellulose esters [177]. In principle, pH-dependent strategies should reliably deliver drugs to the colon; however, this is often not the case in practice [178]. Within the GI tract, many factors (such as the intake of food, the presence of disease, or differences in physiology) can cause pH to deviate from population averages [179, 180]. The resulting inter- and intra-individual variability in pH can cause pH-sensitive formulation coatings to release drug cargo before the colon or not at all [178].

Time-dependent strategies for colonic drug delivery utilise the average time it takes dosage forms to pass from the upper to lower GI tract. Whole GI transit time is very variable, and likely affected by microbiome composition; however, healthy averages of around 27–28 h have been reported [116, 181]. Time-based formulation approaches typically use a coating that remains intact within the stomach and enters a lag phase upon entering the small intestines. This lag phase should be equivalent to or slightly longer than small intestinal transit, resulting in drug release within the colon [182, 183]. Problems with these designs arise when patients' GI transit do not reflect population averages. Numerous patient-specific factors can alter GI transit time, including biological sex, obesity, mental state, and diet/medication changes [111, 184, 185]. Age is also a key cause of transit time variability. Older individuals often have longer GI transit times due to alterations in neuromuscular function, polypharmacy, reduced fibre ingestion, and lower levels of physical activity [186–188]. For this reason, colonic targeting mechanisms based solely on average GI transit may not be the most suitable choice for elderly patients.

Enzyme-dependent methods for delivering drugs to the colon harness the metabolic potential of the microbiome. Film coatings based on polysaccharides are designed to be selectively degraded by microbiota after reaching the bacteria-dense colonic environment. More than 50% of colonic bacteria have the ability to produce amylases; therefore, starch-based coatings should be reliably digested within the colon [189]. Other polysaccharides commonly incorporated into bacteria-sensitive systems include pectin and chitosan [190–192]. An in vivo study has compared the ability of pH-dependent (Eudragit® S) and starch-based (amylose) coatings to

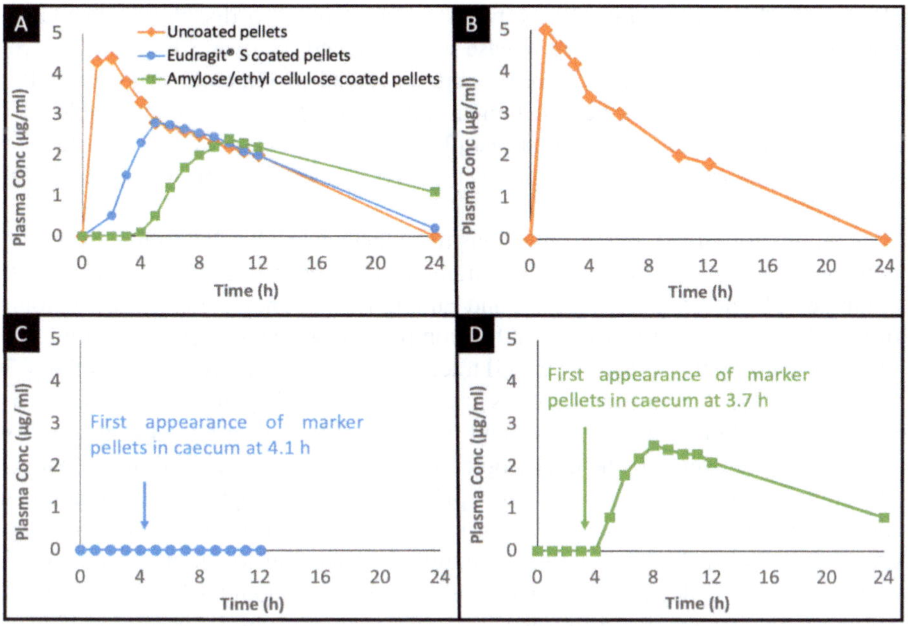

Fig. 3 (**a**) Mean plasma theophylline levels after administration of uncoated, Eudragit®S-coated pellets, and amylose/ethyl cellulose-coated pellets to eight healthy male subjects. **b**, **c**, and **d** show examples of plasma concentration time profiles observed in single subjects: (**b**) uncoated pellets, (**c**) Eudragit® S-coated pellets, and (**d**) amylose/ethyl cellulose-coated pellets. (Reprinted with permission from [193])

target the colon (Fig. 3) [193]. Whilst the amylose coating achieved the best results, it does not completely guarantee reliable colonic drug delivery across patient populations. For example, colonic dysbiosis (e.g. after antibiotic exposure) may limit individuals' ability to completely degrade polysaccharide coatings, especially if colonic transit is reduced (e.g. due to diarrhoea).

3.1.2 Advanced Colonic Targeting

Due to their variability, formulation techniques relying solely on pH, time, or enzyme-dependent mechanisms for colonic drug delivery can be improved upon. A possible resolution is the combination of multiple features, allowing the unreliability of single-trigger mechanisms to be overcome. All permutations of pH, time, and enzyme-sensitive combinations are theoretically possible. One example of pH and time-dependent combination is the Multi Matrix System® (or MMX™) that embeds drug molecules within small lipophilic matrices, which are covered by an enteric coating based on Eudragit® L and S. The enteric coating is dissolved at neutral pH (typically within the distal small intestine), and drug release is subsequently controlled from the underlying matrix. Investigations of the MMX™ system have shown variable-release kinetics between different in vitro dissolution media. In phosphate buffer (pH 7.4), drug release was sustained for over 8 hours with zero-order kinetics, though this differed to that observed in Krebs bicarbonate buffer (pH 7.4) [194, 195]. Further, in human studies, drug PK was comparable to a conventional formulation based on just pH-dependent release. Whilst the MMX™ system shows promise, it has not yet proven itself superior to single-trigger mechanisms.

Combination of pH and enzyme-sensitive mechanisms has led to the successful commercialisation of two coating technologies in recent years. The first, Phloral®, combines a blend of resistant starch and Eudragit® S to enable dual-trigger colonic drug delivery [196]. The Eudragit® S component provides coating stability and physical integrity through the upper GI tract. Upon entering the colon, the Eudragit® S dissolves, and the resistant starch is gradually degraded by enzymes produced by regional microbiota [197]. This system has proved itself as a fail-safe method for colonic drug delivery, as the dual-trigger system overcomes variabilities in patient physiology. For example, if a patient's pH deviates from population average, then the enzyme-sensitive component will still allow site-specific drug release. Similarly, if a patient has a dysbiotic colonic microbiome, leading to failure of the starch component to be degraded, then the pH mechanism should still function well. Phloral® has demonstrated its superior colonic targeting in delivering probiotics, faecal microbiota for treatment of recurrent *C. difficile* infection, and GRP84 and FFAR4 agonists for treatment of obesity [140, 198, 199]. The second newly developed technology, OPTICORE™, builds on the Phloral® design to further optimise colonic drug delivery. OPTICORE™, standing for OPTImized COlonic RElease, formulates an inner alkaline coat (including a neutral enteric polymer) underneath an external Phloral® coating (Fig. 4). Depending on application, a hydroxypropyl methylcellulose layer may separate the alkaline coat from the inner drug core to prevent an acidic drug reacting with the alkaline layer. As the outer Phloral® coat begins to dissolve or be degraded, the inner alkaline layer is dissolved,

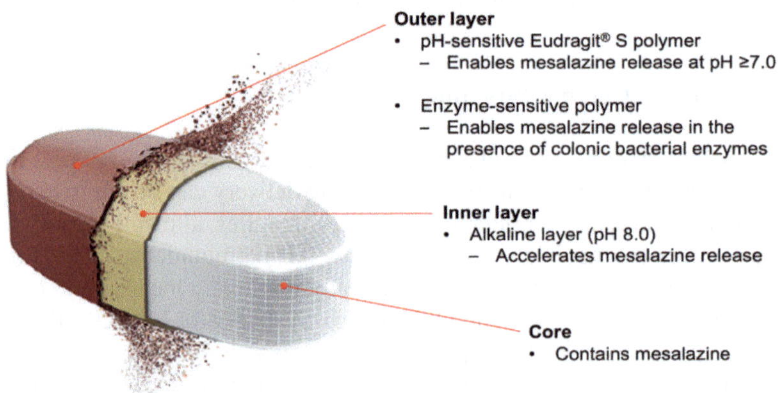

Fig. 4 Graphical illustration of the OPTICORE™ system that can delivery any drug reliably to the colon via combination of pH and enzyme-dependent technologies. (Reproduced with permission from [204])

leading to an environment with elevated pH, buffer capacity, and ionic strength at the inside surface of the Phloral® layer. This micro-environment expedites drug release by rapidly ionising the Eudragit® S portion of the Phloral® coat, resulting in faster dissolution [200, 201]. OPTICORE™ is marketed for the colonic delivery of mesalazine for treatment of ulcerative colitis [202]. The technology allows colonic delivery of 1.6 g of mesalazine, the highest single oral dosage form in the world, facilitating patient compliance through the reduction of dosage frequency. This advantage is especially convenient for elderly patients, who may struggle with complicated polypharmacy regimens [203].

3.2 Novel Microbiome Therapeutics

As a result of the continuous improvements in socio-economic conditions, the elderly population worldwide is constantly rising. An estimate of the United Nation report stated that the number of older persons (age > 65 years) will more than double in the next three decades, projected to increase from 703 million in 2019 to over 1.5 billion in 2050 [205]. As mentioned, ageing can carry a wide range of pathologies, consequently reducing the quality of elders' lives. Since diet is a predominant mediator of the composition and functions of the gut microbiota, scientific communities are studying how microbiota-targeted dietary interventions may prevent the onset of such pathologies, promoting both health span and life span. The presence and abundance of specific microbial strains in the gut are demonstrated to be adjustable through specific dietary modifications. In the last decades, the consumption of fat and proteins in the Western world diet has increased, whilst complex polysaccharides have decreased. Compared to the gut microbiome composition of rural populations of Africa and South America, it is evident how diet and lifestyle strongly influence gut microbiota. For example, Westernized subjects present a reduced microbial diversity with a specific decrease in fibre-degrading bacteria strains, and specifically it has been observed that a high-fat diet has an adverse impact on the abundance of *A. muciniphila* and *Lactobacillus*, both health-promoting strains. On the contrary, a high-fibre diet is highly recommended to restore SCFA-producing bacteria, especially the production of butyrate to reinforce the gut barrier, prevent the colonisation and proliferation of pathogens, and mitigate inflammaging [206]. Moreover, there is evidence that the Mediterranean diet, rich in vegetables, legumes, fruits, nuts, olive oil, and fish, in the elderly population promotes cognitive functions and reduces the risk of frailty and the inflammatory status [207].

In addition to diet modifications to counteract the consequences of a dysbiotic microbiota, over the past two decades, numerous studies have been conducted to investigate the effectiveness of non-invasive treatments such as the consumption of beneficial microorganisms (probiotics), substrates utilised by microorganisms to promote their growth (prebiotics), or a combination of the two (synbiotics) in the elderly population [208].

3.2.1 Probiotics

Since ancient times, fermented food made of living microbes have been consumed to maintain or restore health. The first modern probiotic came from the Nobel laurate Élie Metchnikoff, who, at the beginning of 1900s, hypothesized that regular ingestion of *Lactobacillus* could displace pathogenic intestinal microorganisms, thus increasing longevity and healthy living [209]. Today, probiotics are defined, by the World Health Organization, as live microorganisms that confer a health benefit when administered to the host at defined doses [210]. Their beneficial effects on the host's health status comprise improvements of the gut barrier functions, immunomodulation, production of neurotransmitters, and resistance against potential pathogens [209]. Traditionally, the discovery of probiotics relied on the identification of bacteria strains which are enriched in healthy individuals' guts and altered in a dysbiotic state, conveying a clear benefit to human health. This approach led to the formulation of probiotics principally containing *Lactobacillus*, *Bifidobacterium*, and other lactic acid-producing bacteria (LAB), followed by *E. coli* Nissle 1917, *Saccharomyces*, *Bacillus* spp., *Enterococcus*, *Weissella* spp., and, more recently, *Akkermansia muciniphila* [211, 212]. Moreover, as research on the human microbiota and its functions is intensifying, and as genome sequencing has become more affordable, new potential probiotic discoveries are under development. The isolation and characterisation of new bacteria strains from the human gut, such as *Roseburia intestinalis*, *Faecalibacterium prausnitzii*, *Eubacterium* spp., *Bacteroides* spp., and *A. muciniphila*, could represent the so-called next-generation probiotics [212]. These strains, compared to traditional probiotic strains, require strategic formulation development to overcome challenges associated with aerobic conditions and growth media to be successfully delivered as oral formulations, potentially with colonic targeting [212]. There are data supporting the beneficial effects of probiotics supplementation. Most of the clinical trials in human studies, still today, focus on probiotics containing gram-positive *Lactobacillaceae* and *Bifidobactericeae*.

One clinical study evaluated the effectiveness of *Bifidobacterium longum* and *Lactobacillus helveticus* in ameliorating immune response in an elderly cohort of over 75 years of age, as well as in aged mice. These bacterial strains were administered in the form of a biscuit for 30 days and led to a significant increase in regulatory T cells, B lymphocytes, and natural killer activity compared to the placebo control group [213]. Another controlled study was conducted in 96 elderly patients residing in nursing homes, who consumed fermented yogurt with *Lactobacillus delbrueckii* ssp. *bulgaricus* every morning. These subjects had an increase in total IgA levels in saliva and also in influenza A virus-bound IgA levels after consumption of the study yogurt, suggesting a possible prevention of the influenza A virus in elderly subjects with a weakened immune system [214].

Recently, Wang et al. evaluated the effects of oral probiotics as a prophylaxis for cognitive impairment in more than 100 elderly patients scheduled for orthopaedic or colorectal surgery. The probiotic capsule contained *Bifidobacterium longum*, *Lactobacillus acidophilus*, and *Enterococcus faecalis* and significantly reduced the incidence of postoperative impairment in these subjects and decreased the plasma levels of IL6 and cortisol, compared to the placebo group, possibly via the limitation of the stress response and peripheral inflammation [215]. Moreover, many experimental studies showed how probiotics can also promote longevity. Great attention was given to the nematode *Caenorhabditis elegans*, which seems to be an appropriate model for anti-ageing research, because it experienced life-extending effects from probiotics supplementation, in addition to an increased innate immune response, improved resistance to oxidative stress, and modulated serotonin signalling [39].

Probiotics have gained an exceptional popularity among the population; however, their proof of efficacy in ameliorating age-related disorders is still conflicting among the medical/scientific communities [211]. Unfortunately, to date, most probiotic strains have been chosen on the basis of their viability, stability, and overall ease of production and long-term storage, thus, to confer a strong technological advantage rather than to convey clear beneficial effects. Moreover, even if general probiotic use is considered to be safe, they should be consumed with caution in some aged patients since they could encourage excessive immune stimulation, gastrointestinal adverse effects, or unfavourable metabolic profiles that may prevent gut probiotic colonisation [39]. Furthermore, as the beneficial probiotics' properties are likely to be strain specific, it should not be assumed that the effects of one strain are automatically true for others without scientific confirmation [39].

3.2.2 Prebiotics

The first cases of management of gut microbiota with prebiotics have been the consumption of fermented carbohydrates to modulate levels of resident *lactobacilli* and *bifidobacteria* to obtain health effects. Thanks to technological advancements, in recent years, a wider variety of prebiotics has been determined, and new bacterial strain targets have been identified. Dietary prebiotics are defined as fermented ingredients that are selectively utilised by the host microorganisms conferring benefit health [210]. Prebiotics are complex non-digestible substances within the small intestine that are submitted to fermentation when reaching the colon. In the colon, prebiotics stimulate specific bacterial growth either directly or indirectly through cross-feeding interactions – whereby metabolic products from some microorganisms can be exploited as growth substrates from others [216].

Most prebiotic substances in the market today include carbohydrates of low digestibility rich in galactooligosaccharides (GOS) and fructooligosaccharides (FOS) because of their ability to promote the proliferation of specific groups of bacteria, especially *lactobacilli*, bifidobacteria, or both [217]. To assess the potential effect of these molecules on the gut microflora of elderly persons, Vulevic et al. [218] administered a GOS prebiotic mixture to 44 elderly subjects for 10 weeks in a placebo-controlled, crossover study. At the end of the study period, predominant bacterial groups were quantified, and several beneficial strains were increased, particularly bifidobacteria, at the expense of less beneficial bacterial strains. In the analysis of blood samples, they found a reduction in the production of pro-inflammatory cytokines and increase of the anti-inflammatory interleukin-10, phagocytosis, and NK cell activity. Thus, these prebiotic substrates resulted in positive outcomes on both the microbiome composition and the immune response [218].

Resistant starch (RS) is another carbohydrate well known for its multiple beneficial effects [219]. In the gut, it increases the butyrate production, thus contributing to the maintenance of epithelial integrity, regulating inflammation and influencing gene expression in colonic epithelial cells [220]. Since the elderly diet is often lacking resistant starch, Alfa et al. investigated whether the supplementation of RS could be an effective prebiotic to restore a compromised ageing gut microbiota [221]. In their prospective study, they collected the stools of >70 year-old elderly individuals who were given a daily administration of *MSPrebiotic®*, microgranules of resistant starch, for a 3-month period. The daily consumption of *MSPrebiotic®* was significantly associated with an increase in *Bifidobacterium*; a decrease of the abundance of proteobacteria, in particular *E. coli*/*Shigella* (compared to what was observed at baseline); and a slight increase in the butyrate levels versus placebo group [221].

Whilst the prebiotic substances dominating the market today are galactans and fructans, the ambition to stimulate the proliferation of a wider group of beneficial microbes within the gut has pushed the investigation of new prebiotic compounds. Most of the known compounds are derived from plants, but there is an increasing interest in animal-derived substances, such as oligosaccharides rich in milk, yeast-based substrates, and non-carbohydrates that could reach the colon intact and be utilised by resident microbes [216]. Concerning the human milk oligosaccharides (HMOs), they are complex glycans present in high concentrations in human milk. It was observed that in infants, only 1–2% of these molecules are absorbed, whilst the majority reaches the large intestine to provide selective nutritive substrates to bifidobacteria. They also modulate the immune system and prevent pathogen colonisation [222]. This suggests that HMOs supplementation could represent a valuable strategy to promote the growth of beneficial bacteria also in the gut of adults and elderly people. However, the impact of HMOs on the gut functions of adult patients currently lacks in vivo confirmation.

3.2.3 Synbiotics

When the effects of fermentable substrates and live microorganisms are blended, this creates a synbiotic. In this context, the living microorganisms can be autochthonous, i.e. resident or colonising the host, and allochthonous microorganisms, which are externally applied even if transiently present (e.g. probiotics). Synbiotics can be either complementary or synergistic: the first are formulated as an combination of a prebiotic and probiotic (or multiple of each), which must have their own already established beneficial effects, working independently to elicit one or more health benefits. A synergistic symbiotic is composed of live microorganisms in coadministration with a fermentable substrate that enhances its functionality. Synergistic synbiotics work together (not independently), and neither one of the components needs to be confirmed probiotics or prebiotics; however, they must demonstrate that the combined effect is greater than the effect of a single component [210].

Comparatively to prebiotics and probiotics, there are a limited number of studies examining the effects of synbiotics in older populations. The benefits on the microbiome observed with synbiotics are similar to those of pro- and prebiotics and include increases in the faecal abundance of lactobacilli and bifidobacteria and improved mucosal integrity and butyrate production [208]. Furthermore, they showed beneficial outcomes for conditions such as high fasting glucose, lipid profiles related to obesity, postoperative infections, and some inflammatory markers [223].

In this regard, Cicero et al. [224] investigated the efficacy of the supplementation of synbiotics in the pathogenesis of metabolic syndrome in the a cohort of 60 elderly patients. They were given a formula of *Lactobacillus plantarum*, *Lactobacillus acidophilus*, and *Lactobacillus reuteri* with both inulin and FOS for 60 days, compared to a placebo group that received only maltose as a control. The results showed a significant improvement in many parameters associated with metabolic syndrome, such as waist circumference, total cholesterol, high-density lipoprotein cholesterol, triglycerides, low-density lipoprotein cholesterol, and tumour necrosis factor. Moreover, mean arterial pressure and fasting plasma glucose were significantly reduced suggesting that the treatment with this synbiotic formula of lactobacilli and prebiotics ameliorated the prevalence of metabolic syndrome, some cardiovascular risk factors, and markers of insulin resistance [224]. Although some studies conducted with synbiotics have not demonstrated clear health benefits (e.g. for irritable bowel syndrome [225], renal dysfunction [226], or non-alcoholic fatty liver disease [227]), the future of synbiotics will likely be determined by novel microbial strains and substrates with potential interventions in the gastrointestinal tract and non-gastrointestinal targets, including the female urogenital tract, oral cavity, nasopharyngeal tract, and skin.

3.2.4 Faecal Microbiota Transplantation

Currently, dietary interventions – diet modifications or supplement administration – are the most utilised approaches for modulating the host gut microbiota. In the recent years, faecal microbiota transplantation (FMT) or bacteriotherapy has acquired relevant popularity in treating particular disorders of gastrointestinal nature (e.g. IBD and IBS) and also non-gastrointestinal nature [228]. Precisely, researchers demonstrated that multifactorial diseases such as metabolic, behavioural, autoimmune, and neurological diseases can be associated with an abnormal microbiome composition. Thus, FMT, which refers to the transplantation of liquid filtrate faeces from healthy donors into the gastrointestinal tract of the patient, could be a promising treatment to restore the balance of the intestinal ecosystem and treat the diseases associated with a dysbiotic gut [229]. The FMT therapeutic approach is the first recommended treatment for recurrent or relapsing *C. difficile* infections when the patient has a severe morbidity and does not respond to conventional antibiotics or hospitalisation [230]. Clinical manifestations of *C. difficile* infection vary from mild watery diarrhoea to possibly lethal conditions such as pseudomembranous colitis [229]. To date, there are no data to suggest that *C. difficile* infection causes different symptoms in the elderly compared to other patients. However, it is well documented that increasing age (> 70 years) increases frequency of complications, mortality, and risk of recurrent disease [231]. This disproportion in the older population may be due to immunosenescence, multimorbidity, IBD, chronic liver disease, and prolonged or multiple hospitalisations. However, the most common risk factor implicated in the pathogenesis of *C. difficile* infections is antibiotic use (specifically clindamycin, cephalosporins, fluoroquinolones, or combinations of antibiotics) that leads to gut dysbiosis and allows *C. difficile* to flourish [232]. Although current research mainly focuses on *C. difficile* infections, scientists have been looking into FMT as a potential promising treatment for other pathologies such as metabolic syndrome [233], IBD [234], and even neurodegenerative disorders (e.g. Alzheimer's and Parkinson's diseases) [215].

Despite the increasing general acceptance for the therapeutic use of FMT, it remains a treatment with a potential risk of infections for some patients due to the variable nature of donor stools and the limited understanding of a healthy gut microbiota composition. Few long-term studies have been conducted to assess the safety of this treatment, and more likely microbiota-based medicines with defined mixtures of microorganisms will still be the preferred products for microbiome remodelling [235].

4 Conclusion

After more than a decade of an increase in interest in the field microbiota, it is evident how significant changes in the gut microbiota occur with age and in turn how the microbial community can affect ageing. Indeed, ageing can be a healthy process, as seen in long-living people who live healthily or can be associated with age-related pathological conditions, frailty, weakened immune system, or inflammaging. Ageing is a multifactorial progression; changes in dietary habits, lifestyle, hygiene, and drug administration shape the establishment and development of the gut microbiota. In particular, we have outlined the disruptive impact that various common drugs can have on the microbial composition and microbial diversity of elderly patients and on the contrary, the effect of the microflora on the bioavailability, both biotransformation and bioaccumulation, of medications. In this regard, given the complexity of the medicine-microbiome relationship, it is important to rely and push future research on technological advancements and machine learning algorithms to predict and to elucidate the mechanisms responsible for the interactions.

Another potentially determining factor that shapes the microbiota composition is diet, and that's why a direct modulation of the gut microbiome with probiotics, prebiotics, synbiotics, and other interventions may be applied not only in treating specific age-related pathologies but also ameliorating ageing per se. In the older patients, the microbial targeted supplements seem to be effective in attenuating the occurrence of critical conditions such as frailty and opportunistic strains' infections (*C. difficile*); thus, a microbiota restauration therapy should be made early and often.

References

1. Barko PC, McMichael MA, Swanson KS, Williams DA. The gastrointestinal microbiome: a review. J Vet Intern Med. 2018;32(1):9–25.
2. Sender R, Fuchs S, Milo R. Are we really vastly outnumbered? Revisiting the ratio of bacterial to host cells in humans. Cell. 2016;164(3):337–40.
3. Valdes AM, Walter J, Segal E, Spector TD. Role of the gut microbiota in nutrition and health. BMJ. 2018;361:k2179.
4. Poretsky R, Rodriguez RL, Luo C, Tsementzi D, Konstantinidis KT. Strengths and limitations of 16S rRNA gene amplicon sequencing in revealing temporal microbial community dynamics. PLoS One. 2014;9(4):e93827.
5. Dieterich W, Schink M, Zopf Y. Microbiota in the gastrointestinal tract. Med Sci (Basel). 2018;6(4)
6. Rinninella E, Raoul P, Cintoni M, Franceschi F, Miggiano GAD, Gasbarrini A, et al. What is the healthy gut microbiota composition? A changing ecosystem across age, environment, diet, and diseases. Microorganisms. 2019;7(1)
7. Donaldson GP, Lee SM, Mazmanian SK. Gut biogeography of the bacterial microbiota. Nat Rev Microbiol. 2016;14(1):20–32.
8. Jandhyala SM, Talukdar R, Subramanyam C, Vuyyuru H, Sasikala M, Nageshwar RD. Role of the normal gut microbiota. World J Gastroenterol. 2015;21(29):8787–803.
9. Louis P, Flint HJ. Formation of propionate and butyrate by the human colonic microbiota. Environ Microbiol. 2017;19(1):29–41.
10. Roediger WE. Role of anaerobic bacteria in the metabolic welfare of the colonic mucosa in man. Gut. 1980;21(9):793–8.
11. Chambers ES, Preston T, Frost G, Morrison DJ. Role of gut microbiota-generated short-chain fatty acids in metabolic and cardiovascular health. Curr Nutr Rep. 2018;7(4):198–206.
12. Wilson KA, Han Y, Zhang M, Hess JP, Chapman KA, Cline GW, et al. Inter-relations between 3-hydroxypropionate and propionate metabolism in rat liver: relevance to disorders of propionyl-CoA metabolism. Am J Physiol Endocrinol Metab. 2017;313(4):E413–E28.
13. Macpherson AJ, Uhr T. Induction of protective IgA by intestinal dendritic cells carrying commensal bacteria. Science. 2004;303(5664):1662–5.
14. Zhang X, Han Y, Huang W, Jin M, Gao Z. The influence of the gut microbiota on the bioavailability of oral drugs. Acta Pharm Sin B. 2021;11(7):1789–812.
15. Javdan B, Lopez JG, Chankhamjon P, Lee YJ, Hull R, Wu Q, et al. Personalized mapping of drug metabolism by the human gut microbiome. Cell. 2020;181(7):1661–79. e22
16. Sun C, Chen L, Shen Z. Mechanisms of gastrointestinal microflora on drug metabolism in clinical practice. Saudi Pharm J. 2019;27(8):1146–56.
17. Klünemann M, Andrejev S, Blasche S, Mateus A, Phapale P, Devendran S, et al. Bioaccumulation of therapeutic drugs by human gut bacteria. Nature. 2021;597(7877):533–8.
18. Mackie et al. 1999. Mackie et al.-1999-Developmental microbial ecology of the neonatal ga.pdf.

19. Fanaro et al. 2003. Fanaro et al.-2003-Intestinal microflora in early infancy compositio.pdf.
20. Aagaard K, Ma J, Antony KM, Ganu R, Petrosino J, Versalovic J. The placenta harbors a unique microbiome. Sci Transl Med. 2014;6(237):237ra65.
21. Cao et al. 2014. Cao et al.-2014-Placental Microbiome and Its Role in Preterm Birth.pdf.
22. Romano-Keeler J, Weitkamp JH. Maternal influences on fetal microbial colonization and immune development. Pediatr Res. 2015;77(1–2):189–95.
23. Mishra A, Lai GC, Yao LJ, Aung TT, Shental N, Rotter-Maskowitz A, et al. Microbial exposure during early human development primes fetal immune cells. Cell. 2021;184(13):3394–409. e20
24. Dominguez-Bello MG, Costello EK, Contreras M, Magris M, Hidalgo G, Fierer N, et al. Delivery mode shapes the acquisition and structure of the initial microbiota across multiple body habitats in newborns. Proc Natl Acad Sci USA. 2010;107(26):11971–5.
25. Jakobsson HE, Abrahamsson TR, Jenmalm MC, Harris K, Quince C, Jernberg C, et al. Decreased gut microbiota diversity, delayed Bacteroidetes colonisation and reduced Th1 responses in infants delivered by caesarean section. Gut. 2014;63(4):559–66.
26. Dogra S, Sakwinska O, Soh SE, Ngom-Bru C, Bruck WM, Berger B, et al. Dynamics of infant gut microbiota are influenced by delivery mode and gestational duration and are associated with subsequent adiposity. MBio. 2015;6(1)
27. Rutayisire E, Huang K, Liu Y, Tao F. The mode of delivery affects the diversity and colonization pattern of the gut microbiota during the first year of infants' life: a systematic review. BMC Gastroenterol. 2016;16(1):86.
28. Asakuma S, Hatakeyama E, Urashima T, Yoshida E, Katayama T, Yamamoto K, et al. Physiology of consumption of human milk oligosaccharides by infant gut-associated bifidobacteria. J Biol Chem. 2011;286(40):34583–92.
29. Murk W, Risnes KR, Bracken MB. Prenatal or early-life exposure to antibiotics and risk of childhood asthma: a systematic review. Pediatrics. 2011;127(6):1125–38.
30. Cox LM, Blaser MJ. Antibiotics in early life and obesity. Nat Rev Endocrinol. 2015;11(3):182–90.
31. Ungaro R, Bernstein CN, Gearry R, Hviid A, Kolho K-L, Kronman MP, et al. Antibiotics associated with increased risk of new-onset crohn's disease but not ulcerative colitis: a meta-analysis. Am J Gastroenterol. 2014;109(11):1728–38.
32. Laursen MF, Zachariassen G, Bahl MI, Bergstrom A, Host A, Michaelsen KF, et al. Having older siblings is associated with gut microbiota development during early childhood. BMC Microbiol. 2015;15:154.
33. Tun HM, Konya T, Takaro TK, Brook JR, Chari R, Field CJ, et al. Exposure to household furry pets influences the gut microbiota of infant at 3-4 months following various birth scenarios. Microbiome. 2017;5(1):40.
34. Lawley TD, Walker AW. Intestinal colonization resistance. Immunology. 2013;138(1):1–11.
35. Nagpal R, Mainali R, Ahmadi S, Wang S, Singh R, Kavanagh K, et al. Gut microbiome and aging: physiological and mechanistic insights. Nutr Healthy Aging. 2018;4(4):267–85.
36. Backhed F, Roswall J, Peng Y, Feng Q, Jia H, Kovatcheva-Datchary P, et al. Dynamics and stabilization of the human gut microbiome during the first year of life. Cell Host Microbe. 2015;17(5):690–703.
37. Derrien M, Alvarez AS, de Vos WM. The gut microbiota in the first decade of life. Trends Microbiol. 2019;27(12):997–1010.
38. Arumugam M, Raes J, Pelletier E, Le Paslier D, Yamada T, Mende DR, et al. Enterotypes of the human gut microbiome. Nature. 2011;473(7346):174–80.
39. Vaiserman AM, Koliada AK, Marotta F. Gut microbiota: a player in aging and a target for anti-aging intervention. Ageing Res Rev. 2017;35:36–45.
40. Biagi E, Nylund L, Candela M, Ostan R, Bucci L, Pini E, et al. Through ageing, and beyond: gut microbiota and inflammatory status in seniors and centenarians. PLoS One. 2010;5(5):e10667.

41. Odamaki T, Kato K, Sugahara H, Hashikura N, Takahashi S, Xiao JZ, et al. Age-related changes in gut microbiota composition from newborn to centenarian: a cross-sectional study. BMC Microbiol. 2016;16:90.
42. Claesson MJ, Cusack S, O'Sullivan O, Greene-Diniz R, de Weerd H, Flannery E, et al. Composition, variability, and temporal stability of the intestinal microbiota of the elderly. Proc Natl Acad Sci USA. 2011;108(Suppl 1):4586–91.
43. Claesson MJ, Jeffery IB, Conde S, Power SE, O'Connor EM, Cusack S, et al. Gut microbiota composition correlates with diet and health in the elderly. Nature. 2012;488(7410):178–84.
44. Mitchell et al. 2017. Mitchell et al.-2017-reduced intestinal motility, mucosal barrier funct.pdf.
45. Mitchell EL, Davis AT, Brass K, Dendinger M, Barner R, Gharaibeh R, et al. Reduced intestinal motility, mucosal barrier function, and inflammation in aged monkeys. J Nutr Health Aging. 2017;21(4):354–61.
46. Hollister EB, Gao C, Versalovic J. Compositional and functional features of the gastrointestinal microbiome and their effects on human health. Gastroenterology. 2014;146(6):1449–58.
47. Kim M, Benayoun BA. The microbiome: An emerging key player in aging and longevity. Transl Med Aging. 2020;4:103–16.
48. Zhang C, Li S, Yang L, Huang P, Li W, Wang S, et al. Structural modulation of gut microbiota in life-long calorie-restricted mice. Nat Commun. 2013;4:2163.
49. Lee J, Venna VR, Durgan DJ, Shi H, Hudobenko J, Putluri N, et al. Young versus aged microbiota transplants to germ-free mice: increased short-chain fatty acids and improved cognitive performance. Gut Microbes. 2020;12(1):1–14.
50. Flemer B, Gaci N, Borrel G, Sanderson IR, Chaudhary PP, Tottey W, et al. Fecal microbiota variation across the lifespan of the healthy laboratory rat. Gut Microbes. 2017;8(5):428–39.
51. Cianci R, Franza L, Massaro MG, Borriello R, De Vito F, Gambassi G. The interplay between immunosenescence and microbiota in the efficacy of vaccines. Vaccines (Basel). 2020;8(4)
52. Burcelin R, Serino M, Chabo C, Blasco-Baque V, Amar J. Gut microbiota and diabetes: from pathogenesis to therapeutic perspective. Acta Diabetol. 2011;48(4):257–73.
53. Westfall S, Lomis N, Kahouli I, Dia SY, Singh SP, Prakash S. Microbiome, probiotics and neurodegenerative diseases: deciphering the gut brain axis. Cell Mol Life Sci. 2017;74(20):3769–87.
54. Maier L, Pruteanu M, Kuhn M, Zeller G, Telzerow A, Anderson EE, et al. Extensive impact of non-antibiotic drugs on human gut bacteria. Nature. 2018;555(7698):623–8.
55. Imhann F, Bonder MJ, Vich Vila A, Fu J, Mujagic Z, Vork L, et al. Proton pump inhibitors affect the gut microbiome. Gut. 2016;65(5):740–8.
56. Kim J, Lee H, An J, Song Y, Lee CK, Kim K, et al. Alterations in gut microbiota by statin therapy and possible intermediate effects on hyperglycemia and hyperlipidemia. Front Microbiol. 2019;10:1947.
57. Dias AM, Cordeiro G, Estevinho MM, Veiga R, Figueira L, Reina-Couto M, et al. Gut bacterial microbiome composition and statin intake-a systematic review. Pharmacol Res Perspect. 2020;8(3):e00601.
58. Rogers MAM, Aronoff DM. The influence of non-steroidal anti-inflammatory drugs on the gut microbiome. Clin Microbiol Infect. 2016;22(2):178. e1–e9
59. Flowers SA, Evans SJ, Ward KM, McInnis MG, Ellingrod VL. Interaction between atypical antipsychotics and the gut microbiome in a bipolar disease cohort. Pharmacotherapy. 2017;37(3):261–7.
60. Flowers SA, Baxter NT, Ward KM, Kraal AZ, McInnis MG, Schmidt TM, et al. Effects of atypical antipsychotic treatment and resistant starch supplementation on gut microbiome composition in a cohort of patients with bipolar disorder or schizophrenia. Pharmacotherapy. 2019;39(2):161–70.
61. Elvers KT, Wilson VJ, Hammond A, Duncan L, Huntley AL, Hay AD, et al. Antibiotic-induced changes in the human gut microbiota for the most commonly prescribed antibiotics in primary care in the UK: a systematic review. BMJ Open. 2020;10(9):e035677.

62. Becattini S, Taur Y, Pamer EG. Antibiotic-induced changes in the intestinal microbiota and disease. Trends Mol Med. 2016;22(6):458–78.
63. Beloosesky Y, Nenaydenko O, Gross Nevo RF, Adunsky A, Weiss A. Rates, variability, and associated factors of polypharmacy in nursing home patients. Clin Interv Aging. 2013;8:1585–90.
64. Luna RA, Foster JA. Gut brain axis: diet microbiota interactions and implications for modulation of anxiety and depression. Curr Opin Biotechnol. 2015;32:35–41.
65. Cevenini E, Invidia L, Lescai F, Salvioli S, Tieri P, Castellani G, et al. Human models of aging and longevity. Expert Opin Biol Ther. 2008;8(9):1393–405.
66. Schneeberger M, Everard A, Gomez-Valades AG, Matamoros S, Ramirez S, Delzenne NM, et al. Akkermansia muciniphila inversely correlates with the onset of inflammation, altered adipose tissue metabolism and metabolic disorders during obesity in mice. Sci Rep. 2015;5:16643.
67. Santoro A, Ostan R, Candela M, Biagi E, Brigidi P, Capri M, et al. Gut microbiota changes in the extreme decades of human life: a focus on centenarians. Cell Mol Life Sci. 2018;75(1):129–48.
68. Vinarov Z, Abdallah M, Agundez JAG, Allegaert K, Basit AW, Braeckmans M, et al. Impact of gastrointestinal tract variability on oral drug absorption and pharmacokinetics: an UNGAP review. Eur J Pharm Sci. 2021;162:105812.
69. Stillhart C, Vučićević K, Augustijns P, Basit AW, Batchelor H, Flanagan TR, et al. Impact of gastrointestinal physiology on drug absorption in special populations – an UNGAP review. Eur J Pharm Sci. 2020;147:105280.
70. Mangoni AA, Jackson SHD. Age-related changes in pharmacokinetics and pharmacodynamics: basic principles and practical applications. Br J Clin Pharmacol. 2004;57(1):6–14.
71. Fadda HM. The route to palatable fecal microbiota transplantation. AAPS PharmSciTech. 2020;21(3):114.
72. Yang S, Li X, Yang F, Zhao R, Pan X, Liang J, et al. Gut microbiota-dependent marker TMAO in promoting cardiovascular disease: inflammation mechanism, clinical prognostic, and potential as a therapeutic target. Front Pharmacol. 2019;10(1360)
73. Anderson G, Seo M, Berk M, Carvalho AF, Maes M. Gut permeability and microbiota in parkinson's disease: role of depression, tryptophan catabolites, oxidative and nitrosative stress and melatonergic pathways. Curr Pharm Des. 2016;22(40)
74. McCoubrey LE, Elbadawi M, Orlu M, Gaisford S, Basit AW. Harnessing machine learning for development of microbiome therapeutics. Gut Microbes. 2021;13(1):1–20.
75. Scheline RR. Metabolism of foreign compounds by gastrointestinal microorganisms. Pharmacol Rev. 1973;25(4):451.
76. Sousa T, Paterson R, Moore V, Carlsson A, Abrahamsson B, Basit AW. The gastrointestinal microbiota as a site for the biotransformation of drugs. Int J Pharm. 2008;363(1):1–25.
77. Basit AW, Newton JM, Lacey LF. Susceptibility of the H2-receptor antagonists cimetidine, famotidine and nizatidine, to metabolism by the gastrointestinal microflora. Int J Pharm. 2002;237(1):23–33.
78. Zimmermann M, Zimmermann-Kogadeeva M, Wegmann R, Goodman AL. Mapping human microbiome drug metabolism by gut bacteria and their genes. Nature. 2019;570(7762):462–7.
79. Chankhamjon P, Javdan B, Lopez J, Hull R, Chatterjee S, Donia MS. Systematic mapping of drug metabolism by the human gut microbiome. bioRxiv. 2019;2019:538215.
80. Yadav V, Gaisford S, Merchant HA, Basit AW. Colonic bacterial metabolism of corticosteroids. Int J Pharm. 2013;457(1):268–74.
81. Yadav V, Varum F, Bravo R, Furrer E, Basit AW. Gastrointestinal stability of therapeutic anti-TNF alpha IgG1 monoclonal antibodies. Int J Pharm. 2016;502(1–2):181–7.
82. Wang J, Yadav V, Smart AL, Tajiri S, Basit AW. Stability of peptide drugs in the colon. Eur J Pharm Sci. 2015;78:31–6.

83. Wang J, Yadav V, Smart AL, Tajiri S, Basit AW. Toward oral delivery of biopharmaceuticals: an assessment of the gastrointestinal stability of 17 peptide drugs. Mol Pharm. 2015;12(3):966–73.
84. Sousa T, Yadav V, Zann V, Borde A, Abrahamsson B, Basit AW. On the colonic bacterial metabolism of Azo-bonded prodrugs of 5-aminosalicylic acid. J Pharm Sci. 2014;103(10):3171–5.
85. Coombes Z, Yadav V, McCoubrey LE, Freire C, Basit AW, Conlan RS, et al. Progestogens are metabolized by the gut microbiota: implications for colonic drug delivery. Pharmaceutics. 2020;12(8)
86. Yadav V, Mai Y, McCoubrey LE, Wada Y, Tomioka M, Kawata S, et al. 5-Aminolevulinic acid as a novel therapeutic for inflammatory bowel disease. Biomedicine. 2021;9(5)
87. Tannergren C, Borde A, Borestrom C, Abrahamsson B, Lindahl A. Evaluation of an in vitro faecal degradation method for early assessment of the impact of colonic degradation on colonic absorption in humans. Eur J Pharm Sci. 2014;57:200–6.
88. McCoubrey LE, Basit AW. Addressing drug–microbiome interactions: the role of healthcare professionals. Pharm J. 2022;
89. Fung TC, Vuong HE, Luna CDG, Pronovost GN, Aleksandrova AA, Riley NG, et al. Intestinal serotonin and fluoxetine exposure modulate bacterial colonization in the gut. Nat Microbiol. 2019;4(12):2064–73.
90. Oliphant K, Allen-Vercoe E. Macronutrient metabolism by the human gut microbiome: major fermentation by-products and their impact on host health. Microbiome. 2019;7(1):91.
91. McCoubrey LE, Gaisford S, Orlu M, Basit AW. Predicting drug-microbiome interactions with machine learning. Biotechnol Adv. 2021;107797
92. Clarke G, Sandhu KV, Griffin BT, Dinan TG, Cryan JF, Hyland NP. Gut reactions: breaking down xenobiotic-microbiome interactions. Pharmacol Rev. 2019;71(2):198–224.
93. Yadav V, Varum F, Bravo R, Furrer E, Bojic D, Basit AW. Inflammatory bowel disease: exploring gut pathophysiology for novel therapeutic targets. Transl Res. 2016;176:38–68.
94. Merchant HA, Liu F, Orlu Gul M, Basit AW. Age-mediated changes in the gastrointestinal tract. Int J Pharm. 2016;512(2):382–95.
95. Taggart AJ, McDermott BJ, Roberts SD. The effect of age and acetylator phenotype on the pharmacokinetics of sulfasalazine in patients with rheumatoid arthritis. Clin Pharmacokinet. 1992;23(4):311–20.
96. Okuda H, Ogura K, Kato A, Takubo H, Watabe T. A possible mechanism of eighteen patient deaths caused by interactions of sorivudine, a new antiviral drug, with oral 5-fluorouracil prodrugs. J Pharmacol Exp Ther. 1998;287(2):791–9.
97. Nakayama H, Kinouchi T, Kataoka K, Akimoto S, Matsuda Y, Ohnishi Y. Intestinal anaerobic bacteria hydrolyse sorivudine, producing the high blood concentration of 5-(E)-(2-bromovinyl)uracil that increases the level and toxicity of 5-fluorouracil. Pharmacogenet Genomics. 1997;7(1)
98. Gupta E, Lestingi TM, Mick R, Ramirez J, Vokes EE, Ratain MJ. Metabolic fate of irinotecan in humans: correlation of glucuronidation with diarrhea. Cancer Res. 1994;54(14):3723.
99. Takasuna K, Hagiwara T, Hirohashi M, Kato M, Nomura M, Nagai E, et al. Involvement of β-glucuronidase in intestinal microflora in the intestinal toxicity of the antitumor camptothecin derivative irinotecan hydrochloride (CPT-11) in rats. Cancer Res. 1996;56(16):3752.
100. Haiser HJ, Seim KL, Balskus EP, Turnbaugh PJ. Mechanistic insight into digoxin inactivation by Eggerthella lenta augments our understanding of its pharmacokinetics. Gut Microbes. 2014;5(2):233–8.
101. Haiser HJ, Gootenberg DB, Chatman K, Sirasani G, Balskus EP, Turnbaugh PJ. Predicting and manipulating cardiac drug inactivation by the human gut bacterium Eggerthella lenta. Science. 2013;341(6143):295–8.
102. van Kessel SP, Frye AK, El-Gendy AO, Castejon M, Keshavarzian A, van Dijk G, et al. Gut bacterial tyrosine decarboxylases restrict levels of levodopa in the treatment of Parkinson's disease. Nat Commun. 2019;10(1):310.

103. van Kessel SP, El Aidy S. Contributions of gut bacteria and diet to drug pharmacokinetics in the treatment of parkinson's disease. Front Neurol. 2019;10:1087.
104. Zimmermann M, Zimmermann-Kogadeeva M, Wegmann R, Goodman AL. Separating host and microbiome contributions to drug pharmacokinetics and toxicity. Science. 2019;363(6427):eaat9931.
105. Guo Y, Crnkovic CM, Won K-J, Yang X, Lee JR, Orjala J, et al. Commensal gut bacteria convert the immunosuppressant tacrolimus to less potent metabolites. Drug Metab Dispos. 2019;47(3):194.
106. Lee JR, Muthukumar T, Dadhania D, Taur Y, Jenq RR, Toussaint NC, et al. Gut microbiota and tacrolimus dosing in kidney transplantation. PLoS One. 2015;10(3):e0122399.
107. Campbell C, McKenney PT, Konstantinovsky D, Isaeva OI, Schizas M, Verter J, et al. Bacterial metabolism of bile acids promotes generation of peripheral regulatory T cells. Nature. 2020;
108. Apte SP. Macromolecular excipients exert biological effects via their microbial fermentation absorbable products. J Excip Food Chem. 2016;7(1)
109. Chen K, Zhao H, Shu L, Xing H, Wang C, Lu C, et al. Effect of resveratrol on intestinal tight junction proteins and the gut microbiome in high-fat diet-fed insulin resistant mice. Int J Food Sci Nutr. 2020;71(8):965–78.
110. Hatton GB, Yadav V, Basit AW, Merchant HA. Animal farm: considerations in animal gastrointestinal physiology and relevance to drug delivery in humans. J Pharm Sci. 2015;104(9):2747–76.
111. Madla CM, Gavins FKH, Merchant HA, Orlu M, Murdan S, Basit AW. Let's talk about sex: differences in drug therapy in males and females. Adv Drug Deliv Rev. 2021;113804
112. Keohane DM, Ghosh TS, Jeffery IB, Molloy MG, O'Toole PW, Shanahan F. Microbiome and health implications for ethnic minorities after enforced lifestyle changes. Nat Med. 2020;26(7):1089–95.
113. An R, Wilms E, Masclee AAM, Smidt H, Zoetendal EG, Jonkers D. Age-dependent changes in GI physiology and microbiota: time to reconsider? Gut. 2018;67(12):2213–22.
114. Jostins L, Ripke S, Weersma RK, Duerr RH, McGovern DP, Hui KY, et al. Host-microbe interactions have shaped the genetic architecture of inflammatory bowel disease. Nature. 2012;491(7422):119–24.
115. Takashima S, Tanaka F, Kawaguchi Y, Usui Y, Fujimoto K, Nadatani Y, et al. Proton pump inhibitors enhance intestinal permeability via dysbiosis of gut microbiota under stressed conditions in mice. Neurogastroenterol Motil. 2020;32(7)
116. Roager HM, Hansen LB, Bahl MI, Frandsen HL, Carvalho V, Gobel RJ, et al. Colonic transit time is related to bacterial metabolism and mucosal turnover in the gut. Nat Microbiol. 2016;1(9):16093.
117. González-Sarrías A, Miguel V, Merino G, Lucas R, Morales JC, Tomás-Barberán F, et al. The gut microbiota ellagic acid-derived metabolite urolithin A and its sulfate conjugate are substrates for the drug efflux transporter breast cancer resistance protein (ABCG2/BCRP). J Agric Food Chem. 2013;61(18):4352–9.
118. Ridlon JM, Kang DJ, Hylemon PB, Bajaj JS. Bile acids and the gut microbiome. Curr Opin Gastroenterol. 2014;30(3):332–8.
119. Enright EF, Griffin BT, Gahan CGM, Joyce SA. Microbiome-mediated bile acid modification: role in intestinal drug absorption and metabolism. Pharmacol Res. 2018;133:170–86.
120. Caroli-Bosc F-X, Iliadis A, Salmon L, Macheras P, Montet A-M, Bourgeon A, et al. Ursodeoxycholic acid modulates cyclosporin A oral absorption in liver transplant recipients. Fundam Clin Pharmacol. 2000;14(6):601–9.
121. Enright EF, Joyce SA, Gahan CGM, Griffin BT. Impact of gut microbiota-mediated bile acid metabolism on the solubilization capacity of bile salt micelles and drug solubility. Mol Pharm. 2017;14(4):1251–63.
122. Einarsson K, Nilsell K, Leijd B, Angelin B. Influence of age on secretion of cholesterol and synthesis of bile acids by the liver. N Engl J Med. 1985;313(5):277–82.

123. Kim IH, Kisseleva T, Brenner DA. Aging and liver disease. Curr Opin Gastroenterol. 2015;31(3):184–91.
124. McLachlan AJ, Pont LG. Drug metabolism in older people – a key consideration in achieving optimal outcomes with medicines. J Gerontol Ser A. 2012;67A(2):175–80.
125. Butler JM, Begg EJ. Free drug metabolic clearance in elderly people. Clin Pharmacokinet. 2008;47(5):297–321.
126. Montagner A, Korecka A, Polizzi A, Lippi Y, Blum Y, Canlet C, et al. Hepatic circadian clock oscillators and nuclear receptors integrate microbiome-derived signals. Sci Rep. 2016;6(1):20127.
127. Gandhi AS, Guo T, Shah P, Moorthy B, Chow DSL, Hu M, et al. CYP3A-dependent drug metabolism is reduced in bacterial inflammation in mice. Br J Pharmacol. 2012;166(7):2176–87.
128. Björkholm B, Bok CM, Lundin A, Rafter J, Hibberd ML, Pettersson S. Intestinal microbiota regulate xenobiotic metabolism in the liver. PLoS One. 2009;4(9):e6958.
129. Liu Y, Yang K, Jia Y, Shi J, Tong Z, Fang D, et al. Gut microbiome alterations in high-fat-diet-fed mice are associated with antibiotic tolerance. Nat Microbiol. 2021;6(7):874–84.
130. Peterson VL, Richards JB, Meyer PJ, Cabrera-Rubio R, Tripi JA, King CP, et al. Sex-dependent associations between addiction-related behaviors and the microbiome in outbred rats. EBioMedicine. 2020;55:102769.
131. Donini LM, Savina C, Cannella C. Nutrition in the elderly: role of fiber. Arch Gerontol Geriatr. 2009;49(Suppl 1):61–9.
132. National Health Service. How to get more fibre into your diet Online: NHS Eat Well; 2018. Available from: https://www.nhs.uk/live-well/eat-well/how-to-get-more-fibre-into-your-diet/
133. Asnicar F, Berry SE, Valdes AM, Nguyen LH, Piccinno G, Drew DA, et al. Microbiome connections with host metabolism and habitual diet from 1,098 deeply phenotyped individuals. Nat Med. 2021;27:321–32.
134. McCoubrey LE, Elbadawi M, Orlu M, Gaisford S, Basit AW. Machine learning uncovers adverse drug effects on intestinal bacteria. Pharmaceutics. 2021;13(7)
135. Dethlefsen L, Huse S, Sogin ML, Relman DA. The pervasive effects of an antibiotic on the human gut microbiota, as revealed by deep 16S rRNA sequencing. PLoS Biol. 2008;6(11):e280.
136. Baquero F, Levin BR. Proximate and ultimate causes of the bactericidal action of antibiotics. Nat Rev Microbiol. 2021;19(2):123–32.
137. Uzan-Yulzari A, Turta O, Belogolovski A, Ziv O, Kunz C, Perschbacher S, et al. Neonatal antibiotic exposure impairs child growth during the first six years of life by perturbing intestinal microbial colonization. Nat Commun. 2021;12(1):443.
138. Fouhy F, Guinane CM, Hussey S, Wall R, Ryan CA, Dempsey EM, et al. High-throughput sequencing reveals the incomplete, short-term recovery of infant gut microbiota following parenteral antibiotic treatment with ampicillin and gentamicin. Antimicrob Agents Chemother. 2012;56(11):5811–20.
139. Gao H, Shu Q, Chen J, Fan K, Xu P, Zhou Q, et al. Antibiotic exposure has sex-dependent effects on the gut microbiota and metabolism of short-chain fatty acids and amino acids in mice. mSystems. 2019;4(4):e00048–19.
140. Allegretti JR, Fischer M, Sagi SV, Bohm ME, Fadda HM, Ranmal SR, et al. Fecal microbiota transplantation capsules with targeted colonic versus gastric delivery in recurrent clostridium difficile infection: a comparative cohort analysis of high and lose dose. Dig Dis Sci. 2019;64(6):1672–8.
141. Balsells E, Shi T, Leese C, Lyell I, Burrows J, Wiuff C, et al. Global burden of Clostridium difficile infections: a systematic review and meta-analysis. J Glob Health. 2019;9(1):010407.
142. National Institute for Health and Care Excellence. Antimicrobial stewardship: systems and processes for effective antimicrobial medicine use. United Kingdom; 2015.

143. Mulder M, Radjabzadeh D, Kiefte-de Jong JC, Uitterlinden AG, Kraaij R, Stricker BH, et al. Long-term effects of antimicrobial drugs on the composition of the human gut microbiota. Gut Microbes. 2020;12(1):1795492.
144. Langdon A, Crook N, Dantas G. The effects of antibiotics on the microbiome throughout development and alternative approaches for therapeutic modulation. Genome Med. 2016;8(1):39.
145. Jernberg C, Löfmark S, Edlund C, Jansson JK. Long-term impacts of antibiotic exposure on the human intestinal microbiota. Microbiology. 2010;156(Pt 11):3216–23.
146. Rodrigues RR, Greer RL, Dong X, Kn DS, Gurung M, Wu JY, et al. Antibiotic-induced alterations in gut microbiota are associated with changes in glucose metabolism in healthy mice. Front Microbiol. 2017;8:2306.
147. Hagan T, Cortese M, Rouphael N, Boudreau C, Linde C, Maddur MS, et al. Antibiotics-driven gut microbiome perturbation alters immunity to vaccines in humans. Cell. 2019;178(6):1313–28. e13
148. Cussotto S, Walsh J, Golubeva AV, Zhdanov AV, Strain CR, Fouhy F, et al. The gut microbiome influences the bioavailability of olanzapine in rats. EBioMedicine. 2021:66.
149. Baillargeon J, Holmes HM, Lin Y-L, Raji MA, Sharma G, Kuo Y-F. Concurrent use of warfarin and antibiotics and the risk of bleeding in older adults. Am J Med. 2012;125(2):183–9.
150. Nayak RR, Alexander M, Deshpande I, Stapleton-Gray K, Rimal B, Patterson AD, et al. Methotrexate impacts conserved pathways in diverse human gut bacteria leading to decreased host immune activation. Cell Host Microbe. 2021;29(3):362–77 e11.
151. Vieira-Silva S, Falony G, Belda E, Nielsen T, Aron-Wisnewsky J, Chakaroun R, et al. Statin therapy is associated with lower prevalence of gut microbiota dysbiosis. Nature. 2020;581(7808):310–5.
152. Wu H, Esteve E, Tremaroli V, Khan MT, Caesar R, Mannerås-Holm L, et al. Metformin alters the gut microbiome of individuals with treatment-naive type 2 diabetes, contributing to the therapeutic effects of the drug. Nat Med. 2017;23(7):850–8.
153. Prattichizzo F, Giuliani A, Mensà E, Sabbatinelli J, De Nigris V, Rippo MR, et al. Pleiotropic effects of metformin: shaping the microbiome to manage type 2 diabetes and postpone ageing. Ageing Res Rev. 2018;48:87–98.
154. Freedberg DE, Toussaint NC, Chen SP, Ratner AJ, Whittier S, Wang TC, et al. Proton pump inhibitors alter specific taxa in the human gastrointestinal microbiome: a crossover trial. Gastroenterology. 2015;149(4):883–5. e9
155. Jackson MA, Verdi S, Maxan ME, Shin CM, Zierer J, Bowyer RCE, et al. Gut microbiota associations with common diseases and prescription medications in a population-based cohort. Nat Commun. 2018;9(1):2655.
156. Chassaing B, Van de Wiele T, De Bodt J, Marzorati M, Gewirtz AT. Dietary emulsifiers directly alter human microbiota composition and gene expression ex vivo potentiating intestinal inflammation. Gut. 2017;66(8):1414–27.
157. Jiang Z, Zhao M, Zhang H, Li Y, Liu M, Feng F. Antimicrobial emulsifier-glycerol monolaurate induces metabolic syndrome, gut microbiota dysbiosis, and systemic low-grade inflammation in low-fat diet fed mice. Mol Nutr Food Res. 2018;62(3)
158. Gorkiewicz G, Thallinger GG, Trajanoski S, Lackner S, Stocker G, Hinterleitner T, et al. Alterations in the colonic microbiota in response to osmotic diarrhea. PLoS One. 2013;8(2):e55817.
159. Uebanso T, Kano S, Yoshimoto A, Naito C, Shimohata T, Mawatari K, et al. Effects of consuming xylitol on gut microbiota and lipid metabolism in mice. Nutrients. 2017;9(7)
160. Sato T, Kusuhara S, Yokoi W, Ito M, Miyazaki K. Prebiotic potential of L-sorbose and xylitol in promoting the growth and metabolic activity of specific butyrate-producing bacteria in human fecal culture. FEMS Microbiol Ecol. 2017;93(1)
161. Suez J, Korem T, Zeevi D, Zilberman-Schapira G, Thaiss CA, Maza O, et al. Artificial sweeteners induce glucose intolerance by altering the gut microbiota. Nature. 2014;514(7521):181–6.

162. Zhang C, Jiao S, Wang ZA, Du Y. Exploring effects of chitosan oligosaccharides on mice gut microbiota in in vitro fermentation and animal model. Front Microbiol. 2018;9
163. Reker D, Blum SM, Steiger C, Anger KE, Sommer JM, Fanikos J, et al. "Inactive" ingredients in oral medications. Sci Transl Med. 2019;11(483):eaau6753.
164. Mai Y, Ashiru-Oredope DAI, Yao Z, Dou L, Madla CM, Taherali F, et al. Boosting drug bioavailability in men but not women through the action of an excipient. Int J Pharm. 2020;587
165. Mai Y, Dou L, Madla CM, Murdan S, Basit AW. Sex-dependence in the effect of pharmaceutical excipients: polyoxyethylated solubilising excipients increase oral drug bioavailability in male but not female rats. Pharmaceutics. 2019;11(5)
166. Reker D, Shi Y, Kirtane AR, Hess K, Zhong GJ, Crane E, et al. Machine learning uncovers food- and excipient-drug interactions. Cell Rep. 2020;30(11):3710–6. e4
167. Ahmadmehrabi S, Tang WHW. Gut microbiome and its role in cardiovascular diseases. Curr Opin Cardiol. 2017;32(6):761–6.
168. Huang L, Wang T, Wu Q, Dong X, Shen F, Liu D, et al. Analysis of microbiota in elderly patients with acute cerebral infarction. PeerJ. 2019;7:e6928.
169. Koeth RA, Wang Z, Levison BS, Buffa JA, Org E, Sheehy BT, et al. Intestinal microbiota metabolism of l-carnitine, a nutrient in red meat, promotes atherosclerosis. Nat Med. 2013;19(5):576–85.
170. Ghyselinck J, Verstrepen L, Moens F, Van Den Abbeele P, Bruggeman A, Said J, et al. Influence of probiotic bacteria on gut microbiota composition and gut wall function in an in-vitro model in patients with Parkinson's disease. Int J Pharm X. 2021;3:100087.
171. Defaye M, Gervason S, Altier C, Berthon JY, Ardid D, Filaire E, et al. Microbiota: a novel regulator of pain. J Neural Transm (Vienna). 2020;127(4):445–65.
172. Larroya-Garcia A, Navas-Carrillo D, Orenes-Pinero E. Impact of gut microbiota on neurological diseases: diet composition and novel treatments. Crit Rev Food Sci Nutr. 2019;59(19):3102–16.
173. McConnell EL, Fadda HM, Basit AW. Gut instincts: explorations in intestinal physiology and drug delivery. Int J Pharm. 2008;364(2):213–26.
174. Kararli TT. Comparison of the gastrointestinal anatomy, physiology, and biochemistry of humans and commonly used laboratory animals. Biopharm Drug Dispos. 1995;16(5):351–80.
175. Ibekwe VC, Fadda HM, McConnell EL, Khela MK, Evans DF, Basit AW. Interplay between intestinal pH, transit time and feed status on the in vivo performance of pH responsive ileo-colonic release systems. Pharm Res. 2008;25(8):1828–35.
176. Basit A. Advances in colonic delivery. Drugs. 2005;65:14.
177. Maderuelo C, Lanao JM, Zarzuelo A. Enteric coating of oral solid dosage forms as a tool to improve drug bioavailability. Eur J Pharm Sci. 2019;138:105019.
178. Ibekwe VC, Liu F, Fadda HM, Khela MK, Evans DF, Parsons GE, et al. An investigation into the in vivo performance variability of pH responsive polymers for ileo-colonic drug delivery using gamma scintigraphy in humans. J Pharm Sci. 2006;95(12):2760–6.
179. Hatton GB, Madla CM, Rabbie SC, Basit AW. All disease begins in the gut: influence of gastrointestinal disorders and surgery on oral drug performance. Int J Pharm. 2018;548(1):408–22.
180. Hatton GB, Madla CM, Rabbie SC, Basit AW. Gut reaction: impact of systemic diseases on gastrointestinal physiology and drug absorption. Drug Discov Today. 2019;24(2):417–27.
181. Tian H, Chen Q, Yang B, Qin H, Li N. Analysis of gut microbiome and metabolite characteristics in patients with slow transit constipation. Dig Dis Sci. 2020;
182. Gazzaniga A, Maroni A, Sangalli ME, Zema L. Time-controlled oral delivery systems for colon targeting. Expert Opin Drug Deliv. 2006;3(5):583–97.
183. Maroni A, Zema L, Cerea M, Foppoli A, Palugan L, Gazzaniga A. Erodible drug delivery systems for time-controlled release into the gastrointestinal tract. J Drug Deliv Sci Technol. 2016;32:229–35.
184. Wang YT, Mohammed SD, Farmer AD, Wang D, Zarate N, Hobson AR, et al. Regional gastrointestinal transit and pH studied in 215 healthy volunteers using the wireless motility

capsule: influence of age, gender, study country and testing protocol. Aliment Pharmacol Ther. 2015;42(6):761–72.
185. Delgado-Aros S, Camilleri M, Garcia MA, Burton D, Busciglio I. High body mass alters colonic sensory-motor function and transit in humans. Am J Physiol Gastrointest Liver Physiol. 2008;295(2):G382–G8.
186. Broad J, Kung VWS, Palmer A, Elahi S, Karami A, Darreh-Shori T, et al. Changes in neuromuscular structure and functions of human colon during ageing are region-dependent. Gut. 2019;68(7):1210.
187. O'Mahony D, O'Leary P, Quigley EMM. Aging and intestinal motility. Drugs Aging. 2002;19(7):515–27.
188. Graff J, Brinch K, Madsen JL. Gastrointestinal mean transit times in young and middle-aged healthy subjects. Clin Physiol. 2001;21(2):253–9.
189. Macfarlane GT, Englyst HN. Starch utilization by the human large intestinal microflora. J Appl Bacteriol. 1986;60(3):195–201.
190. Bourgeois S, Laham A, Besnard M, Andremont A, Fattal E. In vitro and in vivo evaluation of pectin beads for the colon delivery of beta-lactamases. J Drug Target. 2005;13(5):277–84.
191. Gadalla HH, Soliman GM, Mohammed FA, El-Sayed AM. Development and in vitro/in vivo evaluation of Zn-pectinate microparticles reinforced with chitosan for the colonic delivery of progesterone. Drug Deliv. 2016;23(7):2541–54.
192. McConnell EL, Murdan S, Basit AW. An investigation into the digestion of chitosan (noncrosslinked and crosslinked) by human colonic bacteria. J Pharm Sci. 2008;97(9):3820–9.
193. McConnell EL, Short MD, Basit AW. An in vivo comparison of intestinal pH and bacteria as physiological trigger mechanisms for colonic targeting in man. J Control Release. 2008;130(2):154–60.
194. Fadda HM, Merchant HA, Arafat BT, Basit AW. Physiological bicarbonate buffers: stabilisation and use as dissolution media for modified release systems. Int J Pharm. 2009;382(1–2):56–60.
195. Goyanes A, Hatton GB, Merchant HA, Basit AW. Gastrointestinal release behaviour of modified-release drug products: dynamic dissolution testing of mesalazine formulations. Int J Pharm. 2015;484(1–2):103–8.
196. Varum F, Cristina Freire A, Fadda HM, Bravo R, Basit AW. A dual pH and microbiota-triggered coating (Phloral(TM)) for fail-safe colonic drug release. Int J Pharm. 2020;583:119379.
197. Ibekwe VC, Khela MK, Evans DF, Basit AW. A new concept in colonic drug targeting: a combined pH-responsive and bacterially-triggered drug delivery technology. Aliment Pharmacol Ther. 2008;28(7):911–6.
198. Dodoo CC, Wang J, Basit AW, Stapleton P, Gaisford S. Targeted delivery of probiotics to enhance gastrointestinal stability and intestinal colonisation. Int J Pharm. 2017;530(1):224–9.
199. Peiris M, Aktar R, Reed D, Cibert-Goton V, Zdanaviciene A, Halder W, et al. Decoy bypass for appetite suppression in obese adults: role of synergistic nutrient sensing receptors GPR84 and FFAR4 on colonic endocrine cells. Gut. 2021;
200. Kararli TT, Kirchhoff CF, Truelove JE. Ionic strength dependence of dissolution for eudragit S-100 coated pellets. Pharm Res. 1995;12(11):1813–6.
201. Varum FJ, Veiga F, Sousa JS, Basit AW. Mucoadhesive platforms for targeted delivery to the colon. Int J Pharm. 2011;420(1):11–9.
202. Varum F, Cristina Freire A, Bravo R, Basit AW. OPTICORE, an innovative and accurate colonic targeting technology. Int J Pharm. 2020;583:119372.
203. Smaje A, Weston-Clark M, Raj R, Orlu M, Davis D, Rawle M. Factors associated with medication adherence in older patients: a systematic review. Aging Med. 2018;1(3):254–66.
204. Awad A, Madla CM, McCoubrey LE, Ferraro F, Gavins FKH, Buanz A, et al. Clinical translation of advanced colonic drug delivery technologies. Adv Drug Deliv Rev. 2022;181:114076.
205. United Nations, Department of Economic and Social Affairs, Population Division. World population ageing 2019: highlights (ST/ESA/SER.A/430); 2019.
206. Ctoi AF, Corina A, Katsiki N, Vodnar DC, Andreicut AD, Stoian AP, et al. Gut microbiota and aging-a focus on centenarians. Biochim Biophys Acta Mol basis Dis. 2020;1866(7):165765.

207. Ghosh TS, Rampelli S, Jeffery IB, Santoro A, Neto M, Capri M, et al. Mediterranean diet intervention alters the gut microbiome in older people reducing frailty and improving health status: the NU-AGE 1-year dietary intervention across five European countries. Gut. 2020;69(7):1218–28.
208. Ale EC, Binetti AG. Role of probiotics, prebiotics, and synbiotics in the elderly: insights into their applications. Front Microbiol. 2021;12:631254.
209. Sanchez B, Delgado S, Blanco-Miguez A, Lourenco A, Gueimonde M, Margolles A. Probiotics, gut microbiota, and their influence on host health and disease. Mol Nutr Food Res. 2017;61(1)
210. Swanson KS, Gibson GR, Hutkins R, Reimer RA, Reid G, Verbeke K, et al. The international scientific Association for Probiotics and Prebiotics (ISAPP) consensus statement on the definition and scope of synbiotics. Nat Rev Gastroenterol Hepatol. 2020;17(11):687–701.
211. Veiga P, Suez J, Derrien M, Elinav E. Moving from probiotics to precision probiotics. Nat Microbiol. 2020;5(7):878–80.
212. O'Toole PW, Marchesi JR, Hill C. Next-generation probiotics: the spectrum from probiotics to live biotherapeutics. Nat Microbiol. 2017;2:17057.
213. Finamore A, Roselli M, Donini L, Brasili DE, Rami R, Carnevali P, et al. Supplementation with Bifidobacterium longum Bar33 and lactobacillus helveticus Bar13 mixture improves immunity in elderly humans (over 75 years) and aged mice. Nutrition. 2019;63–64:184–92.
214. Yamamoto Y, Saruta J, Takahashi T, To M, Shimizu T, Hayashi T, et al. Effect of ingesting yogurt fermented with lactobacillus delbrueckii ssp. bulgaricus OLL1073R-1 on influenza virus-bound salivary IgA in elderly residents of nursing homes: a randomized controlled trial. Acta Odontol Scand. 2019;77(7):517–24.
215. Wang H, Yang F, Zhang S, Xin R, Sun Y. Genetic and environmental factors in Alzheimer's and Parkinson's diseases and promising therapeutic intervention via fecal microbiota transplantation. NPJ Parkinsons Dis. 2021;7(1):70.
216. Cunningham M, Azcarate-Peril MA, Barnard A, Benoit V, Grimaldi R, Guyonnet D, et al. Shaping the future of probiotics and prebiotics. Trends Microbiol. 2021;29(8):667–85.
217. Macfarlane GT, Steed H, Macfarlane S. Bacterial metabolism and health-related effects of galacto-oligosaccharides and other prebiotics. J Appl Microbiol. 2008;104(2):305–44.
218. Vulevic J, Drakoularakou A, Yaqoob P, Tzortzis G, Gibson GR. Modulation of the fecal microflora profile and immune function by a novel trans-galactooligosaccharide mixture (B-GOS) in healthy elderly volunteers. Am J Clin Nutr. 2008;88(5):1438–46.
219. DeMartino P, Cockburn DW. Resistant starch: impact on the gut microbiome and health. Curr Opin Biotechnol. 2020;61:66–71.
220. Yang X, Darko KO, Huang Y, He C, Yang H, He S, et al. Resistant starch regulates gut microbiota: structure, biochemistry and cell signalling. Cell Physiol Biochem. 2017;42(1):306–18.
221. Alfa MJ, Strang D, Tappia PS, Graham M, Van Domselaar G, Forbes JD, et al. A randomized trial to determine the impact of a digestion resistant starch composition on the gut microbiome in older and mid-age adults. Clin Nutr. 2018;37(3):797–807.
222. Šuligoj T, Vigsnæs LK, Abbeele PVD, Apostolou A, Karalis K, Savva GM, et al. Effects of human milk oligosaccharides on the adult gut microbiota and barrier function. Nutrients. 2020;12(9)
223. Coman V, Vodnar DC. Gut microbiota and old age: modulating factors and interventions for healthy longevity. Exp Gerontol. 2020;141:111095.
224. Cicero AFG, Fogacci F, Bove M, Giovannini M, Borghi C. Impact of a short-term synbiotic supplementation on metabolic syndrome and systemic inflammation in elderly patients: a randomized placebo-controlled clinical trial. Eur J Nutr. 2021;60(2):655–63.
225. Ford AC, Quigley EM, Lacy BE, Lembo AJ, Saito YA, Schiller LR, et al. Efficacy of prebiotics, probiotics, and synbiotics in irritable bowel syndrome and chronic idiopathic constipation: systematic review and meta-analysis. Am J Gastroenterol. 2014;109(10):1547–61. quiz 6, 62

226. Firouzi S, Haghighatdoost F. The effects of prebiotic, probiotic, and synbiotic supplementation on blood parameters of renal function: a systematic review and meta-analysis of clinical trials. Nutrition. 2018;51-52:104–13.
227. Hadi A, Mohammadi H, Miraghajani M, Ghaedi E. Efficacy of synbiotic supplementation in patients with nonalcoholic fatty liver disease: a systematic review and meta-analysis of clinical trials: synbiotic supplementation and NAFLD. Crit Rev Food Sci Nutr. 2019;59(15):2494–505.
228. Zeng W, Shen J, Bo T, Peng L, Xu H, Nasser MI, et al. Cutting edge: probiotics and fecal microbiota transplantation in immunomodulation. J Immunol Res. 2019;2019:1603758.
229. Gupta S, Allen-Vercoe E, Petrof EO. Fecal microbiota transplantation: in perspective. Ther Adv Gastroenterol. 2016;9(2):229–39.
230. Cohen SH, Gerding DN, Johnson S, Kelly CP, Loo VG, McDonald LC, et al. Clinical practice guidelines for Clostridium difficile infection in adults: 2010 update by the society for healthcare epidemiology of America (SHEA) and the infectious diseases society of America (IDSA). Infect Control Hosp Epidemiol. 2010;31(5):431–55.
231. Asempa TE, Nicolau DP. Clostridium difficile infection in the elderly: an update on management. Clin Interv Aging. 2017;12:1799–809.
232. De Roo AC, Regenbogen SE. Clostridium difficile infection: an epidemiology update. Clin Colon Rectal Surg. 2020;33(2):49–57.
233. de Groot PF, Frissen MN, de Clercq NC, Nieuwdorp M. Fecal microbiota transplantation in metabolic syndrome: history, present and future. Gut Microbes. 2017;8(3):253–67.
234. Tan P, Li X, Shen J, Feng Q. Fecal microbiota transplantation for the treatment of inflammatory bowel disease: an update. Front Pharmacol. 2020;11:574533.
235. Giles EM, D'Adamo GL, Forster SC. The future of faecal transplants. Nat Rev Microbiol. 2019;17(12):719.

Tailoring Vaccines for Older Individuals: Aging of the Immune System and the Impact on Vaccine Efficacy

Shazia Bashir, Maria Wilson, Diane Ashiru-Oredope, and Sudaxshina Murdan

Contents

Joint first authors: Shazia Bashir and Maria Wilson contributed equally to this work

---

S. Bashir
Department of Pharmaceutics, UCL School of Pharmacy, London, UK

King's College London, London, UK

M. Wilson · S. Murdan (✉)
Department of Pharmaceutics, UCL School of Pharmacy, London, UK
e-mail: s.murdan@ucl.ac.uk

D. Ashiru-Oredope
Department of Pharmaceutics, UCL School of Pharmacy, London, UK

UK Health Security Agency, London, UK

M. Orlu, F. Liu (eds.), *Pharmaceutical Formulations for Older Patients*, AAPS Advances in the Pharmaceutical Sciences Series 51,
https://doi.org/10.1007/978-3-031-35811-1_9

Abstract Tailoring vaccines to the older population (>65 yrs) is an emerging subject that has gained momentum over the past decade. The vaccine efficacy of a formulation administered to young adults is often much higher than in older people. Therefore, there has been a clarion call for vaccines which show improved effectiveness in older individuals.

There is substantial evidence indicating that immunosenescence, which is aging of the immune system, causes dysfunction of the immune system and the associated components, which consequently impacts vaccine response. Therefore, vaccines need to be tailored with specific strategies designed to boost immune responses in older people, in order to achieve increased vaccine efficacy. Although several vaccine formulations have already been designed showing improved vaccine efficacy in older people, research into many more strategies is necessary and ongoing, to further enhance protection in the aging population.

Keywords Immunosenescence · Inflammaging · Immune system · Vaccine formulations · Tailored approach · Older individuals

1 Introduction

Population aging is a global phenomenon, where every country in the world is experiencing growth in the size and proportion of the elderly. In 2019, there were 703 million individuals aged 65 years or over, worldwide. It has been predicted that the number of elderly people will double to 1.5 billion in 2050 and that one in six individuals will be aged 65 years or over [1]. The number of people above age 80 is growing even faster [2]. Providing care for this aging population presents a major challenge as older people are more susceptible to disease, which are more severe and associated with poorer outcomes than in younger individuals [3].

Aging is a complex biological process, associated with a general decline in physiological functions and profound alterations to the immune system. The age-related dysregulation of the immune system, known as *immunosenescence*, causes impaired responses to infections, reduces vaccine efficacy and results in negative clinical outcomes in older adults [4–7] (Fig. 1). Thus, the older population more often succumbs to vaccine-preventable diseases such as influenza [8, 9], where vaccine efficacy is 17–53% in the elderly versus 70–90% in healthy adults [10] and mortality rate is higher. Another example is invasive pneumococcal disease, the incidence of which peaks at <2 years and > 85 years.

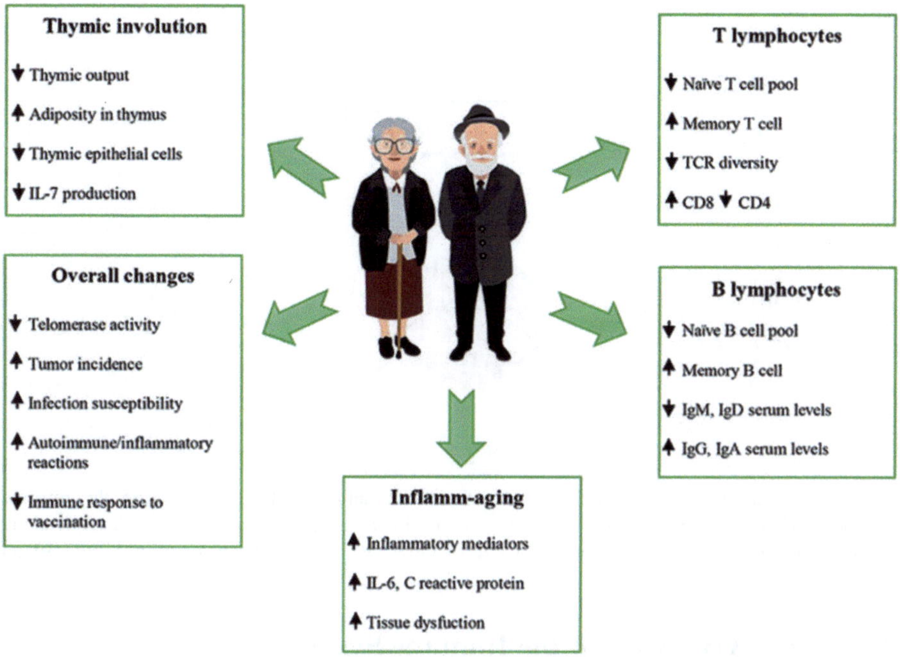

Fig. 1 A summary of the main effects of immunosenescence in the older population that cause a decline in vaccine response. The main changes are due to several factors, including dysregulation of B and T cell functions leading to increased vulnerability to infectious diseases and frailty in old people. (Image reproduced from [7])

Infections are the leading causes of death and morbidity in adults over 65 years old [11]. Prevention of infections is thus very important to ensure healthy aging and to reduce morbidity. Over the last decade, there has been substantial research assessing vaccine efficacy and effectiveness among different age groups [12–16], following the observation of varying levels of vaccine efficacy among different age groups [17, 18]. While many factors influence the immune responses to vaccinations, as shown in Fig. 2, a vaccinee's age is one that is not amenable to improvement by the vaccine developer.

A meta-analysis of the efficacy and effectiveness of influenza vaccines conducted by Osterholm et al. [19] revealed that standard-dose influenza vaccines do not provide adequate protection to individuals over 65 years of age [19]. The authors concluded that to control the impact of the disease on the elderly, influenza vaccines should be upgraded and made more efficacious [19]. In 2009, the Food and Drug Administration (FDA) approved the first influenza vaccine in the USA tailored to older people [20]; the high-dose trivalent inactivated influenza vaccine contained four times the standard antigen dose and elicited higher immune responses [21].

In this chapter, the current knowledge on how multi-faceted characteristics of the immune system are affected by aging will be discussed and how this understanding can be used to tailor existing vaccines to improve efficacy by developing novel strategies to boost the aging immune system.

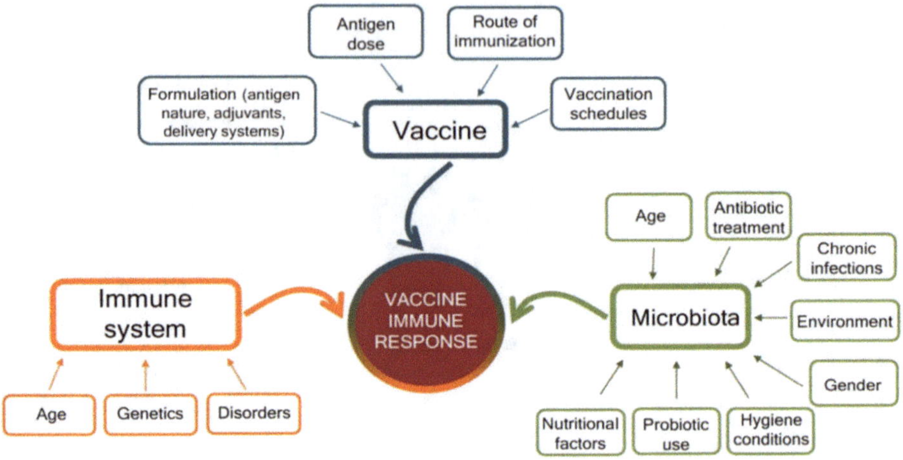

Fig. 2 Vaccination outcome is the result of a combination of factors affecting the immune response. This includes: the immune system of the host (age, genetics, disorders), vaccine formulation (type of antigen, adjuvant, route of delivery), and the host gut microbiota. (Image reproduced from [28])

2 A Brief Overview of the Immune System

2.1 *The Host Defence*

The immune system refers to a collection of cells and proteins that function to protect the skin, respiratory passages, gastrointestinal tract and other areas from 'foreign' antigens, such as bacteria, fungi, parasites, viruses, cancer cells and toxins [22]. In broad terms, the host immune defence mechanism consists of three levels of protection, (1) anatomical and physiological barriers, (2) innate immunity and (3) adaptive immunity.

Anatomical and physiological barriers provide the crucial first line of defence against pathogens. These barriers include intact skin, mucocilliary clearance mechanisms, low stomach pH and bacteriolytic lysozyme in tears, saliva and other secretions [23, 24].

2.2 *The Innate Immune System*

If invading pathogens manage to bypass the first line of defence, then the innate immune system is triggered and will augment and support anatomical and physiological barrier functions in an attempt to combat pathogen invasion. The innate immune system is an antigen-independent (non-specific) defence mechanism that is used by the host immediately or within hours of encountering an antigen. The innate immune response has no immunological memory, and therefore, it is unable to recognise or 'memorise' the same pathogen should the body be exposed to it in the future. Numerous cells are involved in the innate immune response such as phagocytes (macrophages and neutrophils), dendritic cells, mast cells, basophils, eosinophils, natural killer (NK) cells and lymphocytes (B and T cells). Phagocytes are sub-divided into two main cell types: neutrophils and macrophages. Both cells share a similar function: to engulf (phagocytose) microbes. In addition to their phagocytic properties, neutrophils contain granules that, when released, assist in the elimination of pathogenic microbes. Unlike neutrophils (which are short-lived cells), macrophages are long-living cells that not only play a role in phagocytosis but are also involved in antigen presentation to T cells [25].

2.3 The Adaptive Immune System

Adaptive immunity, on the other hand, is antigen-dependent and antigen-specific and, therefore, involves a lag time between exposure to the antigen and maximal response. The hallmark of adaptive immunity is the capacity for memory, which enables the host to mount a more rapid and efficient immune response upon subsequent exposure to the antigen.

The adaptive immune system involves a tightly regulated interplay between antigen-presenting cells (APCs) and T and B lymphocytes (Fig. 3). This pathway facilitates pathogen-specific immunologic effector pathways, generation of immunologic memory and regulation of host immune homeostasis. Lymphocytes develop and are activated within a series of lymphoid organs (spleen, bone marrow and thymus) comprising the lymphatic system. The activation mechanism generates a diverse repertoire of receptor specificities capable of recognising components of all potential pathogens. Innate and adaptive immunity are not mutually exclusive mechanisms of host defence but rather are complementary, with defects in either system resulting in host vulnerability [25, 26].

3 The Effect of Aging on the Immune System

3.1 Immunosenescence and Inflammaging

Immunosenescence and inflammaging are hallmark characteristics of the remodelling process that the immune system undergoes with age. Immunosenescence is the sum of all the age-related alterations that occur with the immune system, which make the elderly more susceptible to infection and decrease their immune response towards pathogens. It involves changes in the repertoire and functional capacity of both the innate and adaptive immune systems [6].

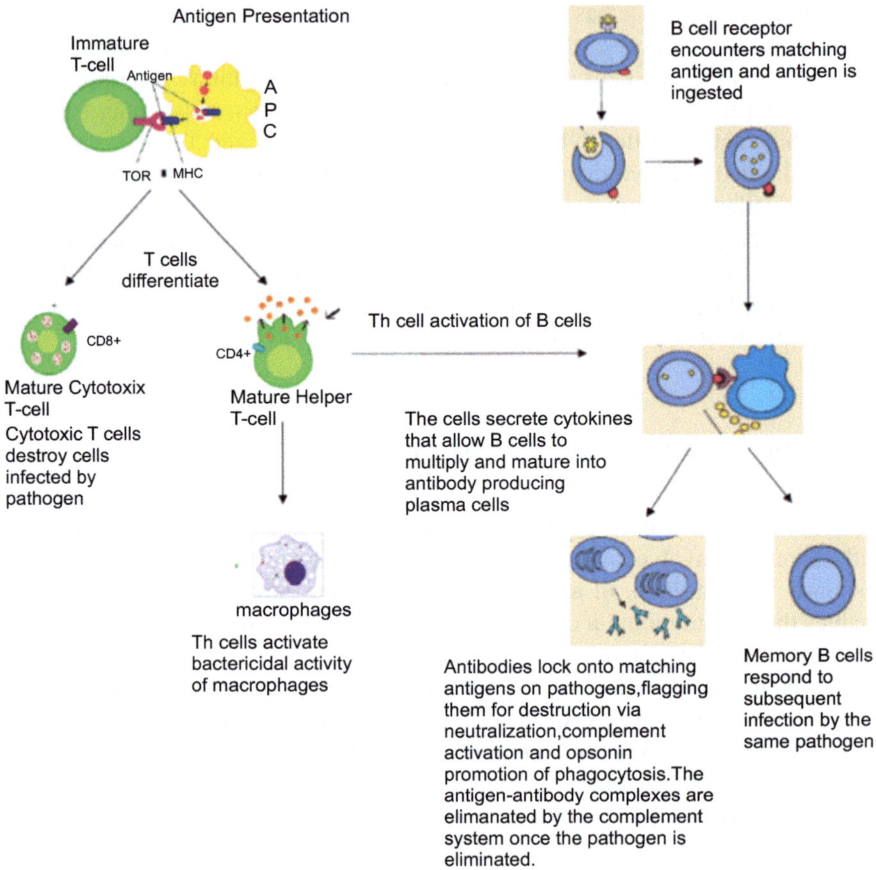

Fig. 3 The activation of T and B cells when exposed to foreign or non-self antigens. T cells express a unique antigen-binding receptor on the cell surface called T cell receptor (TCR) and require the action of APCs (usually dendritic cells, B cells and macrophages) to recognise a specific antigen. The surface of APCs express cell-surface proteins known as major histocompatibility complex (MHC). MHC are classified as class I, which are found on all nucleated cells, or class II which are only found on certain cells such as B cells, dendritic cells and macrophages. The MHC complex displays fragments of antigen when a cell is infected with a pathogen or has phagocytosed foreign protein. T cells are activated when they encounter an APC that has digested an antigen and is displaying antigen fragments bound to its MHC molecules. The MHC-antigen complex activates the TCR and the T cell secretes cytokines which further control the immune response. T cells are stimulated to differentiate into either cytotoxic T cells ($CD8^+$ cells), T-helper (Th) cells ($CD4^+$ cells) or regulatory T cells (Treg). Cytotoxic T cells are mainly involved in the destruction of cells infected by foreign agents. Upon resolution of the infection, most effector cells die and are cleared by phagocytes. However, a few of these cells are retained as memory cells that can quickly differentiate upon encounter with the same antigen. Treg will limit and supress the immune system to prevent excessive immune reactions to self-antigens and auto-immune reactions. B cells are involved in the anti-body or humoral response (whereas T cells dominate the cellular immune response). B cells can recognise antigens directly. After maturation, they leave the bone marrow expressing a unique antigen-binding receptor on their membrane. Upon activation by a foreign antigen, B cells will produce antibodies. They undergo proliferation and differentiation into plasma cells or memory B cells. Memory B cells remain long after the infection has cleared and continue to express antigen-binding receptors. The plasma cells are short-lived and undergo apoptosis, once the stimulating agent that induced the immune response, is eliminated. (Image reproduced from [22])

Outlined in Tables 1 and 2 are a number of age-related changes, which occur in specific cells of the innate and the adaptive immune systems, respectively. Moreover, these cells will undergo senescence, resist apoptosis and eventually cease to replicate, adapting a senescence-associated secretory phenotype (SASP), which affects chemical signalling. As a downstream consequence of dysregulation between stimulatory and inhibitory signals, a prolonged state of sterile, low-grade inflammation, referred to as inflammaging, is commonly observed in frail individuals [27, 28]. Moreover, Giovanni et al. [29] demonstrated that chronic inflammation, including excessive levels of proinflammatory cytokines, such as interleukin 6 (IL 6), tissue necrosis factor-α (TNF-α) and C- reactive protein (CRP), are associated with mortality risks in older people [29]. While acute inflammation is necessary for pathogen elimination and healing of injuries, a chronic, elevated baseline level inflammatory state is not desirable as this disrupts immune function, makes the older people more susceptible to diseases and adversely affects responses to vaccines [30, 31].

As with many other aspects of aging, the impact on immune response is an area that warrants more intensive study. The following sections summarise current knowledge regarding the effects of aging on the various components of the immune

Table 1 A summary of the age-related changes to components and systems of the innate immune system

Immune cell	Changes due to aging	References
Toll-like receptors (TLRs)	Expression and function of TLRs on surface of monocytes, dendritic cells and neutrophils, decline with age	[33, 35–37]
Monocytes	Changes in transcriptional and functional levels according to age, reducing production of cytokines and interferons	[38, 39]
Macrophages	Aging in macrophages influences many processes including TLR signalling, polarisation, phagocytosis and wound repair	[8, 36, 40, 41]
Dendritic cells	Impairment of mDCs and pDCs due to aging causing ineffective antigen presentation	[5, 42–44]
Neutrophils	Functional impairment affecting recruitment of neutrophils to site of infection leading to delayed wound healing and susceptibility to pathogens	[44, 45]
Natural killer (NK) cells	Decreased NK cell activity. A shift in NK cell differentiation leading to negative recruitment of other cells during the immune response that impairs signal amplification	[46, 48, 49]
Eosinophils	Declining eosinophil 'effector' function in the airways for both humans and mice	[50–53]
Mast cells	An increase in the population of mast cells in human organs and organs of other mammals and vertebrates, but with reduced degranulation and migration	[58–61]
Complement system	Increased levels of terminal pathway components such as C1q, C3, C4, C5, C8, C9. Decrease in factor B and D. Dysfunction in signalling cascade leading to increased risk of developing AMD, AD and PD	[76–78, 81–83]

Table 2 A summary of the age-related changes to components of adaptive immunity

Immune cell	Changes due to aging	References
T cells	Decrease in thymic output with age leading to decreased number of naive T helper and cytotoxic T lymphocytes causing reduced immune response to new pathogens	[6, 85–87, 89–91]
B cells	Decline in B cell production in the bone marrow. Limited diversity in B cell repertoire, an increase in antigen-experienced cells but a reduced output of naïve B cells	[92–94, 97–99]
Lymph nodes	Organisation of T and B cell areas becomes less distinct, impairing lymphocyte homeostasis and migration of immune cells during an immune response	[106–112]
Bone marrow	Decrease of haematopoietic tissue, resulting in decreased B and T cell lymphopoiesis	[96, 101–104]

system in an attempt to fully understand the complexity of the immunosenescence process and its impact on vaccine immune responses.

3.2 *Immunosenescence of the Innate Immune System*

While the relatively nonspecific innate response is considered to be less affected by immunosenescence than the adaptive immune system, age-related changes do affect the composition, phenotype and function of innate immune system cells and are discussed below.

3.2.1 Toll-Like Receptors (TLRs)

TLRs are highly conserved receptors that are expressed on innate immune cells, such as monocytes, macrophages and dendritic cells. They can recognise a variety of stimuli, including pathogen-associated molecular patterns (PAMPs) such as bacterial lipoproteins, lipopolysaccharides and microbial DNA and RNA. TLR receptor signalling has a crucial role in vaccination by linking innate and adaptive immune responses [32]. It has been found that cell surface expression of members of the TLR family (receptors and co-effector molecules) do not show a consistent age-dependent change across immune system models. There are, however, impaired downstream signalling events, including inhibition of positive and activation of negative modulators of TLR-induced signalling events and altered cytokine production [33, 34]. The expression and function of TLRs in monocytes [35], dendritic cells [36] and neutrophils [37] are reported to decline with advancing age in humans, although not all TLR expressions decrease in the older groups. For example, surface expression of TLR-1 on monocytes was found to decrease significantly with age, while surface TLR2 expression on monocytes was unchanged;

in addition, age-associated decrease in TLR-4 surface expression was also observed [35]. In contrast, studies using murine models show that aging do not affect the expression levels of TLR [34]; however, most studies include healthy subjects (either human or murine) and the presence of comorbidities could possibly reveal further insight into the defects of age-associated TLR expression and function [35]. An age-dependent decrease in TLR function in human DCs has been linked with poor antibody responses to influenza immunisation [36].

3.2.2 Macrophages and Monocytes

Macrophages and monocytes express a broad range of pattern recognition receptors (PRRs) such as TLRs, which, when triggered, activate an inflammatory signalling cascade. Macrophages and monocytes play a crucial role in the immune response by clearing infectious agents and subsequently cleaning the debris caused by inflammation and infection. De Maeyer and Chambers recently presented some additional effects of aging on macrophages and monocytes including decreased efferocytosis (the process by which apoptotic cells are removed) and reduced immune resolution, both of which directly affect immune function [38]. For example, in response to TLR agonists, human monocyte subsets were found to have different transcriptional or functional levels according to age, and this difference induced alterations in surface molecule expression, causing reduced production of interferons and cytokines such as IL-1β [39].

Macrophage function is also altered by age. The expression of TLR on macrophages is reduced in older humans and mice [40, 41]. In the older population, the role of macrophages in promoting chronic activation of senescence markers such as IL-1β and SASP cytokines, as well as inflammatory cytokine production, has also been highlighted [36]. IFN-γ plays an important role in macrophage activation. It has been shown that macrophages from old mice expressed 50% less MHC class II molecules on cell surfaces of older mice in comparison to the young, upon stimulation with IFN-γ. Panda et al. also noted a reduction in the number of CD68+ macrophages, which is usually abundant in the bone marrow [36]. These findings, therefore, demonstrate that aging in macrophages influences many processes including TLR signalling, polarisation, phagocytosis and wound repair [8].

3.2.3 Dendritic Cells

Dendritic cells are APCs and are responsible for the capture and processing of antigens for presentation to T cells. Two principal types of dendritic cells are recognised: myeloid (mDCs) and plasmacytoid (pDCs). Activated mDCs produce mainly IL-12, while pDCs produce IFN-α in response to viral infections and IL-4 in response to parasites [42]. Currently, knowledge about the effects of aging on dendritic cells from various tissues in humans is limited, and while there are some discrepancies in the changes in the dendritic cell population with age, the impairment of all subsets of dendritic cell function due to aging is widely reported [5, 43, 44]. Significant reduction in the total number of dendritic cells and of mDCs in healthy people over 60 years of age compared with younger individuals has been reported [42]. Consequently, the facilitating of signal transduction to T cells required for efficient T-cell-dependent antibody response is affected, resulting from ineffective antigen presentation by dendritic cells.

3.2.4 Neutrophils

Neutrophils circulate in the bloodstream and are the first cells to migrate to an infection site [45]. Aging affects the recruitment of neutrophils, which, in turn, disrupts internal signalling pathways, which can lead to delayed wound healing and high levels of susceptibility to invasive pathogens such as methicillin-resistant Staphylococcus aureus [44, 46, 47].

3.2.5 Natural Killer (NK) Cells

With age, significant impairments have been reported in the two main mechanisms by which NK cells confer host protection: direct cytotoxicity and the secretion of immunoregulatory cytokines and chemokines [48]. In elderly subjects, decreased NK cell activity has been shown to be associated with an increased incidence and severity of viral infection, highlighting the clinical implications that age-associated changes in NK cell biology have on the health of older adults [49]. As a marker of NK maturity, an increased percentage of CD57+ cells suggest a shift towards a more mature circulating NK pool that occurs with age. With regard to NK cell subsets, studies have shown that NK cells undergo a shift in differentiation with aging, which results in an increase in the ratio of $CD56^{dim}$ to $CD56^{bright}$ cell line. This promotes greater cytotoxic activity as the $CD56^{dim}$ functions in that capacity, while $CD56^{bright}$ plays a more regulatory role [48]. Furthermore, as a consequence of this change, $CD56^{dim}$ cells begin to express the $CD57^{+}$ surface marker and ignore cytokine signals. This negatively affects recruitment of other cells during the immune response as the production and release of chemokines IL-2 and IL-12, important for signal amplification, are dependent on the regulatory function of $CD56^{+}$ cells [46].

3.2.6 Eosinophils and Basophils

Both eosinophils and basophils reside primarily in the connective tissue underlying the epithelium of the respiratory and gastrointestinal tract. Once activated, they secrete a range of granule-derived proteins as well cytokines, leukotrienes and prostaglandins. Most studies reported in the literature have assessed age-related changes with reference to asthmatic patients. For example, a study found that airway eosinophilia (excessive eosinophils in the lungs) was comparable in all ages of asthmatic subjects; however, there were age-related changes in peripheral blood eosinophil 'effector' functions [50]. This indicated the possibility of differences in the manifestation of diseases because of age-related changes in inflammatory cell function [51]. In animal studies, aged mice displayed greater airway eosinophilia in response to antigen challenge in comparison to younger mice [52]. Conversely, airway hyper-responsiveness decreased in elderly mice, suggesting that eosinophil effector function declines [53].

Basophils are involved in inflammatory and immune responses; for example, they can promote humoral memory responses, drive IgE-dependent dermatitis and shift an immune response towards Th-2-dominated inflammation [54]. There is very little understanding of the effects of aging on basophil numbers and function. In animal models, basophils were more abundant in the bone marrow and spleens of 19-month-old mice compared to 4-month-old mice, and aged basophils in the bone marrow and spleen tended to express less CD200R3 and more CD123 [55]. In addition, microbiota transfers from young and old mice to germ-free recipients showed the role of aged microbiota on the activation of bone marrow-derived basophils [55].

3.2.7 Mast Cells

A key set of immune cells, mast cells also undergo age-related changes. Despite their established role in promoting inflammation in allergic reactions, it is known that mast cells play a critical role in cross-talk between the innate and the adaptive immune system [56]. Mast cells are innate immune cells effective in immunosurveillance in the skin, responding to a broad range of stimuli via a wide range of surface receptors, including the high affinity receptor for IgE [57]. The main age-associated changes in mast cells include an increase in the population in the organs of humans and other mammals and vertebrates [58–61] but a reduced degranulation and migration [58].

Mast cells exhibit a greater density (4.5 fold) near mesenteric lymph vessels (MLV), than in mesenteric tissue remote to MLVs [62]. They have the ability to influence lymphatic contractility, mediated by histamine. It has been reported that mast cells can produce, store and release, upon activation, numerous bio-active and vasoactive mediators that continuously modulate the surrounding tissues [56, 63]. Aging affects the functional status of lymphatic mast cells, causing a state of chronic basal mast cell activation and inability to react to acute stimuli [62]. Consequently, it has been suggested that there is an age-associated deterioration in the effective transportation of immune cells through the lymphatic system in cases of acute inflammatory stimulation [62].

Studies in mice suggest that mast cells could potentially be used as a target for adjuvants in vaccines delivered intradermally [64]. Further studies confirm that the c48/80 mast cell activator can safely and effectively stimulate mast cells to release granules containing TNF-α, which promotes draining of DCs into the lymph nodes

[65]. This could be a potential target for the elderly population as the number of mast cells in the skin increases with age, despite what is believed to be a compensatory mechanism for age-associated change in function [58]. However, more in-depth studies need to be done to determine what mechanisms could be explored to exploit the adjuvant characteristics and balance the responses that would be obtained so that health and safety are maintained.

3.2.8 The Complement System

The complement system consists of more than 40 serum proteins and cell membrane bound receptors [66] that form an important branch of the innate immune system and also play a key role in the enhancement of adaptive immunity by facilitating B and T cell responses that help eliminate pathogens [67, 68]. The complement proteins remain as inactive zymogens until they react with one another in an enzymatic cascade, resulting in an amplified response to foreign antigens (Fig. 3). Activation of the complement system causes induction of various anti-microbial effects such as neutralisation of pathogens, regulation of the inflammatory effect, chemotaxis and enhancement of the adaptive immune response. To minimise damage to host cells from the complement system, host cell surfaces have regulatory proteins that inhibit a complement cascade [69]. There are three distinct pathways through which complement can be activated, but they all result in the activation and enzymatic cleavage of C3 (the most abundant complement protein found in the body). Enzymatic cleavage of C3 initiates the formation of the activation products, C3a, C3b, C5a and C5b-9 (a membrane attack complex, MAC) [70].

There are three distinct pathways through which complement can be activated:

1. The classical pathway can be initiated by the binding of C1q, the first protein of the complement cascade, to immune complexes consisting of IgM or IgG that recognise and bind to the surface of foreign antigens to form an antibody-antigen complex. Through cleavage of molecular structures on the antibody-antigen complex, C3a and C3b are released. C3b is an opsin and will amplify the complement cascade and facilitate phagocytosis as well as initiate a cascade that results in the formation of the membrane attack complex (MAC), which forms a pore on the surface of the pathogen leading to cell lysis [71, 72].
2. The lectin pathway is activated when either mannose-binding lectin (MBL) or Ficolin (both pattern recognition receptors) detects mannose or N-acetyl glucosamine containing structures of carbohydrates on the surface of bacteria, viruses, parasites or yeast. Both MBL and Ficolin exist in the serum as complex structures called the MBL-associated proteins (MASPs) [73, 74].
3. The alternate pathway, unlike the classical and lectin pathways, is continuously active at low levels as surveillance for the presence of pathogens. Healthy host cells are resistant to this low level of activity, and proteins within this pathway can bind to a wide range of suitable sites that are not specifically detected by the other complement pathways [75].

Dysfunction of the complement system has been implicated in aging-related diseases including age-related macular degeneration, Alzheimer's disease and Parkinson's disease [76–78]. The most common cause of irreversible blindness in Western populations is age-related macular degeneration, characterised by deposits in the form of lesions (drusen), of cellular debris, lipids and various protein components, a number of which belong to the innate immune system, in the inner layers of the retina. Immunohistochemistry studies identified deposition of complement proteins as well as a number of inflammatory mediators in drusen, believed to be caused by an inadequately regulated complement cascade that significantly contributes to the pathophysiology of age-related macular disease [79, 80].

There are very few studies that have assessed the effect of aging on the complement system. Overall, it has been found that in healthy older subjects, there are higher levels of complement components such as C1q, C3, C4, C5, C9 and haemolytic complement activity compared to the young [81]. During infections, younger people will show a dramatic increase in their complement levels; however, this is not seen in the elderly population because of an already higher base level [82]. A study that analysed the complement proteins involved in the three different pathways in healthy older Caucasian individuals found significant increase in activity of the CP and AP of the complement system. In line with this enhanced functional activity, the levels of terminal pathway complement proteins, C5, C8 and C9 were also raised with age [83]. The levels of Factor D [83] and Factor B [81], both essential components of AP, showed decreased levels in the elderly. The mechanism and significance of these age-related changes are not fully understood; however, it has been suggested that the increase in levels of terminal pathway components with age could be a mechanism to compensate for the impaired clearance of pathogens and apoptotic cells due to lower cellular immunity [83, 84].

3.3 *Immunosenescence of the Adaptive Immune System*

If pathogens are able to overcome innate defence mechanisms, then the adaptive immune system is initiated. Dendritic cells capture antigens from pathogens, mature, differentiate and migrate to regional draining lymph nodes to stimulate antigen-specific T and B cells. Following antigen-specific clonal expansion of B and T cells, the invading pathogen or the pathogen-infected cells are removed by specific antibody and T cells.

The following is an overview of the changes that have been observed in the adaptive immune system during aging, with a focus on alterations that might impact immune responses to infection or vaccination. Table 2 provides a summary of the details described.

3.3.1 T Cells

T lymphocytes originate from haematopoietic stem cells within the bone marrow. Some of these cells subsequently become lymphoid progenitor cells that leave the bone marrow and travel to the thymus via the blood. Once in the thymus, T lymphocytes undergo a selection process in which the majority of developing T cells (called thymocytes) will not survive. Each T lymphocyte has a T cell receptor (TCR), which is specific to a particular antigen. T lymphocytes that survive thymic selection will mature and leave the thymus. When naïve T lymphocytes encounter a specific antigen, they will proliferate and differentiate into one of several effector T lymphocyte subsets [26].

The primary lymphoid organs, namely, the bone marrow and the thymus, undergo significant changes with age, which adversely affect the duration of immunity and vaccine efficacy and effectiveness [85]. Thymic involution, which refers to the atrophy of the thymus, begins from infancy and continues throughout life, causing a decrease in the thymic output of approximately 3% each year [6, 86]. Thus, the number of naive T helper and cytotoxic T lymphocytes produced in an older individual is significantly lower than that of a younger person. Instead, there is a higher ratio of primed memory T lymphocytes for specific antigens that were encountered during an individual's lifetime. This shift means that the aging adaptive immune system's response to new pathogens especially for viruses like influenza, which constantly undergo antigenic drift, is impaired. Despite having an increased repertoire of primed memory cells, over time, the elderly often lose their immunity towards certain antigens they had previously been immune to. Further, having a reduced number of naive CD4+ T lymphocytes suggests that inadequate help will be provided for B cell activation and proliferation and antibody production. Upon antigenic encounter, the naive repertoire of CD4+ T cells of aged individuals displays reduced proliferative potential and cytokine secretion and shows inadequate effector responses towards vaccines [87].

Additionally, T lymphocytes, though CD8+ to a greater extent than CD4+, have been known to senesce with age, which limits the number of cell cycles that mature T cells undergo since their telomeres shorten overtime. Once a particular length is reached, usually under 12 kilobase pairs, the lymphocyte is no longer able to replicate since telomerase (the enzyme that lengthens telomeres) activity declines as well. This process appears to contribute to cell surface receptor modification, where certain receptors such as CD27 and CD28, typically found in abundance in young people, appear to diminish in older adults [6, 88]. CD28 is a co-stimulatory receptor required for heightened CD4 + T cell activation through its ability to stimulate cytokine production. Studies show that lack of CD28 results in poor vaccine responses and increased susceptibility to infections. A reduction in T regulatory populations, including CD45RA+ and CCR7+ cells, also contributes to a decline in immune function [6, 89, 90]. It is clear that the TCR diversity brought on by aging affects the functional capacity of T lymphocytes impairing immune function and, by extension, vaccine responses in the elderly [91].

3.3.2 B Cells

Aging in mice was found to reduce the number of naïve B cells and plasma cells while increasing the population of CD27+ memory B cells. Human peripheral B cell percentages and numbers significantly decrease with age, and although B lymphopoiesis is active throughout life, there is a decline in B cell production in the bone marrow in aged groups [92–94].

With regards to the B cell repertoire, there is contradicting evidence which could be due to the complexity of B cell lineage diversity [95]. Several studies report that like T cells, memory B lymphocytes accumulate, surface receptor diversity decreases and receptor binding affinity is reduced with age [27, 87, 95]. This strongly suggests the decreased potential of B cells to respond to novel antigens, proliferate and differentiate into plasma cells, producing antibodies upon antigen encounter. However, other researchers have reported that the number of naive cells remained unchanged but also noted a decrease in the number of plasma cells as well as memory B cells in the bone marrow and peripheral blood with age. This inconsistency, according to Pritz et al., is possibly due to different markers being associated with B cell subsets [96]. Further, aging B cells have been reported to promote elevated levels of proinflammatory cytokines, thus contributing to inflammaging [27].

Diversity in B cell repertoire is essential for an effective immune response, given that B cells provide a variety of specific antibodies. Many elderly individuals are known to have limited diversity in B cell repertoire, potentially contributing to susceptibility to infections, and an inability to raise an efficient response to vaccines. Studies have indicated that aging may cause significant changes in the selection process during affinity maturation of B cells [97] and also that the B cell repertoire is often less diverse in the elderly with evidence of non-pathogenic clonal expansions [98], where the bone marrow output of naïve B cells is reduced. Since the overall numbers of B cells remain the same, this indicates a reduced input of new B cells to the overall population. Therefore, a greater proportion of antigen-experienced cells indicates evidence of clonality, leading to a repertoire that mainly consists of antigen-experienced cells [98]. This loss of diversity is likely to be correlated with poor vaccine responses against many pathogens [99].

A number of naive B cells are known to deplete along with the B cell and antibody diversity and isotype class-switching capacity. However, there seems to be an increase in the effector B cell count. An investigation of the humoral responses to the pneumococcal vaccine in old and young people revealed that although the IgG levels in the elderly were comparable, IgA and IgM levels were much lower. Spectratyping analysis conducted 1 week and then 1 month after vaccination showed that while baseline titres were restored in the young after 1 month, the elderly had significantly lower antibody serum levels for both periods [100].

3.3.3 The Bone Marrow

The bone marrow undergoes significant change in composition with age, and being the place where the hematopoietic stem cell precursors for all immune cells are derived, it is likely that this is a contributor to the weakened immune response observed in the elderly [96]. However, the intricacies are still to be fully understood. However, studies have shown that the amount of haematopoietic tissues decreases in the bone marrow with age and is replaced by fat tissue [101]. The haematopoietic stem cell (HSC) niche is a region of the bone marrow in which HSC resides. Under normal, healthy physiological conditions, HSCs have a high potential for self-renewal and differentiation. However, this capability has been found to be affected by age. Aged HSCs have an impaired ability to proliferate and develop due to shortened teleomeres, which play a vital role in preserving the integrity of genetic information [102]. As a result, B and T lymphopoiesis is decreased [103]. Further, age-related decline of naïve T-cells is more pronounced in the bone marrow than in the peripheral blood [104].

3.3.4 The Lymph Nodes

While age-related changes in primary lymphoid organs (bone marrow and, in particular, the thymus, which involutes in the first third of life) have long been appreciated, changes affecting aging secondary lymphoid organs and, in particular, aging lymph nodes have been less well characterised [105]. In general, lymph nodes in both mice and humans become smaller and less cellular with aging [106]. Similar to thymic involution, histological studies of the lymph nodes show that the organisation is less distinct (especially between T and B cell areas) [107], with an accumulation of adipocytes [106] and signs of fibrosis [108]. This has the potential to impair both lymphocyte homeostasis and the migration of immune cells during the course of an immune response [105, 109]. Fibroblastic reticular cells (FRCs) comprise a large portion of the lymph node stromal network, creating channels for chemokine transport and promoting DC and T cell migration [110]. The number of FRCs declines with age, resulting in subsequent disorganisation of this transport network in the lymph node. Similar disorganisation between T and B cell areas occurs in the aging spleen [111]. Decreased lymph node swelling has also been noted in older mice following viral infection, indicating a reduced influx of immune cells to the lymph nodes [109, 112].

4 The Mucosal Immune System

As discussed above, the first line of defence against invading pathogens due to ingestion or inhalation is anatomical and physiological barriers. Aside from mounting an effective immune response against food-borne pathogens, the mucosal immune system must also recognise the harmless antigens associated with food and commensal microorganisms and generate immunological tolerance against them [113]. Secretory IgA (SIgA) is the predominating immunoglobulin along mucosal surfaces, and SIgA antibodies are generated in response to subsequent colonisation of pathogens in gastrointestinal, respiratory and genito-urinary mucosal tissues and can confer protection at these sites [114]. Signs of mucosal senescence first appear in the gut immune system [115]. A study on oral and parenteral vaccination in naturally ageing mice showed that age-associated decrease in antigen-specific immune responses occurs earlier in the mucosal immune system than in systemic immune system [116]. Although many factors have the potential to influence vaccine immunogenicity and therefore vaccine effectiveness, increasing evidence from clinical studies and animal models now suggests that the composition and function of the gut microbiota are crucial factors modulating immune responses to vaccination [117].

Alterations in the mucosal immune system occur in advanced aging, which results in a failure of induction of SIgA antibodies for protection from infectious diseases [118]. Most studies on the effects of aging on mucosal immunity focus primarily on the gut-associated lymphoid tissue (GALT) [119]. It has been shown that GALT-mediated immune responses are more susceptible to aging than are lymphoid tissues involved in peripheral immunity [118]. This effect coincides with age-related increases in the incidence and severity of gastrointestinal infections, tumours and inflammatory diseases and reduction in the vaccination efficacy [118, 120, 121] (Fig. 4).

4.1 An Overview of Mucosal Immunity

Mucosal surfaces cover the pulmonary, nasal, oral, gastric and genitourinary tracts. They share a common cellular organisation composed of a semi-permeable epithelial layer that sits on top of a basement layer underneath, which lies loose connective tissue termed the lamina propria, which contains blood vessels, lymphatics, immune cells and other components. The mucosal immune system of higher mammals is sophisticated and consists of an integrated network of tissues, lymphoid and mucous membrane-associated cells that are collectively known as the mucosa-associated lymphoid tissue (MALT). The lymphoid tissue exists as aggregated, organised immune tissue within the mucosa and are associated with local immune responses. The three major regions of MALT are (i) the gut-associated lymphoid tissue (GALT), including the well-characterised Peyer's patches, consisting of organised lymphoid nodules found in the small intestine; (ii) the nasal-associated lymphoid tissue (NALT), consisting of Waldeyer's ring of tonsils and adenoids; and (iii) the bronchus-associated lymphoid tissue (BALT). In addition, colonic patches, caecal patches and isolated lymphoid follicles (ILFs) also form part of the MALT. The most abundant antibody isotype secreted across mucous membranes is secretory immunoglobulin A (SIgA) [122] and provides mucosal immune protection as a

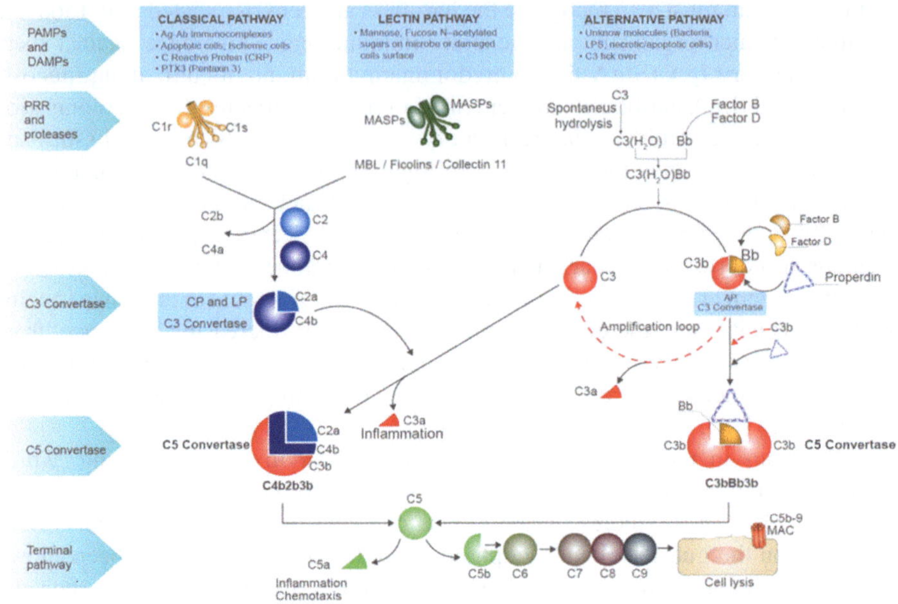

Fig. 4 A schematic overview of the complement system. The classical pathway is initiated when the C1q domain binds to the Fc domain of the antigen immunocomplex. This activates C1r which subsequently cleaves C1s which in turn cleaves C2 to C2b and C4 to C4a leading to the formation of C4b2a (C3 convertase). This complex cleaves C3 to C3a and C3b. C3a acts as a recruiter of inflammatory cells (anaphlotoxin), and C3b binds to the C4b2a complex to form C5 convertase (C4b2a3b). The C5 convertase initiates the formation of the membrane attack complex (MAC) that inserts into cell membranes, forming functional pores, causing cell lysis. The initiating molecules for the lectin pathway are collectins (MBL, ficolin and collectin-11), which are multimeric lectin complexes. These bind to specific carbohydrate patterns on the surface of pathogens or damaged cells that are recognised as uncommon to the host, forming MBL-associated proteins (MASPs). MASP1 and MASP2 cleaves C2 and C4 to form the C3 convertase complex, similar to the CP pathway. Activation of the AP can be induced by spontaneous hydrolysis of C3 into C3(H2O) which resembles C3b in structure. If C3b binds to a nearby surface that is incapable of inactivating it (such as bacteria/yeast cells or damaged host cells), this then leads to the amplification of the AP. The presence of complement regulators on healthy cells ensures the spontaneous hydrolysis of C3 is kept in check. C3(H2O) can bind factor B which is then cleaved by factor D, generating Ba and Bb and the initial AP C3 convertase (C3(H2O)Bb). This complex can cleave C3 to C3a and C3b. C3b is able to create new C3 convertase (C3bBb) in the presence of factors B and D with the protein properdin stabilising C3bBb, thus acting as an 'amplification loop' for other pathways (CP and LP) as well as the AP. Properdin stabilisation also occurs in the formation of the C5 convertase complex (C3bBb3b) which is then involved in the formation of C5b-9 (MAC). (Image reproduced from Franzin et al. [66])

result of the ability to interact with polymeric Ig (pIgR), an antibody transporter expressed on the basolateral surface of epithelial cells [123]. The induction and maintenance of an antigen specific mucosal immune response in the intestine is a multistep process (Fig. 5). In the intestines, SIgA promotes immune tolerance by entrapping dietary antigens and microorganisms in the mucus and downregulating

Fig. 5 The key stages of the induction and maintenance of mucosal immunity in the gut

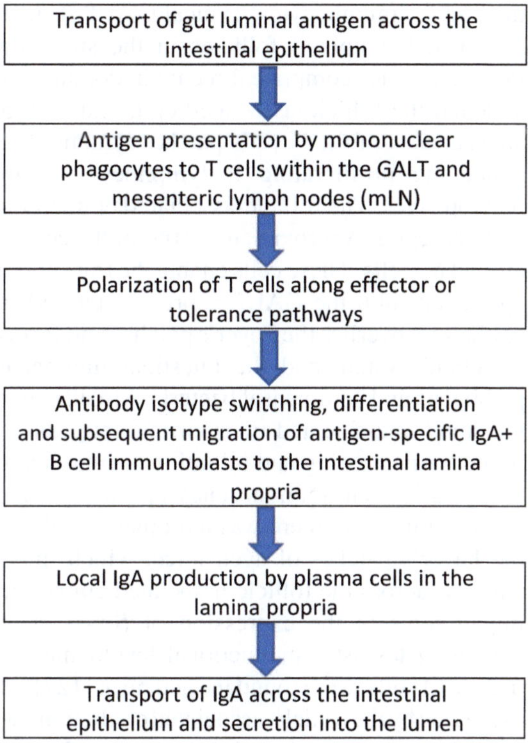

the expression of proinflammatory bacterial epitopes on commensal bacteria and consequently plays a key role in managing and regulating bacterial communities within the gastrointestinal system [123, 124].

4.2 *Immunosenescence of Mucosal Immunity*

Below is a review, summarising the current understanding of how the function of the mucosal immune system in the intestine is affected by aging, particularly the effects on the innate immune system and IgA production by B cells. A summary of the reported age-related changes to components and physiological structures involved in mucosal immunity is shown in Table 2.

4.2.1 M Cells

GALT, which consists of both isolated and aggregated lymphoid follicles [125], is the site where antigen recognition and mucosal immune responses are initiated. GALT is one of the largest lymphoid organs in the body, containing up to 70% of the body's immunocytes. Typical GALT structures can be seen in Peyer's patches, aggregated lymphoid follicles in the small intestinal mucosa. Morphologically, Peyer's patches comprise three main domains: the follicular area, the parafollicular area and the follicle-associated epithelium [125]. The follicular and parafollicular areas consist of the Peyer's patches lymphoid follicle, which has a germinal centre containing proliferating B lymphocytes, follicular dendritic cells (FDC) and macrophages. The follicle-associated epithelium is a one-cell-thick layer composed of enterocytes. Approximately 10% of the cells consist of specialised epithelial cells termed M-cells. These cells enable the transportation of antigens across the intestinal epithelium into the GALT, a process termed transcytosis. The follicle-associated epithelium overlies the Peyer's patches and forms the interface between the intestinal lymphoid system and the intestinal luminal environment, preventing access to pathogens and commensal bacteria. M-cells are continuously sampling the lumen of the small intestine and transporting antigens to the underlying mucosal lymphoid tissue for processing and initiation of immune responses [126].

By using an immunosenescent mouse model ($\geq$ 18 months old), it was demonstrated that there was a dramatic decline in the density of mature M cells in the Peyer's patches of aged mice, which impeded the transcytosis of particulate antigens across the follicle-associated epithelium [127]. In the same study, specific impairments in the expression of Spi-B, a transcription factor responsible for mediating downstream functional development of immature M cells into functionally mature M cells, were found in aged mice [127]. The chemokine CCL20 is specifically expressed by the follicle-associated epithelium, where it mediates the chemoattraction of CCR6-expressing lymphocytes and leukocytes towards the intestinal epithelium. CCL20-CCR6-stimulation has also been suggested to influence M-cell maturation as their density is reduced in CCR6-deficient mice [128, 129]. The expression of CCL20 by the follicle-associated epithelium was found to be reduced in aged mice [130]. As a consequence of the reduced CCL20 expression, the chemoattraction of certain populations of mononuclear phagocytes and lymphocytes, including those with apparent 'M cell inducing' potential towards the follicle-associated epithelium, was similarly reduced [127]. This implies that the effects of ageing on M cell status will impact M cell-targeted vaccines in the elderly [131, 132].

4.2.2 Goblet Cells

Goblet cells are mucin-secreting cells in the gut epithelium. They may also provide passages for the delivery of low-molecular-weight soluble antigens to CD103+ MNP in the lamina propria [133]. Goblet cell dysfunction has been associated with the development of intestinal inflammation [134], but whether aging influences their density, function and/or antigen sampling via goblet-cell associated antigen passages is uncertain. Increased goblet cell density has been reported in the intestines of 12-month-old rats, but no differences were observed in the follicle-associated epithelium of aged mice [119].

4.2.3 Epithelial Cell Barrier Integrity

Both animal and human studies demonstrate an age-dependent increase in the permeability of the gastrointestinal barrier [135]. An intact intestinal epithelium provides a crucial protective barrier between the contents of the gut lumen, including pathogens, and host tissues. Intercellular tight junctions at the apical lateral membrane link neighbouring epithelial cells together and help determine paracellular permeability. Data from two independent studies have demonstrated the expression of molecular components of tight-junctions including zonula occludens-1 (ZO-1), junctional adhesion molecules (JAMs) and occludins, is decreased in the intestinal epithelia of aged rats [136] and baboons [137]. These studies suggest that aging is associated with significant intestinal barrier dysfunction. The precise implications for mucosal immunity are uncertain. Disruption to tight junctions in the intestinal epithelium may lead to increased paracellular permeability to luminal antigens and/or proinflammatory stimuli such as bacterial endotoxins. Whether this apparent dysregulation in intestinal permeability contributes to the low-level chronic inflammation often observed in the elderly (inflammaging) remains to be determined [138].

Disruption of the intestinal barrier can also be caused by other underlying conditions associated with old age. Studies conducted on young and old mice found that cerebral arterial occlusion increased gut permeability, allowing translocation of bacteria across the gut epithelium. It was quickly resolved in the young, but the older mice developed sepsis [139]. Whether alterations in gut permeability will have direct implications on mucosal vaccine efficacy remains to be seen.

4.2.4 Mucosal T Lymphocytes

The human gastrointestinal tract is a major reservoir of total body lymphocytes (approx. 60%) [140] and a region of high antigenic exposure. Mucosal T cells are particularly reliant on signals and cues from the microbiota for development and differentiation [141]. Together with the innate branch of the gut immune system, these effector lymphocytes form the first line of defence against invading pathogens and play a crucial role in maintaining barrier integrity. The T lymphocytes in these compartments are significantly different from conventional T cells in terms of ontogeny, phenotype and function, thus indicating highly specialised adaptive immune responses that are influenced by the external gut environment [142]. Although most T cells originate from the thymus and migrate to the gut, it is well-established that the intestine functions as a region for lymphocyte development [143].

There are three defined anatomical compartments of the mammalian intestine that contain mucosal T lymphocytes. The GALT, including Peyer's patches and mesenteric lymph nodes, possess T cells in organised lymphoid structures, and in the other two compartments, T lymphocytes are scattered among nonlymphoid cell types. These include T cells distributed within the lamina propria known as lamina propria lymphocytes (LPL), which lie between the villi or crypts away from the epithelial layer, and intraepithelial lymphocytes (IEL), which reside below the tight junctions of the epithelial cells – also termed 'epithelial compartments' on the other side of the basement membrane from the lamina propria. IEL subsets have been of particular interest in immunology as their development and function have been shown to be significantly influenced by the gut microbiota [141, 142, 144, 145]. Most IEL subsets develop via extrathymic pathways. Studies have shown that the transfer of pluripotent stem cells into mice devoid of the thymus demonstrates clear evidence of the development and proliferation of IELs in murine intestines [146].

Mucosal T cells are heterogeneous in phenotype and function and are classified into two general subsets based on their T cell receptor (TCR) and coreceptor expression. Mucosal T cells in the lamina propria are generally populated with 'type a' T cells that express TCRαβ together with CD4 or CD8αβ as TCR coreceptors. 'Type b' mucosal T cells are more prevalent in the mucosal epithelium and express either TCRαβ or TCRγδ and typically also CD8αα homodimers and lack expression of the conventional TCR coreceptors CD4 and CD8αβ. Peyer's patches contain $CD4^+$ or $CD8^+$ cells together with the TCRαβ. Studies that have assessed the impact of commensal bacteria on the intestinal mucosal immune system have compared the intestines of animals that are germ-free (gnotobiotic) with those that possess intestinal bacteria but are devoid of pathogens (specific pathogen free, SPF). It has been found that the cellular content of the mucosal immune system is shaped by the presence of the commensal intestinal bacterial flora. For example, areas of Peyer's patches containing T cells are diminished with a generally reduced lymphocyte cellularity in germ-free conditions [124]. The T cell component of the lamina propria (largely composed of CD4+ (ab TCR)) lymphocytes is also considerably reduced and restored to normal levels by conventionalisation (when animals acquire a normal intestinal flora) [147–149]. The heterologous populations of the IEL group, comprising cells that use the ϒδTCR and those that use the αβTCR, have displayed no difference in αβTCR CD8 αα, whereas other components of the group have been reduced in germ-free rodents [147, 148, 150, 151].

Immunosenescence of the mucosal immune system in terms of vaccine efficacy in the elderly has not been as extensively studied, in comparison to the age-related decline of the systemic immune system. However, research in the past decade on the impact of the gut microbiota on aspects of mucosal immunity has begun to unravel the complex symbiotic and dynamic relationship that exists between commensal intestinal flora and components of the mucosal-associated lymphatic system, which can aid the exploitation of this knowledge to produce effective vaccines for the elderly.

Information on age-related changes to the mammalian immune system is scarce; however, alterations to mucosal T cells in the murine intestinal compartment have been reported in several studies. The mucosal immune response of aged mice (12–14 months, or 2 yrs) was compared to young adult mice (6–8wks), and it was found that mice older than 1 year showed reduced antigen response to immunisation with three once-weekly oral doses of ovalbumin and a native cholera toxin as adjuvant [116]. There were significantly higher levels of CD4+ T cells from both the spleen and Peyer's patches in young mice in comparison to the aged mice, demonstrating impaired gastrointestinal associated immune response. It has also been noted that the GALT of mice shows signs of age-related decline earlier than that of the nasal-associated lymphatic immune response [152]. Further evidence of immunosenescence at the intestinal mucosal surface, particularly with regard to IEL subsets, has been revealed in several studies [153, 154]. For example, Santiago et al. determined mucosal T cell levels in the GALT to understand the mechanism underpinning the age-related decline in oral tolerance (immune hypo-responsiveness to fed antigens) in aged mice (12–24 months). It was attributed to several factors, namely, the diminished frequency of TCRϒδ+ as well as a lower frequency of TCRαβ+CD8αα + IEL in 24-month-old mice. These IEL subsets and the associated TCRs are involved in gut regulatory activities, indicating disruption in gut homeostasis during the aging process [155]. In contrast, elevated levels of $CD4^+CD25^+Foxp3^+$ and $CD4^+CD25^+LAP^+$ cells were found in mesenteric lymph nodes, Peyer's patches and lamina propria of older mice and were possibly upregulated as a compensatory mechanism to maintain some gut homeostasis in the intestinal mucosa [155].

There are conflicting reports involving human subjects, with regard to mucosal T cell populations and their effector sites. Biopsies of jejunal cells taken from healthy individuals revealed significant immunosenescent effects on intestinal cellular immunity, showing higher IgA plasma cell counts and significantly lower IEL counts in the elderly (>70 yrs) in comparison to the younger control group (25–50 yrs) [156]. In contrast, Beharka et al. examined saliva, blood and intestinal biopsies from young and old healthy subjects and found that IELs did not demonstrate age-related changes [157].

Mucosa-associated invariant T cells (MAIT) have been given much attention since they are found in abundance in mucosal locations such as the gut but also comprise 10% of the peripheral blood lymphocytes in humans [158]. The immunological role of MAIT cells is not fully understood, but the fact that they behave like innate cells and gravitate to the gut in response to a broad range of microbial organisms indicates their involvement in infection or regulatory control of commensal microbes [159]. In old age, MAIT levels have been found to decrease to 10% of the value found during fertile age. Further evidence of a reduction of MAIT cells with age has also been reported by Lee et al., who showed that although circulating MAIT cell levels varied widely (0.19% to 21.7%) in the study subjects, they were significantly lower in older people (age, 61–92 years) compared to younger ones (age, 21–40 years) [160].

4.2.5 Effects on Secretory IgA Responses

There are conflicting reports in the literature regarding the overall quantity of IgA produced by the aged mucosal immune system. Several studies have detected a significant increase in the amount of IgA produced in aged mice. The faecal IgA levels of aged mice have been found to be significantly more elevated in comparison to the young [116, 130, 156]. In contrast, a few studies also reported the absence of age-related differences in the amounts of nonspecific immunoglobulins secreted into the intestinal lumen in vivo or into the medium by cultured duodenal biopsies [116, 156, 161], while other researchers observed that the production of antigen-specific IgA is diminished with aging [118, 162, 163]. However, antigen-specific IgA responses are adversely affected in the elderly and have a lower affinity [162, 164–166]. Antibody-mediated protection is highly dependent on the affinity and isotype of the antibody produced, where quality rather than quantity is determinant for an effective immune response, and a decrease in high affinity antibodies has been detected in the elderly, although the reasons and mechanisms are not quite clear [167] (Table 3).

The B cell repertoire changes with age, and experiments in mice revealed subtle alterations in variable region gene usage and a decrease in the ratio of antibodies reacting with nominal versus autologous antigens [167, 168]. A diverse B-cell repertoire is important to ensure provision of a wide range of antibodies against potential pathogens. This diversity is maintained by rearrangements in the immunoglobulin gene, which produces B cells with unique immunoglobulin genes and different antigen specificities. High-affinity antibodies are produced by B cells because of affinity maturation, which occurs within the germinal centres of lymphoid tissues. In the intestine, immunoglobulin isotype class switching to IgA appears to occur only in the organised lymphoid structures of the GALT such as Peyer's patches and isolated lymphoid follicles [169–171], which contain all the necessary cellular components required to generate IgA-committed B cells, including B cell follicles with germinal centres, T cells and networks of follicular dendritic cells. When a B-cell is activated by an antigen, it enters the germinal centre and proliferates,

Table 3 Summary of the age-related changes to the gastrointestinal tract (GIT) epithelial barrier

Component of the mucosal immune system in the GIT	Changes due to aging	References
M cells	Decline in the density of mature M cells found in aged mice	[127–132]
Goblet cells	Uncertain whether there is a change in function, density or antigen-sampling ability in aged mice or humans	[133, 134]
Epithelial cell barrier integrity	Increase in permeability of the GI barrier. Expression of molecular components of tight junctions is decreased in the intestinal epithelia of rats and baboons	[135–138]
Secretory IgA	Decrease in high affinity antibodies detected in the elderly. Antigen-specific IgA responses are adversely affected and have a lower affinity	[116, 118, 130, 156, 161–171]
Mucosal T cells	Lower levels of CD4+ T cells in aged mice, indicating a reduced immune response in comparison to young adult mice. Diminished frequency and expression of TCR receptors on IEL subsets necessary for gut homeostasis in elderly mice. Conflicting reports in in human studies regarding IEL subsets. Decreased levels of MAIT cells in elderly	[116, 152–159]

and mutations are accumulated within the immunoglobulin gene (termed hypermutation). These mutations attempt to improve the antigen binding affinity of the antibody. Thus, the newly encoded antibody is expressed on the B-cell surface, and negative selection occurs whereby the cell undergoes apoptosis unless it receives rescue signals based on the antigen affinity of the new antibody. If the B-cell receives the appropriate rescue signals, its immunoglobulin gene undergoes further rounds of mutation and selection, ultimately developing into an antibody secreting plasma cell or memory B cell. During this response, the antigen-specific B cell matures and proliferates to produce high-affinity antibodies. The B cells proliferate, and their Ig genes are hypermutated. Only the most effective antibody-producing cells will survive to become memory B cells or plasma cell precursors. Once high-affinity IgA-committed B cells are selected, they then circulate through the bloodstream and lymphatic system to seed the lamina propria with plasma cell precursors, which synthesise dimeric IgA for secretion through the epithelium into the gut lumen [122]. The subsequent migration of IgA^{+} B cell plasmablasts from Peyer's patches and their homing to the intestinal lamina propria have been reported to be reduced in aged rodents [162, 172].

A decrease in high-affinity antibodies in the elderly could be due to a defect in selection or a defect in hypermutation [116, 119, 165]. The hypermutation rate is similar in young and old B cells, but aged cells have increased accumulations of mutations [97]. Furthermore, within Peyer's patches of elderly humans, the process of Ig gene selection within the germinal centre is also decreased [97]. In aged mice, the intestinal IgA repertoire diversity has been shown to be increased [173] and skewed to one that is more reflective of the systemic B-cell pool [130]. Some very old humans, however, display a dramatically reduced diversity in their peripheral blood B-cells, which is associated with frailty [98].

4.3 Immunosenescence of the Microbiota

The human microbiota consists of trillions of microorganisms that are resident, primarily along the gastrointestinal mucosa but also at other sites such as the lungs, vaginal tract and the skin. These populations consist of bacteria, fungi, parasites and archaea, which have evolved over millions of years. Numerous reports show an immense diversity in the human microbiota, depending on body localisation, health and disease status, as well as age [174, 175]. The terms 'microbiota' and 'microbiome' are often used interchangeably, with the former term referring to the specific number of commensal organisms inhabiting the gut and the latter term pertaining to the aggregate of unique genomes from all the microorganisms within the body. Gut microbiota includes 1000–1500 bacterial species in total; however, each individual possesses approximately 160 out this range of species, indicating that the composition of the microbiota is substantially different among individuals and is related to environmental changes and genetic inheritance [176, 177]. Environmental factors play a very important role in the diversity of the gut microbiome. Even mice with the same genotype housed in separate cages within the same facility show differences in microbiota compositions [178].

The symbiotic relationship between microbiota and the host is mutually beneficial. The host provides an important habitat and nutrients to the microbiota, while the gut microbiota supports the development of the metabolic system and the maturation of the intestinal immune system (Fig. 6) by providing beneficial nutrients, e.g. by the synthesis of vitamins [179] and short-chain fatty acids (SCFAs) [180, 181]. Consequently, the interaction between the microbiota and intestinal immune system is critical to maintain mucosal homeostasis.

The gut microbiota development starts at birth, showing instability and low diversity. Environmental factors affecting the diversity of microbiota include the mother's intrauterine microbiota and birth type [182]. During childhood, the microbiota remains unstable until the infant is 2–3 years old and is highly dependent on environmental factors such as introduction of solid food, place of residence, antibiotic exposure and genetic host factors [183–185]. Maturity of the microbiota is established after the age of 3 years, although the composition and abundance of individual microbiota appears to continue to change until adulthood and the personal and healthy core native microbiota remains relatively stable in adulthood [186–189]. A direct correlation between bacterial diversity and healthy status of microbiota has been recognised [190, 191]. Reflecting the age-related alterations in lifestyle and dietary intake, reduced and altered intestinal function, gut morphology and physiology, a senescent immune system and increasing comorbidities, the aging process deeply alters the proportion and composition of the different taxa, leading to; a loss in diversity and proportion of beneficial bacteria, change in the dominant species and increase in some inflammation-causing enteropathogens [187, 192–194]. Comorbidities associated with gut microbiota tend to become more frequent as the host gets older [195, 196], even though it remains unclear whether microbiota alterations are a cause or consequence of host aging.

Intestinal microbes play an essential role in the development and expansion of the gut mucosal and systemic immune function, as described in several studies in germ-free and gnotobiotic animals and treatment with specific microorganisms [197–199]. Constant interaction between the intestinal microbiota and mucosal immune system achieves intestinal homeostasis (Fig. 6). However, once the balance is broken, dysfunction of the intestinal immune system can trigger a variety of diseases. For example, the role of the gut microbiota dysbiosis in inflammatory bowel diseases and Clostridium difficile infection is well known [200–203]. There is also increasing evidence that demonstrates the influence of gut microbiota on far-reaching organs such as the central nervous system and its disorders. The microbiota-gut-brain-axis (MGBA), a bidirectional communication network between gut microbes and their host, is affected, and dysbiosis-induced dysfunction of the MGBA is seen with aging [204].

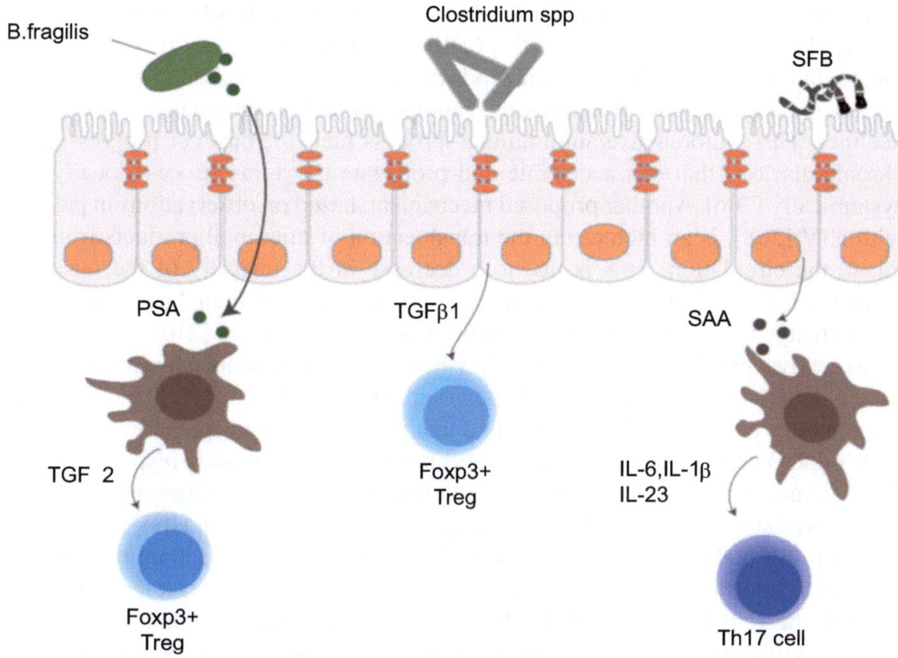

Fig. 6 The interaction of the microbiota with epithelial cells to maintain gut homeostasis. Commensal bacteria such as Clostridium spp. and *Bacteroides fragilis*, and its immunomodulatory molecule PSA (polysaccharide-A) promote the differentiation and expansion of Foxp3+ Treg cells in the gut through various mechanisms. *B. fragilis*, through PSA, can act directly on CD4 T cells by inducing production of TGF-β2, which drives Treg differentiation. In contrast, Clostridium spp. influence Treg differentiation via epithelial cell-derived TGF-β1 whereas microbially derived ATP and epithelium-adhering bacteria such as segmented filamentous bacteria (SFB) stimulate the induction of intestinal Th17 cells by binding to the intestinal epithelium causing the release of serum amyloid A (SAA) that stimulates intestinal antigen presenting cells to secrete the cytokines IL-23, IL-6 and IL-1β which assists in Th 17 activation and differentiation. Antigen-specific Th1 cell differentiation can be promoted by intracellular pathogens such as *Listeria monocytogenes* and *Toxoplasma gondii*. Microbiota antigens are sampled via (1) transepithelial dendrites of dendritic cells (DC), (2) transcytosis through microfold cells (M cell), or (3) goblet cell-associated antigen passages (GAP) and then induce T-cell differentiation in the mesenteric lymph nodes or de novo in the lamina propria. (Image reproduced from [135])

5 Inflammaging

Inflammaging is the long-term result of chronic physiological stimulation of the innate immune system, which can become damaged during ageing. Evidence of increased levels of tumour necrosis factor (TNF), IL-6 and other pro-inflammatory cytokines in the serum of older individuals compared to young, supports this hypothesis [28, 205]. The innate immune system, in particular the monocyte-macrophage network, is thought to be at the centre of inflammaging [206, 207]. Accumulating evidence indicates that the major source of inflammatory stimuli is represented by endogenous/self, misplaced or altered molecules, resulting from damaged or dead cells and organelles (cell debris), recognised by receptors of the innate immune system. While their production is physiological and increases with age, their disposal by the proteasome via autophagy and/or mitophagy progressively declines. This ‘autoreactive/autoimmune’ process fuels the onset or progression of chronic diseases that can accelerate and propagate the aging process locally and systemically [206]. Another proposed mechanism, based on observations in patients with HIV [208], is an increase in the translocation of microbial products from the gut to the circulation. This is due to a decrease in the integrity of the intestinal epithelial barrier [209], which makes it less efficient in containing bacterial growth, and a change in the composition of the gut microbiota with age [210].

Excessive baseline inflammation has been associated with poor responses to vaccination [211]. Investigations into gene signatures predictive of influenza vaccine responses in young and old adults found that pre-vaccination signatures associated with T and B cell function were positively correlated with antibody responses at day 28 post-vaccination, while monocyte- and inflammation-related genes were negatively correlated with antibody responses [212]. Another study on HBV (hepatitis B vaccination) on the elderly demonstrated that more pronounced inflammatory gene expression profiles at baseline predicted a poorer response to vaccination [213]. Overall, these findings support the theory that elevated baseline inflammation may have a significant role in the age-related hypo-responsiveness to vaccination and thus reducing background inflammation might be a promising strategy to enhance vaccine responses [214].

Senescent cells are a major contributor to the inflammaging process, therefore targeting these cells can reduce inflammaging and increase the opportunity to induce protasective immunity in the elderly. Mouse models have been developed where senescent cells can be specifically removed in vivo. These studies have demonstrated that the animals have increased life span, improved fitness and reduced fur loss [215, 216]. ABT263, a specific inhibitor for BCL2 and BCL-x (antiapoptopic pathways), was used in an aged mouse model and resulted in the rejuvenation of hematopoietic stem cells [217]. Another potential therapeutic area is to stimulate the activity of the individuals’ own immune system against senescent cells [214]. Several approaches are being developed, which are aimed at dampening or altering the pre-existing inflammation by using anti-inflammatories as novel adjuvants, such as cytokine blockers (e.g. anti IL-10 and anti-TGF-b) in vaccines [211] (discussed in more detail below).

6 Immunological Mechanisms of Vaccination

6.1 *Vaccine Efficacy and Effectiveness*

Immunogenicity, which is the stimulation of the body's immune system in response to an antigen, is critical for the success of any vaccine. While therapeutic vaccines are used for the treatment of an already existing medical condition or non-infectious disease such as cancer, prophylactic vaccines are used to prime the body's immune system to prevent disease, associated symptoms and proliferation of the disease [218]. Prophylactic vaccines contain small amounts of attenuated antigen, potent enough to trigger an immune response and stimulate the production of memory cells specific to that antigen. The aim is to expose the immune system to the antigen in a safe way so that upon subsequent encounters or exposure to a more virulent form, the body can mount a more rapid and effective immune response [105]. Therefore, vaccines tend to promote a strong adaptive immune response as this confers immunological memory. Substances called adjuvants are sometimes added to vaccines to boost the immune response by either targeting and triggering components of the innate or the adaptive immune system, promoting rapid cellular recruitment, boosting antibody titres and stimulating memory cell proliferation [219].

The most popular marker for quantifying vaccine efficacy is serological testing; however, studies show that this has its limitations. For example, while high levels of antibodies are usually a predictor of a strong immune response, low antibody levels do not necessarily indicate a poor immune response as T cell responses are also key indicators. Hence, more techniques are being used to determine vaccine efficacy, and cellular, innate and cytokine responses are also considered and assessed [220]. As it relates to measuring vaccine effectiveness, post-vaccination studies are critical to determine if the vaccine is impactfully reducing the disease burden in society. It is therefore important to understand the difference between vaccine efficacy and effectiveness. Vaccine efficacy is a measure of the extent of how well a vaccine can work to give positive results under ideal conditions such as those which are being tested during clinical trials [221]. Vaccine effectiveness, on the other hand, is determined by the extent to which the vaccine can provide positive results under real-life conditions and prevent disease and death and/or lower infection rates. This is evaluated after the vaccine becomes widely available for public use [222], and there are still many challenges and opportunities to protect the aging population (Fig. 7). Vaccine efficacy may be used to predict vaccine effectiveness, although there is no guarantee that the prediction will be accurate. However, the information on vaccine effectiveness can provide data for implementing public health policies for the future [223].

Studies that have evaluated vaccine efficacy and effectiveness have revealed that several factors could potentially affect how well vaccines work to protect individuals [117]. Some of these factors have already been considered, and attempts at developing vaccines tailored to the elderly have yielded more promising results than the one-size-fits-all traditional approach [224, 225]. Other than influenza, this has

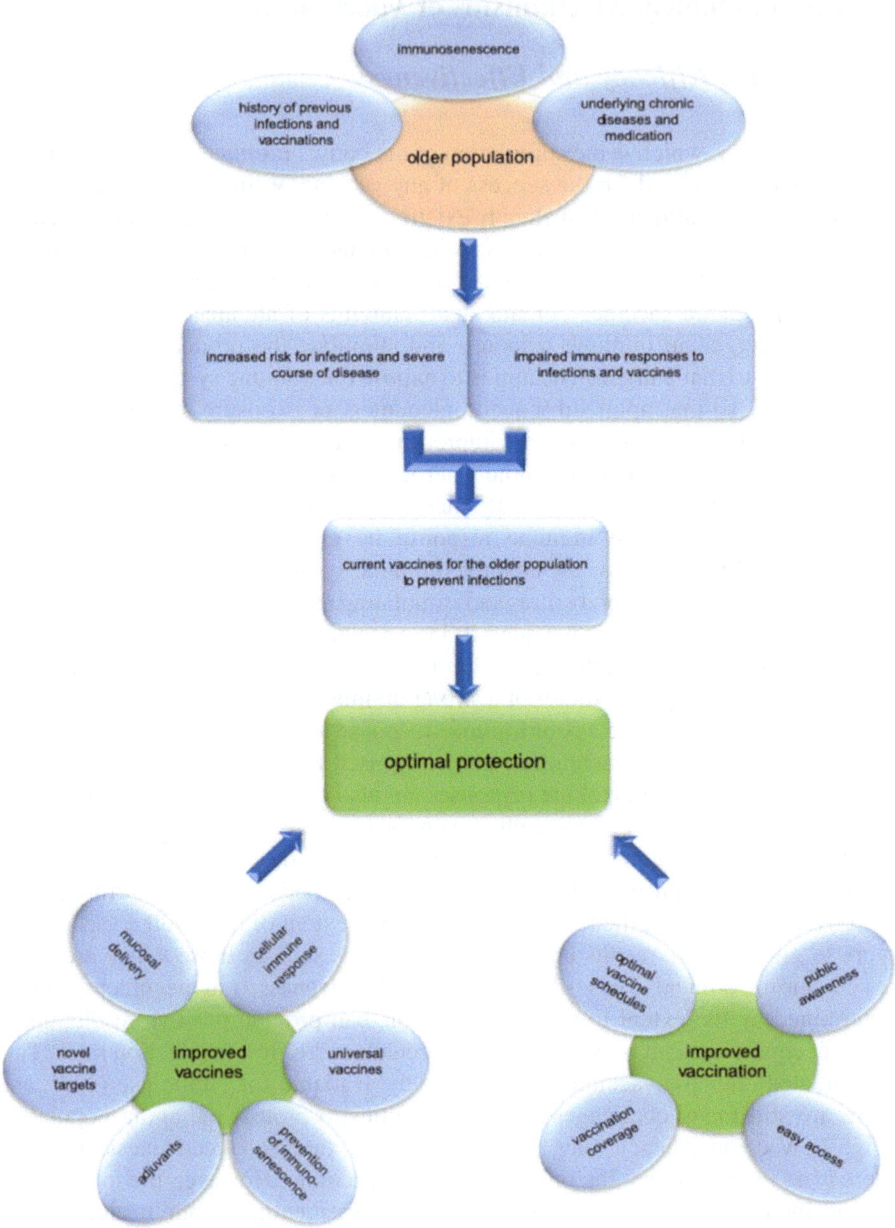

Fig. 7 Challenges and future prespectives on the development of effective vaccines for the elderly. (Image reproduced from [224])

been demonstrated in vaccines recommended for the elderly against herpes zoster and pneumonia [30].

6.2 The Immunisation Process

Stimulation of the immune response begins when the vaccine antigen particles are recognised by pattern recognition receptors including TLRs and nucleotide-binding oligomerisation domain-like receptors, also known as NOD-like receptors, of APCs such as dendritic cells. The APCs have now become activated because of this antigen-receptor interaction and are trafficked and drained into lymph nodes. If the vaccine contains adjuvants, these act to intensify the danger signal by several mechanisms including induction of a depot, cytokine and chemokine release, recruitment of immune cells and promoting the antigen-presenting activity of APCs. The antigen particles are then internalised, processed and presented by APCs via MHC molecules [226].

Naive T lymphocytes recognise and bind to the displayed molecules through their T cell receptors and become activated. Naive B cells also become activated by detecting and binding soluble antigens via their B cell receptor. When CD4+ (helper) T cells bind MHC class II molecules, this can trigger the T cell-dependent pathway of B lymphocyte proliferation, ultimately leading to plasma cell differentiation and production of various antibody isotypes as well as memory B cell proliferation. These specialised plasma cells are responsible for producing antibodies specific to the antigen and are what causes the antibody titre to increase in the bloodstream within 2 weeks of antigen exposure. In addition, some plasma cells migrate to the bone marrow and remain there and continue to produce antibodies for years [227].

Simultaneously, CD8+ (cytotoxic) T cells once activated trigger the proliferation of CD8+ effector and memory T cells. Added to this, neutrophils play an important role in T cell activation, and mast cells heighten the adaptive immune response through the release of pro-inflammatory cytokines and vasoactive mediators. Complement proteins through opsonisation tag antigen particles for recognition by APCs, which further amplifies the B cell effector response. The complement system also facilitates cross-talk between the cells of thc innate immune system and the adaptive immune system.

Despite differences in terms of intensity of immunogenicity triggered in alternative routes of vaccine administration, the mechanism previously outlined seems to be consistent [228–230].

It is evident that there are a number of cells and mediators involved in the functioning of this intricate system regardless of the route of administration of the vaccine. One could appreciate this complexity as some of the components have a synergistic effect on each other and also depend on the surrounding chemical environment to exert a certain phenotypic effect. It follows then that as an individual ages, changes to one or several of these components could limit the immune system's response to a single vaccine formulation.

7 Types of Vaccine

The type of vaccine also has a crucial part to play in the immunogenicity in terms of efficacy (Fig. 8) [231]. More importantly though, the safety of the type of vaccine employed for the elderly is paramount. Hence, vaccines formulated for this age group must strike a balance between these two essentials.

Live-attenuated vaccines can trigger both the cell-mediated and humoral responses and are the most immunogenic but have the least tolerability. Although others such as subunit vaccines are often better tolerated, they require the addition of adjuvants or the dosage increased as a compensatory measure [232]. However, while strong immunogenicity is the aim, safety is usually the drawback of live vaccines as they run the risk of reverting to virulent forms as demonstrated by the polio vaccine, which caused rare cases of vaccine-associated paralytic poliomyelitis and vaccine-derived poliovirus [233]. The implications of this could be detrimental given the vulnerable immune systems of the elderly [226]. Nonetheless, all vaccine types have their pros and cons in terms of safety and efficacy, but what is important is applying strategies to minimise the disadvantages and attain maximum efficacy and safety profiles [234].

8 Vaccines Tailored to the Older Population

A number of vaccines are available for use by individuals of all ages. Certain vaccines are mainly for older people, while others are recommended or are only recommended on a conditional basis or are available in a different formulation or at a different dosage regimen for older people [235]. However, not many are specifically designed for the elderly. There are 15 diseases that adults aged 65 years and above should be vaccinated against. The focus for mass vaccinations in many high-income countries for older adults is on those vaccines which have demonstrated the greatest potential for preventing morbidity and mortality: influenza, pneumococcal infection and herpes zoster. Table 4 gives a summary of the main vaccines administered and highlights those which are recommended and tailored for the elderly.

Prior to the onset of the COVID-19 pandemic in 2019, the world's population over the age of 65 was estimated at 703,000,000 and was projected to grow by almost 17% over the next 30 years [1]. However, studies show that increase in life expectancy is not directly proportional to the health span of the aging population [246]. In an effort to promote healthy aging, vaccination is one of the strategies recommended to address this global issue. Despite this, the potential of this strategy is limited by several factors. It has been noted that elderly individuals have been excluded from clinical trials or included in small numbers [247]. This indicates that this group is not adequately represented in such studies and thus are likely to contribute to reduced effectiveness upon approval. Notably, the aging population has contributed to a reduction in overall vaccine efficacy in some cases [18]. Given

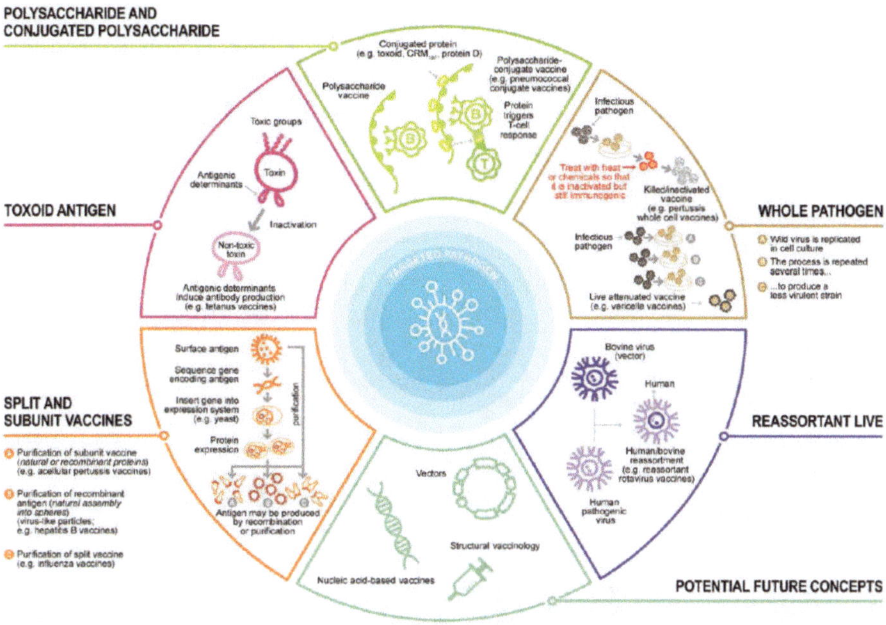

Fig. 8 The difference in immunogenicity and tolerability of different types of vaccines. Live attenuated vaccines contain pathogens that have been weakened to be less virulent than their wild-type counterparts. In their altered form, they can only mimic disease in a mild way, or not at all. Inactivated vaccines contain dead pathogens, so they are safer to use, particularly for the immunocompromised, however immunogenicity and duration of protection is less than live vaccines and they may require additional adjuvants and booster doses to improve protection. Subunit vaccines contain selected fragments of the pathogen e.g. proteins and polysaccharides from the surface of the bacteria, or parts of the virus that may form virus-like particles (VLP). Subunit vaccines are less immunogenic than live and inactivated vaccines because they contain less antigens. Examples of subunit vaccines include tetanus toxoid, inactivated split and subunit seasonal influenza. Some bacteria such as *Clostridium tetanii* cause disease by releasing pathogenic toxins. Vaccines against these diseases contain inactivated toxins (either heat or formaldehyde killed). The inactivated toxins are no longer pathogenic but retain the capability of producing neutralising antibodies. VLP vaccines contain specific viral proteins that can self-assemble to form structures similar to the original virus, but without the infectious element since the viral genome is lacking. However, the native antigenic protein is preserved. Examples of VLPs are the human papilloma virus (HPV) that protects against cervical cancer. (Image reproduced from [230])

the impact of immunosenescence, it is not surprising that the elderly do not respond to vaccines in the same way as young, healthy individuals with robust immune systems. Hence, there is merit in including this population in clinical studies and also studying them as a separate cohort. Furthermore, vaccines tailored to the elderly may be a cost-effective way of supporting and promoting a healthier aging population. Currently, as far as the literature shows, there are two vaccines on the market specifically tailored to the elderly, influenza and herpes zoster [30]. These vaccines, along with the Pneumococcal PCV13 vaccine and the Tdap vaccines, are recommended for the elderly, despite the fact that the latter two were not initially

Table 4 Common vaccines, their dosing recommendations and whether they are tailored to older adults

Vaccine	Type of vaccine	Recommended for the elderly (yes/no)	Tailored for the elderly (yes/no)	Regulatory dosage recommendations for the elderly	References
Tetanus and diphtheria and pertussis (Tdap or td)	Toxoid and protein	Yes	No	1 dose every 10 years with boosters	[236]
Influenza	Inactivated	Yes	Yes: Adjuvant (MF59) added [absent in the standard formulation given to young adults]	1 dose per year	[235, 237]
	Inactivated		Yes: High antigen dose [x4 standard dose]		
	Live attenuated	No	No		
Pneumococcal	(PPSV23) Polysaccharide	Yes	No	1 dose	[235]
Herpes zoster	Live attenuated	No	No	2 doses, 2 to 6 months apart	[235, 238]
	Recombinant subunit	Yes	Yes: Adjuvant (AS01B) added along with VZ virus glycoprotein E [absent in the standard formulation]	2 doses	

Varicella zoster (VZ)	Live attenuated	Yes	No	2 doses recommended if at additional risk	[239]
Hepatitis A	Inactivated	Yes	No	2–3 doses recommended if at additional risk	[235]
Hepatitis B	Protein	Yes	No		
Yellow fever	Live attenuated	Yes	No	Recommended for travel to areas where virus is endemic	[240]
Measles, mumps and rubella (MMR)	Live attenuated	People born before 1957 are presumed to be protected against MMR due to natural infection, and an MMR vaccine is not needed	No	People born before 1957 are presumed to be protected against MMR due to natural infection, and an MMR vaccine is not needed	[241]
Rabies	Inactivated	Yes	No	Recommended soon after exposure	[242]
Polio	Inactivated	Yes	No	Recommended if at additional risk and were not vaccinated as a child	[243]
Typhoid	Inactivated	Yes	No	Recommended if at additional risk	[244]
Haemophilus influenzae type b (Hib)	Polysaccharide conjugate inactivated	No (Hib vaccination is not recommended for most people 5 years or older	No	1–3 doses and only recommended depending on the indication	[245]
Human papillomavirus (HPV)	Recombinant	No	No	NA	[245]
COVID-19	mRNA, viral vectors, inactivated SARS-CoV-2 virus, peptide, protein, conjugate, DNA	Yes	No	Booster doses for older people	https://www.raps.org/news-and-articles/news-articles/2020/3/covid-19-vaccine-tracker, others

formulated specifically for this age group [225]. The specifically formulated vaccines will be discussed below.

8.1 *Influenza*

Substantial research surrounding ways to ease the disease burden of influenza in the aging population has led to a host of vaccine formulations using strategies aimed at boosting immune response in the elderly. To date, the Centers for Disease Control and Prevention (CDC) licensed the adjuvanted trivalent inactivated influenza vaccine (aIIV3), the high-dose trivalent influenza vaccine (HD-IIV3), the recombinant influenza trivalent vaccine (RIV3) and recently the quadrivalent versions of each of those [248].

The HD-IIV3 contains the inactivated split-virus influenza virus at an antigen dose of 60µg per strain, which is four times the dose found in the standard influenza vaccine containing 15µg of haemagglutinin (HA). It contains three antigenic parts, two of which give protection over the influenza A strains H1N1 and H3N2 and one for a B strain. This high dose was tested on individuals over the age of 65 and was successfully able to provide significantly higher antibody responses in which the A strains demonstrated superiority and the B strain non-inferiority. However, while pain was not severe in subjects given the HD-IIV3, they experienced more local reactions including severe swelling and erythema as compared to the subjects given the standard dose (SD) [21]. Notwithstanding its superiority over elderly subjects given the standard dose, the high-dose IIV was unable to mount an immune response comparable to young adults given the SD [249]. Moreover, there have been concerns about the level of protection provided by the HD-IIV3 since seropositive values obtained in the elderly do not necessarily translate to protection against influenza. Chen et al. investigated the cell-mediated responses measuring the interferon gamma (IFN-γ) produced following vaccination. It was found that the young adults consistently produced higher levels of (IFN-γ) [249]. Albeit indicative of a lower T cell response, the results show the importance of this tailored approach in at least being able to trigger higher antibody responses in the elderly population.

It is known that the intradermal (ID) route confers greater immunogenicity because of the plethora of immune cells including DCs embedded in the dermis. Accordingly, the potential of this route as a dose-sparing strategy was explored [250]. Intradermal influenza immunisation was shown to provide enhanced immune responses in old people while maintaining safety [251]. Notably, despite the administration of equal HA doses (15µg per strain), the formulation given intradermally provided greater seroprotection than intramuscular (IM) delivery [251]. The superiority of the intradermal influenza over the intramuscular route was further confirmed in a subsequent study carried out by Chan et al. [252]. Moreover, the results of a subsequent controlled Phase II clinical trial of the HD-IIV3, which compared the intradermal formulation containing the standard dose, high-dose intramuscular and standard-dose IM in older adults, revealed that both the high dose and the intradermal vaccines produced higher immunogenicity than the standard dose. Although the HD-IIV3 triggered a greater immune response than the intradermal formulation, the latter reached a satisfactory standard to be regarded as a suitable strategy for use in this age group [253]. However, the intradermal vaccine elicited more injection site reactions than the intramuscularly administered vaccines. The FDA has only approved intradermal influenza vaccines for use in adults under the age of 65 years [254].

The adjuvanted influenza vaccine was another formulation designed to boost immune responses in elderly people [255]. This formulation containing the MF59 adjuvant oil-in-water emulsion has proved to be more immunogenic than non-adjuvanted formulations and was associated with reduction in hospitalisation and susceptibility to pneumonia [256, 257]. Another adjuvanted influenza subunit vaccine, Aflunov®, containing MF59 and antigen strain from influenza A H5N1, was found to be safe and immunogenic in the elderly when 7.5μg of antigen was administered. Surprisingly, this vaccine was better tolerated in the elderly population than in other age groups [258]. Additionally, it was found to have improved tolerability over the seasonal vaccine formulation containing 15μg of antigen per strain, possibly because of its lower antigen content [258]. To further investigate the most effective strategy for enhancing immune responses in the elderly, a dose ranging study of the MF59 adjuvanted and non-adjuvanted formulations administered via the ID and the IM route was conducted [259]. The ID vaccine formulation was not adjuvanted and contained a lower dose of antigen. The findings demonstrated that both the adjuvanted formulations and the ID formulation significantly improved immunogenicity as compared to the regular seasonal non-adjuvanted formulation administered via the IM route but were associated with more injection site reactions. However, the adjuvanted formulation stimulated a stronger immune response than the ID [260].

These results show that tailored strategies can mount a more robust immune response in the elderly in a sufficiently safe manner. Although more research is warranted to understand the extent of protection, vaccine efficacy and immunogenicity, it is evident that the aging population can benefit from formulations that give them greater protection.

In order to provide better protection to the elderly from rapidly mutating and co-circulating influenza strains, quadrivalent vaccines containing two strains each of influenza A and B were developed [261]. The HD quadrivalent inactivated influenza (IIV4) showed non-inferiority to the IIV3 pre-existing vaccine and was also able to induce adequate immune responses against the additional B strain [262]. Due to safety and improved efficacy of this vaccine over the standard influenza vaccine, it was highly recommended to be given to adults over 65 years old [263].

Recently, a recombinant influenza vaccine became available, and the efficacy in older adults was evaluated. Although it showed greater immunogenicity than the standard dose of IIV4, it must be noted that it contained three times the HA (45μg per strain) than in the standard dose (15μg per strain) [264]. The evidence consistently demonstrates that higher antigen dose elicits a stronger immune response in the aging population. Therefore, this seems to be a reliable technique that could be used to design vaccines in the future for the elderly.

While influenza vaccines are effective at reducing the disease burden in this population, there is some difficulty in properly evaluating its effectiveness [265], due to differences in study design and the fact that the majority of studies are observational in nature and conclusions are often subject to bias [266]. Continuously updated systematic literature of influenza vaccine effectiveness from the Cochrane library emphasised this and reported that older people receiving this vaccine are at lower risk of morbidities, hospitalisations and death, with an approximate 3% reduction with low-to-moderate certainty evidence [267].

Other innovative strategic approaches have been proposed including incorporating cytokine immunomodulators into the inactivated influenza vaccine formulation to boost immune responses [268]. More recently, a novel adjuvant consisting of a single-stranded mRNA has been tested in aged mice and was able to induce T cell activation and IFN-y influx [269].

8.2 Herpes Zoster

Herpes zoster is caused by the same virus that causes varicella zoster (chicken pox) in children, and because it is a latent infection, it has been known to resurface when the immune system is compromised upon aging. To address this, a live attenuated vaccine called Zostavax was formulated with a high antigenic content calculated to be on average 14 times the potency of the varicella formulation for children [270]. It had an efficacy of almost 70% in persons aged 50–59 years, which dropped substantially to approximately 38% in persons over the age of 70 [271]. Additionally, immunocompromised persons were at greater risk of contracting herpes if the live component of the vaccine were to revert to a virulent form [271].

A new herpes zoster vaccine (Shingrix®) was recently approved for older adults. It exploits the recombinant subunit technology, and it contains 50μg of the varicella zoster virus glycoprotein E and the $AS01_B$ adjuvant. Vaccine efficacy was 97% for this age group, of which approximately 25% were over the age of 70. In a similar study focused on people over 50 years, the vaccine efficacy of Shingrix® was 91% in persons aged 70–79 and 89% in individuals aged 80 years or more [272]. The vaccine proved to be superior to the Zostavax® formulation for both safety and efficacy. Moreover, this vaccine formulation provides protection for almost a decade to subjects over 60 years of age as compared to approximately 7 years with Zostavax® [273]. Cellular and humoral immunity were evaluated 9 years after Shingrix® was first administered, and immune responses were still above pre-vaccination baseline levels [273]. Chlibek et al. noted that reduction in adaptive immune responses mainly occurred during the first year of vaccination, but this did not compromise the efficacy of Shingrix® [274]. The vaccine was found to be safe in immunocompromised individuals including in people living with cancer, HIV and kidney transplant [275].

Today Shingrix® is one of the top four vaccines recommended for the elderly by the CDC, while the use of Zostavax® was discontinued in the USA at the end of 2020 [276]. The use of an adjuvant system and a subunit antigen instead of a live attenuated virus boosted immune responses while simultaneously ensuring safe and long-term protection in healthy as well as immunosuppressed older people.

9 Tailored Approaches that can be Applied to Formulate Vaccines for the Elderly

There are several approaches that can be considered to develop more effective vaccine formulations to boost the elderly immune system. This can be achieved by (a) making the vaccine more potent (examples of which have been discussed above), (b) using adjuvants to enhance the immune response and (c) application of immunomodulators and other interventions to alter host immunity. Enhancing the immune response in older people requires careful safety considerations, since there is very little clinical data in real terms, to demonstrate which alteration may be potentially harmful and could trigger excessive unwanted or fatal immune reactions [277]. Figure 9 provides an illustrated example of some interventions that are currently being investigated for improving and enhancing the immune response to vaccine formulations [278], and below is a summary of various approaches and the concepts underpinning the rationale.

9.1 Addition of Adjuvants

9.1.1 Anti-Inflammatory Adjuvants

The use of anti-inflammatory adjuvants has been proposed with the rationale that aging people, especially those with underlying conditions or diseases such as cancer, often suffer from chronic inflammation, which limits their responses to vaccination [211]. Using adjuvants to silence signalling pathways promoting inflammation or broadly hindering inflammation may decrease the elevated baseline inflammation and help increase vaccine response in older people [211].

9.1.2 Cytokine Adjuvants

Attaching cytokines to whole inactivated influenza vaccines induces greater immunogenicity and offers a dose-sparing advantage [268]. Administration of vaccine formulations containing membrane-bound IL-12 and IL-4 immunomodulators in aged mice elicited noteworthy cellular and humoral responses. However, the IL-4 formulation was unable to provide adequate protection to aged mice with one dose,

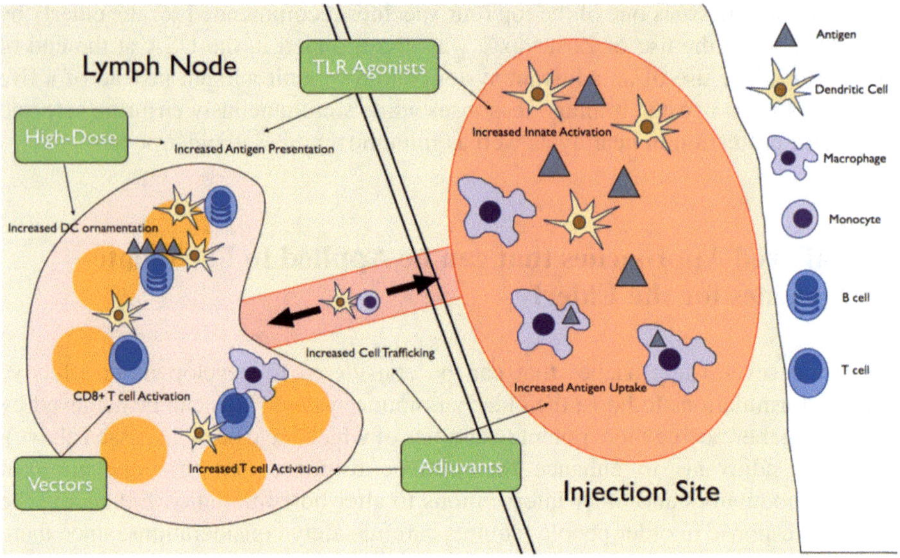

Fig. 9 Current strategies under investigation for the development of novel vaccines to enhance the aging immune system by stimulating various pathways. High-dose vaccines increase antigen presentation and ornamentation of dendritic cells therefore increasing B cell activation over time. Adjuvants can enhance 'immunocompetence' in the environment around the antigen to produce a more robust immune response to increase antigen uptake and presentation. TLR agonists can stimulate and enhance the innate immune system. Finally, viral-vector based vaccines can enhance immunogenicity by stimulating CD8+ T cells and eliminating pathogen-infected cells. (Image reproduced from [277])

while the IL-12 was more successful. This is likely due to IL-12 being able to induce IFN-γ, TNF-α, NK cell activity and Th1 responses. This approach not only offers a more immunogenic strategy than split influenza vaccines, but it also provides protection at a lower dose in contrast to high-dose influenza vaccines that are currently available. Additionally, this can potentially reduce the adverse events associated with the high-dose vaccines [268]. This approach may possibly yield positive results for other vaccines recommended for older people.

9.1.3 Single-Stranded RNA-Based Adjuvant

Preclinical studies on single-stranded RNA (ssRNA) showed potential as an adjuvant for an influenza vaccine (IIV) in older people. In aged mice, the presence of ssRNA in IIV formulation enhanced both cellular and innate immunity specifically IgG1 and IgG2a, but more of the latter, which favours Th1 response pathway. IgA responses were also triggered. The ssRNA was able to overcome the effects of immunosenescence at the genetic level by promoting the expression of CD25 and CD154 in pulmonary cytotoxic T lymphocytes. It also stimulated splenic T helper cells, which release IFN-γ and TNF-α. In the absence of ssRNA, lower neutralising antibody titres and reduced primed T lymphocyte and B lymphocyte responses were found. The extent of T-cell stimulation elicited by the ssRNA in aged mice was close to that found in young mice [269].

9.1.4 Cationic Lipid/DNA Complex Adjuvant

Inclusion of a cationic lipid/DNA complex adjuvant in a trivalent IIV influenza vaccine significantly improved vaccine efficacy in old rhesus macaques animal models [279]. A comparison of the changes in adaptive immune cells population between young and old primates showed a similar profile to that found in humans, indicating that rhesus macaques undergo immunosenescence as well. The viral load and viral replication in the respiratory tract were greatly reduced by the adjuvant, while in the absence of the adjuvant, viral replication was not impeded in the older animals [279].

9.1.5 Flagellin Adjuvant: The TLR5 Activator (TLR Agonist)

Preclinical studies in mice demonstrated that vaccines designed with flagellin, a TLR5 stimulator, linked to the HA peptide of the influenza A strain, protected aged mice, although the response to the high-dose vaccine was inferior to that in young mice. The study further showed that a low vaccine dose could not adequately protect aged mice from infection despite being able to do so in juvenile mice. The performance of this vaccine formulation was found to be limited by immunosenescence of cellular and humoral components often found in older people [280].

A more recent animal study involving flagellin demonstrated that direct interaction with caveolin-1, a structural protein, via TLR5, can potentially improve pneumococcal vaccine responses in older people [281]. TLR5 appeared to be regulated by caveolin-1, which was found in abundance in aged macrophages. In contrast to the previous study carried out by Leng et al., TLR5 activity was proven to be well preserved in the macrophages of aged mice [281]. When caveolin-1 was downregulated, the expression of TLR5 decreased considerably, and the opposite was also true. Hence, when the adjuvant-vaccine complex, inclusive of flagellin and the pneumococcal antigen in the form of a surface protein, activated the pathway involving TLR5 and caveolin-1, immune responses in aged mice increased to a greater extent. Additionally, the mucosal vaccine was able to confer both mucosal and systemic protection against *Streptococcus pneumoniae* [281].

9.2 *Other Approaches*

9.2.1 Trained Immunity

Trained immunity is another target for improving vaccine responses in the elderly. This concept suggests that innate cells such as monocytes and NK cells have immunological memory and are able to respond in an amplified manner to subsequent encounters with a pathogen. The mechanism involves epigenetic alterations, which occur when pattern-recognition receptors and damage-associated molecular patterns initially interact leading to metabolic modifications. These changes cause the upregulation of genes responsible for the activation of key cytokines involved in the immune response. The enzymes involved in this signalling pathway, such as histone acetyltransferase, cause the chromatin to open allowing proinflammatory cytokines to be produced rapidly upon subsequent antigen encounters. Interestingly, this type of memory is long term and lasts beyond the life span of the cells which activated it [282]. Adjuvants that induce trained immunity can potentially be used in vaccine formulations for the elderly. However, this approach may not be appropriate for persons suffering from chronic inflammation as it can augment the condition and is only recommended for persons who have increased susceptibility or are at high risk of contracting infections [282].

9.2.2 Virus-Like Particles Acquired from Plants

Virus-like particles (VLP) obtained from plant sources have shown some promise in animal studies on H1N1 influenza strain. Recent studies have explored the possibilities of VLPs as a promising alternative to the split virion form of the influenza H1N1 strain for better immune responses in the elderly [283, 284]. The study in mice involved comparing VLP to split virions given as a single dose intranasally or intramuscularly. The IM route was superior to the intranasal route; however, although all the juvenile subjects who received either formulation were adequately protected from infection, the VLP group had stronger adaptive immune responses [284]. Additionally, aged mice had balanced humoral and cellular responses; however, only 80% survived lethal challenge, while all the young mice survived. Interestingly, despite lower immune responses, aged mice given VLP intranasally also had a higher chance of surviving than the group receiving the split virus formulation [283]. More recent animal studies demonstrated that in addition to the regular intramuscular administration, booster doses of VLP given via intranasally, and concurrent vaccine administration by the two routes, offer superior protection in aged mice over the split virion formulations [284].

10 Vaccination of Older Adults and Public Health

In addition to designing vaccines specifically for older people to enhance vaccine effectiveness, vaccines must be taken up to be useful. Vaccine availability is unfortunately not fully translated into vaccine usage. Some of the specific challenges to increasing vaccination rates in older adults and for which public/population health approaches can be useful include the following:

- A common misconception is that people often mistakenly believe that vaccines are only for children.
- Regulatory/mandatory requirements for vaccinations. While infants and children have vaccine requirements for school and/or other activities, there are no such requirements for older adults, hence less incentive for vaccination.
- Access to vaccinations – reduced mobility for some older adults, which prevents them from leaving the house, leading to barriers accessing vaccinations.
- Vaccine hesitancy – while vaccine hesitancy can occur across all ages, it is particularly concerning in older adults who can develop severe complications from vaccine-preventable diseases.

For example, with the COVID-19 vaccination programme in the UK, the first Pfizer/BioNTech vaccine was administered to a 90-year-old woman. The vaccination programme prioritised those at the highest risk of serious illness or death as a result of the infection including older people, those considered clinically extremely vulnerable and health- and social care professionals before it was expanded to adults more than 6 months later and then young people, with children 5–12 years old only joining the programme in April 2022. The COVID-19 vaccination programme has thus significantly reduced morbidity and mortality in older adults, and more must be done to increase vaccine uptake for other vaccine-preventable diseases, such as influenza and herpes zoster. For example, World Immunization Week, celebrated in the last week of April, could be exploited by organisations promoting healthy aging to highlight vaccination and increase vaccine uptake in older adults.

11 Conclusion

The population of older individuals is growing at a faster rate than in previous years. Preventative medical approaches such as vaccination is an effective strategy to ensure that older people not only live longer but also healthier and are protected from vaccine-preventable diseases. However, vaccine efficacy is often lower in older people due to intrinsic, age-related immune system remodelling.

Immunosenescence and inflammaging, key hallmarks of aging, involve alterations in the innate and adaptive cell populations and reduced functionality. These alterations are most pronounced in individuals with underlying health conditions, who are therefore most vulnerable. The need for vaccines specifically designed for older people, considering the age-associated immune system dysregulation, has been clear for many years.

The need for more strategic vaccine designs for the aging population is being addressed to some extent. Several innovative potential tailoring strategies are in the pipeline, and research is ongoing to find ways to ameliorate the issue of unsatisfactory vaccine responses and improve vaccine efficacy and effectiveness in the elderly. To date, only two commercially available vaccines, namely, the influenza and herpes zoster vaccine, are tailored to the elderly. Several strategies have been applied to the influenza vaccine to improve vaccine responses in the elderly, such as increasing vaccine dose and use of different antigens and adjuvants. Several innovative technologies are currently in research and development and showing promise for improved vaccine efficacy in this vulnerable group. While the cost of tailoring vaccines is beyond the scope of this review, it is expected that more effective vaccines will lessen the age-related burden on healthcare facilities and ultimately will become the most effective economic option.

It is therefore imperative that research in this field increases significantly in order to combat immunosenscence and inflammaging to advance understanding of how to bolster the immune response to vaccines for the older population, ensuring improved clinical outcomes for the global aging population.

References

1. World Population Aging 2019: Highlights. 2019 New York: United Nations.
2. World Population Prospects 2019: Highlights. 2019: New York.
3. Gavazzi G, Krause KH. Ageing and infection. Lancet Infect Dis. 2002;2(11):659–66.
4. Nikolich-Žugich J. The twilight of immunity: emerging concepts in aging of the immune system. Nat Immunol. 2018;19(1):10–9.
5. Crooke SN, et al. Immunosenescence and human vaccine immune responses. Immun Ageing. 2019;16(1):25.
6. Grubeck-Loebenstein B, et al. Immunosenescence and vaccine failure in the elderly. Aging Clin Exp Res. 2009;21(3):201–9.
7. Aiello A, et al. Immunosenescence and its hallmarks: how to oppose aging strategically? A review of potential options for therapeutic intervention. Front Immunol. 2019;10:2247.
8. Oh SJ, Lee JK, Shin OS. Aging and the immune system: the impact of immunosenescence on viral infection, immunity and vaccine immunogenicity. Immune Netw. 2019;19(6):e37.
9. Disease Burden of Flu. Influenza (Flu) [Web Page] 2022 7/01/2022; Available from: https://www.cdc.gov/flu/about/burden/index.html.
10. Aspinall R, et al. Challenges for vaccination in the elderly. Immun Ageing. 2007;4(1):9.
11. Esme M, et al. Infections in the elderly critically-III patients. Front Med. 2019:6, 118.
12. Mallory RM, et al. A study to evaluate the immunogenicity and shedding of live attenuated influenza vaccine strains in children 24–<48 months of age. Vaccine. 2020;38(5):1001–8.
13. Di Pietrantonj C, et al. Vaccines for measles, mumps, rubella, and varicella in children. Cochrane Database Syst Rev. 2020;4:CD004407.
14. Sah P, et al. Optimizing the impact of low-efficacy influenza vaccines. Proc Natl Acad Sci U S A. 2018;115(20):5151–6.
15. Radke S, et al. Age-specific effectiveness following each dose of acellular pertussis vaccine among infants and children in New Zealand. Vaccine. 2017;35(1):177–83.

16. Michiels B, et al. A systematic review of the evidence on the effectiveness and risks of inactivated influenza vaccines in different target groups. Vaccine. 2011;29(49):9159–70.
17. Gabutti G, et al. Evaluation of efficacy and effectiveness of live attenuated zoster vaccine. J Prev Med Hyg. 2014;55(4):130–6.
18. Russell K, et al. Influenza vaccine effectiveness in older adults compared with younger adults over five seasons. Vaccine. 2018;36(10):1272–8.
19. Osterholm MT, et al. Efficacy and effectiveness of influenza vaccines: a systematic review and meta-analysis. Lancet Infect Dis. 2012;12(1):36–44.
20. FDA Approves High-Dose Flu Vaccine for Seniors. December 28, 2009.; Available from: https://www.medscape.com/viewarticle/714319.
21. Falsey AR, et al. Randomized, double-blind controlled phase 3 trial comparing the immunogenicity of high-dose and standard-dose influenza vaccine in adults 65 years of age and older. J Infect Dis. 2009;200(2):172–80.
22. Warrington R, et al. An introduction to immunology and immunopathology. Allergy, Asthma Clin Immunol. 2011;7(1):S1.
23. Marshall JS, et al. An introduction to immunology and immunopathology. Allergy, Asthma Clin Immunol. 2018;14(2):49.
24. Chaplin DD. Overview of the immune response. J Allergy Clin Immunol. 2010;125(2 Suppl 2):S3–S23.
25. Turvey SE, Broide DH. Innate immunity. J Allergy Clin Immunol. 2010;125(2 Suppl):S24–32.
26. Bonilla FA, Oettgen HC. Adaptive immunity. J Allergy Clin Immunol. 2010;125(2 Suppl):S33–40.
27. Pereira B, Xu X-N, Akbar AN. Targeting inflammation and Immunosenescence to improve vaccine responses in the elderly. Front Immunol. 2020;11:583019.
28. Franceschi C, et al. Inflammaging and anti-inflammaging: a systemic perspective on aging and longevity emerged from studies in humans. Mech Ageing Dev. 2007;128(1):92–105.
29. Giovannini S, et al. Interleukin-6, C-reactive protein, and tumor necrosis factor-alpha as predictors of mortality in frail, community-living elderly individuals. J Am Geriatr Soc. 2011;59(9):1679–85.
30. Ciabattini A, et al. Vaccination in the elderly: the challenge of immune changes with aging. Semin Immunol. 2018;40:83–94.
31. Thomas R, Wang W, Su D-M. Contributions of age-related thymic involution to Immunosenescence and Inflammaging. Immun Ageing. 2020;17(1):2.
32. Dowling JK, Mansell A. Toll-like receptors: the swiss army knife of immunity and vaccine development. Clin Transl Immunol. 2016;5(5):e85.
33. Metcalf TU, et al. Global analyses revealed age-related alterations in innate immune responses after stimulation of pathogen recognition receptors. Aging Cell. 2015;14(3):421–32.
34. Olivieri F, et al. Toll like receptor signaling in "inflammaging": microRNA as new players. Immun Ageing. 2013;10(1):11.
35. van Duin D, et al. Age-associated defect in human TLR-1/2 function. J Immunol. 2007;178(2):970.
36. Panda A, et al. Age-associated decrease in TLR function in primary human dendritic cells predicts influenza vaccine response. J Immunol. 2010;184(5):2518.
37. Montgomery RR, Shaw AC. Paradoxical changes in innate immunity in aging: recent progress and new directions. J Leukoc Biol. 2015;98(6):937–43.
38. De Maeyer RPH, Chambers ES. The impact of ageing on monocytes and macrophages. Immunol Lett. 2021;230:1–10.
39. Metcalf TU, et al. Human monocyte subsets are transcriptionally and functionally altered in aging in response to pattern recognition receptor agonists. J Immunol. 2017;199(4):1405–17.
40. Renshaw M, et al. Cutting edge: impaired toll-like receptor expression and function in aging. J Immunol. 2002;169(9):4697–701.
41. Shaw AC, et al. Dysregulation of human toll-like receptor function in aging. Ageing Res Rev. 2011;10(3):346–53.

42. Liu YJ. IPC: professional type 1 interferon-producing cells and plasmacytoid dendritic cell precursors. Annu Rev Immunol. 2005;23:275–306.
43. Panda A, et al. Human innate immunosenescence: causes and consequences for immunity in old age. Trends Immunol. 2009;30(7):325–33.
44. Solana R, et al. Innate immunosenescence: effect of aging on cells and receptors of the innate immune system in humans. Semin Immunol. 2012;24(5):331–41.
45. Nauseef WM, Borregaard N. Neutrophils at work. Nat Immunol. 2014;15(7):602–11.
46. Crooke SN, et al. Immunosenescence: a systems-level overview of immune cell biology and strategies for improving vaccine responses. Exp Gerontol. 2019;124:110632.
47. Tseng CW, et al. Innate immune dysfunctions in aged mice facilitate the systemic dissemination of methicillin-Resistant S. aureus. PLoS One. 2012;7(7):e41454.
48. Poli A, et al. CD56bright natural killer (NK) cells: an important NK cell subset. Immunology. 2009;126(4):458–65.
49. Hazeldine J, Lord JM. The impact of ageing on natural killer cell function and potential consequences for health in older adults. Ageing Res Rev. 2013;12(4):1069–78.
50. Vignola AM, et al. Effect of age and asthma duration upon elastase and alpha1-antitrypsin levels in adult asthmatics. Eur Respir J. 2003;22(5):795–801.
51. Mathur SK, et al. Age-related changes in eosinophil function in human subjects. Chest. 2008;133(2):412–9.
52. Desai A, Grolleau-Julius A, Yung R. Leukocyte function in the aging immune system. J Leukoc Biol. 2010;87(6):1001–9.
53. Busse PJ, et al. Effect of ageing on pulmonary inflammation, airway hyperresponsiveness and T and B cell responses in antigen-sensitized and -challenged mice. Clin Exp Allergy. 2007;37(9):1392–403.
54. Gurish MF. The expanding universe of the basophil. Blood. 2009;113(12):2616.
55. van Beek AA, et al. Aged mice display altered numbers and phenotype of basophils, and bone marrow-derived basophil activation, with a limited role for aging-associated microbiota. Immun Ageing. 2018;15:1–11.
56. Gri G, et al. Mast cell: an emerging partner in immune interaction. Front Immunol. 2012;3:120.
57. Pilkington SM, et al. Aged human skin accumulates mast cells with altered functionality that localize to macrophages and vasoactive intestinal peptide-positive nerve fibres. Br J Dermatol. 2019;180(4):849–58.
58. Kutukova NA, et al. Mast cells and aging. Adv Gerontol. 2017;7(1):68–75.
59. El-naseery NI, et al. Aging-associated immunosenescence via alterations in splenic immune cell populations in rat. Life Sci. 2020;241:117168.
60. Tsuboi I, et al. Age-related decline of mast cell regeneration in senescence-accelerated mice (SAMP1) after chemical myeloablation due to senescent stromal cell impairment. Exp Biol Med (Maywood). 2012;237(11):1289–97.
61. Karaca T, Yörük M, Uslu S. Age-related changes in the number of mast cells in the avian lymphoid organs. Anat Histol Embryol. 2006;35(6):375–9.
62. Pal S, Meininger CJ, Gashev AA. Aged lymphatic vessels and mast cells in perilymphatic tissues. Int J Mol Sci. 2017;18(5):965.
63. Krystel-Whittemore M, Dileepan KN, Wood JG. Mast cell: a multi-functional master cell. Front Immunol. 2015;6:620.
64. McGowen AL, et al. The mast cell activator compound 48/80 is safe and effective when used as an adjuvant for intradermal immunization with bacillus anthracis protective antigen. Vaccine. 2009;27(27):3544–52.
65. Fang Y, Xiang Z. Roles and relevance of mast cells in infection and vaccination. J Biomed Res. 2016;30(4):253–63.
66. Franzin R, et al. Inflammaging and complement system: a link between acute kidney injury and chronic graft damage. Front Immunol. 2020;11:734.

67. Fearon DT, Carroll MC. Regulation of B lymphocyte responses to foreign and self-antigens by the CD19/CD21 complex. Annu Rev Immunol. 2000;18:393–422.
68. Reid KB, Porter RR. The proteolytic activation systems of complement. Annu Rev Biochem. 1981;50:433–64.
69. Liszewski MK, Post TW, Atkinson JP. Membrane cofactor protein (MCP or CD46): newest member of the regulators of complement activation gene cluster. Annu Rev Immunol. 1991;9:431–55.
70. Barnum SR, Bubeck D, Schein TN. Soluble membrane attack complex: biochemistry and immunobiology. Front Immunol. 2020;11:2891.
71. Brown JS, et al. The classical pathway is the dominant complement pathway required for innate immunity to *Streptococcus pneumoniae* infection in mice. Proc Natl Acad Sci. 2002;99(26):16969–74.
72. Janeway CA Jr, Travers P, Walport M, Shlomchik MJ, et al. The complement system and innate immunity, in Immunobiology: the immune system in health and disease. New York: Garland Science; 2001.
73. Wallis R. Interactions between mannose-binding lectin and MASPs during complement activation by the lectin pathway. Immunobiology. 2007;212(4–5):289–99.
74. Sørensen R, Thiel S, Jensenius JC. Mannan-binding-lectin-associated serine proteases, characteristics and disease associations. Springer Semin Immunopathol. 2005;27(3):299–319.
75. Merle NS, et al. Complement system part I – molecular mechanisms of activation and regulation. Front Immunol. 2015;6:262.
76. Schramm EC, et al. Genetic variants in the complement system predisposing to age-related macular degeneration: a review. Mol Immunol. 2014;61(2):118–25.
77. Reichwald J, et al. Expression of complement system components during aging and amyloid deposition in APP transgenic mice. J Neuroinflammation. 2009;6:35.
78. McGeer EG, Klegeris A, McGeer PL. Inflammation, the complement system and the diseases of aging. Neurobiol Aging. 2005;26(Suppl 1):94–7.
79. Ambati J, Atkinson JP, Gelfand BD. Immunology of age-related macular degeneration. Nat Rev Immunol. 2013;13(6):438–51.
80. Anderson DH, et al. The pivotal role of the complement system in aging and age-related macular degeneration: hypothesis re-visited. Prog Retin Eye Res. 2010;29(2):95–112.
81. Nagaki K, et al. The effect of aging on complement activity (CH50) and complement protein levels. J Clin Lab Immunol. 1980;3(1):45–50.
82. Kalula S, Ross K. Immunosenescence – inevitable or preventable? Curr Allerg Clin Immunol. 2008:21, 126.
83. Gaya da Costa M, et al. Age and sex-associated changes of complement activity and complement levels in a healthy Caucasian population. Front Immunol. 2018:9.
84. Cole DS, Morgan BP. Beyond lysis: how complement influences cell fate. Clin Sci (Lond). 2003;104(5):455–66.
85. Sambhara S, McElhaney JE. Immunosenescence and influenza vaccine efficacy. Curr Top Microbiol Immunol. 2009;333:413–29.
86. Hazeldine J, Lord JM. Innate immunesenescence: underlying mechanisms and clinical relevance. Biogerontology. 2015;16(2):187–201.
87. Allen JC, et al. Understanding immunosenescence and its impact on vaccination of older adults. Vaccine. 2020;38(52):8264–72.
88. Xu W, Larbi A. Markers of T cell senescence in humans. Int J Mol Sci. 2017;18(8):1742.
89. Moro-García M, Alonso-Arias R, Lopez-Larrea C. When aging reaches CD4+ T-cells: phenotypic and functional changes. Front Immunol. 2013;4:107.
90. Jagger A, et al. Regulatory T cells and the immune aging process: a mini-review. Gerontology. 2014;60(2):130–7.
91. Yoshida K, et al. Aging-related changes in human T-cell repertoire over 20years delineated by deep sequencing of peripheral T-cell receptors. Exp Gerontol. 2017;96:29–37.

92. Rossi MI, et al. B lymphopoiesis is active throughout human life, but there are developmental age-related changes. Blood. 2003;101(2):576–84.
93. Frasca D, et al. Aging down-regulates the transcription factor E2A, activation-induced cytidine deaminase, and Ig class switch in human B cells. J Immunol. 2008;180(8):5283–90.
94. Zharhary D. Age-related changes in the capability of the bone marrow to generate B cells. J Immunol. 1988;141(6):1863–9.
95. Scholz JL, et al. A comparative review of aging and B cell function in mice and humans. Curr Opin Immunol. 2013;25(4):504–10.
96. Pritz T, Weinberger B, Grubeck-Loebenstein B. The aging bone marrow and its impact on immune responses in old age. Immunol Lett. 2014;162(1, Part B):310–5.
97. Banerjee M, et al. Age- and tissue-specific differences in human germinal center B cell selection revealed by analysis of IgVH gene hypermutation and lineage trees. Eur J Immunol. 2002;32(7):1947–57.
98. Gibson KL, et al. B-cell diversity decreases in old age and is correlated with poor health status. Aging Cell. 2009;8(1):18–25.
99. Goronzy JJ, Weyand CM. Understanding immunosenescence to improve responses to vaccines. Nat Immunol. 2013;14(5):428–36.
100. Wu Y-C, Kipling D, Dunn-Walters D. Age-related changes in human peripheral blood IGH repertoire following vaccination. Front Immunol. 2012;3:193.
101. Hartsock RJ, Smith EB, Petty CS. Normal variations with aging of the amount of hematopoietic tissue in bone marrow from the anterior iliac crest. A study made from 177 cases of sudden death examined by necropsy. Am J Clin Pathol. 1965;43:326–31.
102. Geiger H, de Haan G, Florian MC. The ageing haematopoietic stem cell compartment. Nat Rev Immunol. 2013;13(5):376–89.
103. Linton PJ, Dorshkind K. Age-related changes in lymphocyte development and function. Nat Immunol. 2004;5(2):133–9.
104. Herndler-Brandstetter D, et al. The impact of aging on memory T cell phenotype and function in the human bone marrow. J Leukoc Biol. 2012;91(2):197–205.
105. Thompson HL, et al. Functional and homeostatic impact of age-related changes in lymph node stroma. Front Immunol. 2017;8:706.
106. Luscieti P, et al. Human lymph node morphology as a function of age and site. J Clin Pathol. 1980;33(5):454–61.
107. Becklund BR, et al. The aged lymphoid tissue environment fails to support naïve T cell homeostasis. Sci Rep. 2016;6:30842.
108. Maiborodin IV, et al. Age-related changes in the structure of lymphoid organs: a review of the literature. Adv Gerontol. 2016;6(4):282–90.
109. Richner JM, et al. Age-dependent cell trafficking defects in draining lymph nodes impair adaptive immunity and control of West Nile virus infection. PLoS Pathog. 2015;11(7):e1005027.
110. Bajénoff M, et al. Stromal cell networks regulate lymphocyte entry, migration, and territoriality in lymph nodes. Immunity. 2006;25(6):989–1001.
111. Aw D, et al. Disorganization of the splenic microanatomy in ageing mice. Immunology. 2016;148(1):92–101.
112. Uhrlaub JL, et al. Dysregulated TGF-β production underlies the age-related vulnerability to chikungunya virus. PLoS Pathog. 2016;12(10):e1005891.
113. Shi N, et al. Interaction between the gut microbiome and mucosal immune system. Mil Med Res. 2017;4:14.
114. Breedveld A, van Egmond M. IgA and FcαRI: pathological roles and therapeutic opportunities. Front Immunol. 2019;10:553.
115. Sato S, Kiyono H, Fujihashi K. Mucosal Immunosenescence in the gastrointestinal tract: a mini-review. Gerontology. 2015;61(4):336–42.
116. Koga T, et al. Evidence for early aging in the mucosal immune system. J Immunol. 2000;165(9):5352–9.

117. Lynn DJ, et al. Modulation of immune responses to vaccination by the microbiota: implications and potential mechanisms. Nat Rev Immunol. 2022;22(1):33–46.
118. Fujihashi K, McGhee JR. Mucosal immunity and tolerance in the elderly. Mech Ageing Dev. 2004;125(12):889–98.
119. Mabbott NA, et al. Aging and the mucosal immune system in the intestine. Biogerontology. 2015;16(2):133–45.
120. Fagiolo U, et al. Humoral and cellular immune response to influenza virus vaccination in aged humans. Aging (Milano). 1993;5(6):451–8.
121. Fujihashi K, Sato S, Kiyono H. Mucosal adjuvants for vaccines to control upper respiratory infections in the elderly. Exp Gerontol. 2014;54:21–6.
122. Macpherson AJ, et al. The immune geography of IgA induction and function. Mucosal Immunol. 2008;1(1):11–22.
123. Turula H, Wobus CE. The role of the polymeric immunoglobulin receptor and secretory immunoglobulins during mucosal infection and immunity. Viruses. 2018;10(5):237.
124. Macpherson AJ, et al. IgA responses in the intestinal mucosa against pathogenic and non-pathogenic microorganisms. Microbes Infect. 2001;3(12):1021–35.
125. Neutra MR, Mantis NJ, Kraehenbuhl J-P. Collaboration of epithelial cells with organized mucosal lymphoid tissues. Nat Immunol. 2001;2(11):1004–9.
126. Neutra MR, Frey A, Kraehenbuhl JP. Epithelial M cells: gateways for mucosal infection and immunization. Cell. 1996;86(3):345–8.
127. Kobayashi A, et al. The functional maturation of M cells is dramatically reduced in the Peyer's patches of aged mice. Mucosal Immunol. 2013;6(5):1027–37.
128. Westphal S, et al. Resistance of chemokine receptor 6-deficient mice to Yersinia enterocolitica infection: evidence of defective M-cell formation in vivo. Am J Pathol. 2008;172(3):671–80.
129. Ebisawa M, et al. CCR6hiCD11c(int) B cells promote M-cell differentiation in Peyer's patch. Int Immunol. 2011;23(4):261–9.
130. McDonald KG, et al. Aging impacts isolated lymphoid follicle development and function. Immun Ageing. 2011;8(1):1.
131. Misumi S, et al. Targeted delivery of immunogen to primate M cells with tetragalloyl lysine dendrimer. J Immunol. 2009;182(10):6061–70.
132. Nochi T, et al. A novel M cell-specific carbohydrate-targeted mucosal vaccine effectively induces antigen-specific immune responses. J Exp Med. 2007;204(12):2789–96.
133. McDole JR, et al. Goblet cells deliver luminal antigen to CD103+ dendritic cells in the small intestine. Nature. 2012;483(7389):345–9.
134. Dvorak AM, Dickersin GR. Crohn's disease: transmission electron microscopic studies. I. Barrier function. Possible changes related to alterations of cell coat, mucous coat, epithelial cells, and Paneth cells. Hum Pathol. 1980;11(5 Suppl):561–71.
135. Parrish AR. The impact of aging on epithelial barriers. Tissue Barriers. 2017;5(4):e1343172.
136. Ren WY, et al. Age-related changes in small intestinal mucosa epithelium architecture and epithelial tight junction in rat models. Aging Clin Exp Res. 2014;26(2):183–91.
137. Tran L, Greenwood-Van Meerveld B. Age-associated remodeling of the intestinal epithelial barrier. J Gerontol Series A. 2013;68(9):1045–56.
138. Franceschi C, et al. Inflamm-aging. An evolutionary perspective on immunosenescence. Ann N Y Acad Sci. 2000;908:244–54.
139. Crapser J, et al. Ischemic stroke induces gut permeability and enhances bacterial translocation leading to sepsis in aged mice. Aging (Albany NY). 2016;8(5):1049–63.
140. Dock J, et al. Distinct aging profiles of CD8+ T cells in blood versus gastrointestinal mucosal compartments. PLoS One. 2017;12(8):e0182498.
141. Smith PM, Garrett WS. The gut microbiota and mucosal T cells. Front Microbiol. 2011;2:111.
142. Cheroutre H. Starting at the beginning: new perspectives on the biology of mucosal T cells. Annu Rev Immunol. 2004;22(1):217–46.

143. Ferguson A, Parrott DM. The effect of antigen deprivation on thymus-dependent and thymus-independent lymphocytes in the small intestine of the mouse. Clin Exp Immunol. 1972;12(4):477–88.
144. Rocha B, Vassalli P, Guy-Grand D. The V beta repertoire of mouse gut homodimeric alpha CD8+ intraepithelial T cell receptor alpha/beta + lymphocytes reveals a major extrathymic pathway of T cell differentiation. J Exp Med. 1991;173(2):483–6.
145. Hayday A, et al. Intraepithelial lymphocytes: exploring the third way in immunology. Nat Immunol. 2001;2(11):997–1003.
146. Lefrançois L, Puddington L. Extrathymic intestinal T-cell development: virtual reality? Immunol Today. 1995;16(1):16–21.
147. Helgeland L, et al. Microbial colonization influences composition and T-cell receptor V beta repertoire of intraepithelial lymphocytes in rat intestine. Immunology. 1996;89(4):494–501.
148. Helgeland L, et al. Regional phenotypic specialization of intraepithelial lymphocytes in the rat intestine does not depend on microbial colonization. Scand J Immunol. 1997;46(4):349–57.
149. Takeuchi M, et al. Age-related changes of T cell subsets in intestinal intraepithelial lymphocytes of mice. Eur J Immunol. 1993;23(6):1409–11.
150. Guy-Grand D, et al. Two gut intraepithelial CD8+ lymphocyte populations with different T cell receptors: a role for the gut epithelium in T cell differentiation. J Exp Med. 1991;173(2):471–81.
151. Umesaki Y, et al. Expansion of alpha beta T-cell receptor-bearing intestinal intraepithelial lymphocytes after microbial colonization in germ-free mice and its independence from thymus. Immunology. 1993;79(1):32–7.
152. Hagiwara Y, et al. Protective mucosal immunity in aging is associated with functional $CD4^+$ T cells in nasopharyngeal-associated Lymphoreticular tissue. J Immunol. 2003;170(4):1754–62.
153. Ishimoto Y, et al. Age-dependent variation in the proportion and number of intestinal lymphocyte subsets, especially natural killer T cells, double-positive CD4+ CD8+ cells and B220+ T cells, in mice. Immunology. 2004;113(3):371–7.
154. Hayashi O, et al. Flow cytometric analysis of age-related changes in intestine intraepithelial lymphocyte subsets and their functional preservation after feeding mice on spirulina. J Med Food. 2009;12(5):982–9.
155. Santiago AF, et al. Aging correlates with reduction in regulatory-type cytokines and T cells in the gut mucosa. Immunobiology. 2011;216(10):1085–93.
156. Arranz E, et al. Immunosenescence and mucosal immunity: significant effects of old age on secretory IgA concentrations and intraepithelial lymphocyte counts. Gut. 1992;33(7):882–6.
157. Beharka AA, et al. Effect of age on the gastrointestinal-associated mucosal immune response of humans. J Gerontol Series A. 2001;56(5):B218–23.
158. Novak J, et al. The decrease in number and change in phenotype of mucosal-associated invariant T cells in the elderly and differences in men and women of reproductive age. Scand J Immunol. 2014;80(4):271–5.
159. Godfrey DI, et al. The biology and functional importance of MAIT cells. Nat Immunol. 2019;20(9):1110–28.
160. Lee OJ, et al. Circulating mucosal-associated invariant T cell levels and their cytokine levels in healthy adults. Exp Gerontol. 2014;49:47–54.
161. Penn ND, et al. Ageing and duodenal mucosal immunity. Age Ageing. 1991;20(1):33–6.
162. Thoreux K, Owen RL, Schmucker DL. Intestinal lymphocyte number, migration and antibody secretion in young and old rats. Immunology. 2000;101(1):161–7.
163. Schmucker DL. Intestinal mucosal immunosenescence in rats. Exp Gerontol. 2002;37(203):197.
164. Taylor LD, Daniels CK, Schmucker DL. Ageing compromises gastrointestinal mucosal immune response in the rhesus monkey. Immunology. 1992;75(4):614–8.
165. Dunn-Walters DK, Banerjee M, Mehr R. Effects of age on antibody affinity maturation. Biochem Soc Trans. 2003;31(2):447–8.

166. Ademokun A, et al. Vaccination-induced changes in human B-cell repertoire and pneumococcal IgM and IgA antibody at different ages. Aging Cell. 2011;10(6):922–30.
167. Nicoletti C, Yang X, Cerny J. Repertoire diversity of antibody response to bacterial antigens in aged mice. III. Phosphorylcholine antibody from young and aged mice differ in structure and protective activity against infection with Streptococcus pneumoniae. J Immunol. 1993;150(2):543–9.
168. Yang X, Stedra J, Cerny J. Relative contribution of T and B cells to hypermutation and selection of the antibody repertoire in germinal centers of aged mice. J Exp Med. 1996;183(3):959–70.
169. Shikina T, et al. IgA class switch occurs in the organized nasopharynx- and gut-associated lymphoid tissue, but not in the diffuse lamina propria of airways and gut. J Immunol. 2004;172(10):6259–64.
170. Barone F, et al. IgA-producing plasma cells originate from germinal centers that are induced by B-cell receptor engagement in humans. Gastroenterology. 2011;140(3):947–56.
171. Knoop K, Newberry R. Isolated lymphoid follicles are dynamic reservoirs for the induction of intestinal IgA. Front Immunol. 2012;3:84.
172. Schmucker DL, et al. Basis for the age-related decline in intestinal mucosal immunity. Clin Dev Immunol. 2003;10(2–4):167–72.
173. Lindner C, et al. Age, microbiota, and T cells shape diverse individual IgA repertoires in the intestine. J Exp Med. 2012;209(2):365–77.
174. Fierer N, et al. The influence of sex, handedness, and washing on the diversity of hand surface bacteria. Proc Natl Acad Sci. 2008;105(46):17994–9.
175. Turnbaugh PJ, et al. A core gut microbiome in obese and lean twins. Nature. 2009;457(7228):480–4.
176. Eckburg PB, et al. Diversity of the human intestinal microbial flora. Science. 2005;308(5728):1635–8.
177. Lee YK, Mazmanian SK. Has the microbiota played a critical role in the evolution of the adaptive immune system? Science. 2010;330(6012):1768–73.
178. Hasegawa M, Inohara N. Regulation of the gut microbiota by the mucosal immune system in mice. Int Immunol. 2014;26(9):481–7.
179. Kau AL, et al. Human nutrition, the gut microbiome and the immune system. Nature. 2011;474(7351):327–36.
180. McDermott AJ, Huffnagle GB. The microbiome and regulation of mucosal immunity. Immunology. 2014;142(1):24–31.
181. Topping DL, Clifton PM. Short-chain fatty acids and human colonic function: roles of resistant starch and nonstarch polysaccharides. Physiol Rev. 2001;81(3):1031–64.
182. Gosalbes MJ, et al. Meconium microbiota types dominated by lactic acid or enteric bacteria are differentially associated with maternal eczema and respiratory problems in infants. Clin Exp Allergy. 2013;43(2):198–211.
183. Koenig JE, et al. Succession of microbial consortia in the developing infant gut microbiome. Proc Natl Acad Sci. 2011;108(Supplement 1):4578–85.
184. Jermy A. Host genes shape the gut microbiota. Nat Rev Microbiol. 2010;8(12):838–9.
185. Palmer C, et al. Development of the human infant intestinal microbiota. PLoS Biol. 2007;5(7):e177.
186. Yatsunenko T, et al. Human gut microbiome viewed across age and geography. Nature. 2012;486(7402):222–7.
187. Salazar N, et al. The human intestinal microbiome at extreme ages of life. Dietary intervention as a way to counteract alterations. Front Genet. 2014:5, 406.
188. Ragonnaud E, Biragyn A. Gut microbiota as the key controllers of "healthy" aging of elderly people. Immun Ageing. 2021;18(1):11.
189. Rinninella E, et al. What is the healthy gut microbiota composition? A changing ecosystem across age, environment, diet, and diseases. Microorganisms. 2019;7(1):14.
190. Lloyd-Price J, Abu-Ali G, Huttenhower C. The healthy human microbiome. Genome Med. 2016;8(1):51.

191. Kong F, et al. Gut microbiota signatures of longevity. Curr Biol. 2016;26(18):R832–3.
192. Claesson MJ, et al. Composition, variability, and temporal stability of the intestinal microbiota of the elderly. Proc Natl Acad Sci. 2011;108(Supplement 1):4586–91.
193. Nagpal R, et al. Ontogenesis of the gut microbiota composition in healthy, full-term, vaginally born and breast-fed infants over the first 3 years of life: a quantitative bird's-eye view. Front Microbiol. 2017;8:1388.
194. Odamaki T, et al. Age-related changes in gut microbiota composition from newborn to centenarian: a cross-sectional study. BMC Microbiol. 2016;16(1):90.
195. Han B, et al. Microbial genetic composition tunes host longevity. Cell. 2017;169(7):1249–1262.e13.
196. Bartosch S, et al. Characterization of bacterial communities in feces from healthy elderly volunteers and hospitalized elderly patients by using real-time PCR and effects of antibiotic treatment on the fecal microbiota. Appl Environ Microbiol. 2004;70(6):3575–81.
197. Belkaid Y, Timothy W. Hand, role of the microbiota in immunity and inflammation. Cell. 2014;157(1):121–41.
198. Rakoff-Nahoum S, et al. Recognition of commensal microflora by toll-like receptors is required for intestinal homeostasis. Cell. 2004;118(2):229–41.
199. Wu HJ, Wu E. The role of gut microbiota in immune homeostasis and autoimmunity. Gut Microbes. 2012;3(1):4–14.
200. Machiels K, et al. A decrease of the butyrate-producing species Roseburia hominis and Faecalibacterium prausnitzii defines dysbiosis in patients with ulcerative colitis. Gut. 2014;63(8):1275–83.
201. Zuo T, Ng SC. The gut microbiota in the pathogenesis and therapeutics of inflammatory bowel disease. Front Microbiol. 2018;9:2247.
202. Theriot CM, et al. Antibiotic-induced shifts in the mouse gut microbiome and metabolome increase susceptibility to clostridium difficile infection. Nat Commun. 2014;5(1):3114.
203. Kho ZY, Lal SK. The human gut microbiome – a potential controller of wellness and disease. Front Microbiol. 2018;9:1835.
204. Honarpisheh P, Bryan RM, McCullough LD. Aging microbiota-gut-brain axis in stroke risk and outcome. Circ Res. 2022;130(8):1112–44.
205. Puzianowska-Kuźnicka M, et al. Interleukin-6 and C-reactive protein, successful aging, and mortality: the PolSenior study. Immun Ageing. 2016;13(1):21.
206. Franceschi C, et al. Inflammaging and 'Garb-aging'. Trends Endocrinol Metabol. 2017;28(3):199–212.
207. Prattichizzo F, et al. Senescence associated macrophages and "macroph-aging": are they pieces of the same puzzle? Aging (Albany NY). 2016;8(12):3159–60.
208. Brenchley JM, et al. Microbial translocation is a cause of systemic immune activation in chronic HIV infection. Nat Med. 2006;12(12):1365–71.
209. Meier J, Sturm A. The intestinal epithelial barrier: does it become impaired with age? Dig Dis. 2009;27(3):240–5.
210. Heintz C, Mair W. You are what you host: microbiome modulation of the aging process. Cell. 2014;156(3):408–11.
211. Alter G, Sekaly RP. Beyond adjuvants: antagonizing inflammation to enhance vaccine immunity. Vaccine. 2015;33(Suppl 2):B55–9.
212. Fourati S, et al. Pre-vaccination inflammation and B-cell signalling predict age-related hyporesponse to hepatitis B vaccination. Nat Commun. 2016;7(1):10369.
213. Bartholomeus E, et al. Transcriptome profiling in blood before and after hepatitis B vaccination shows significant differences in gene expression between responders and non-responders. Vaccine. 2018;36(42):6282–9.
214. Chambers ES, Akbar AN. Can blocking inflammation enhance immunity during aging? J Allergy Clin Immunol. 2020;145(5):1323–31.
215. Baar MP, et al. Targeted apoptosis of senescent cells restores tissue homeostasis in response to Chemotoxicity and aging. Cell. 2017;169(1):132–147.e16.

216. Baker DJ, et al. Naturally occurring p16Ink4a-positive cells shorten healthy lifespan. Nature. 2016;530(7589):184–9.
217. Chang J, et al. Clearance of senescent cells by ABT263 rejuvenates aged hematopoietic stem cells in mice. Nat Med. 2016;22(1):78–83.
218. Schijns V, et al. Modulation of immune responses using adjuvants to facilitate therapeutic vaccination. Immunol Rev. 2020;296(1):169–90.
219. Coffman RL, Sher A, Seder RA. Vaccine adjuvants: putting innate immunity to work. Immunity. 2010;33(4):492–503.
220. Zimmermann P, Curtis N. Factors that influence the immune response to vaccination. Clin Microbiol Rev. 2019;32(2):e00084.
221. Gavi.org. What is the difference between efficacy and effectiveness? [Online] 2020.; Available from: https://www.gavi.org/vaccineswork/what-difference-between-efficacy-and-effectiveness.
222. Weinberg GA, Szilagyi PG. Vaccine epidemiology: efficacy, effectiveness, and the translational research roadmap. J Infect Dis. 2010;201(11):1607–10.
223. Fedson DS. Measuring protection: efficacy versus effectiveness. Dev Biol Stand. 1998;95:195–201.
224. Poland GA, Ovsyannikova IG, Kennedy RB. Personalized vaccinology: a review. Vaccine. 2018;36(36):5350–7.
225. Wagner A, Weinberger B. Vaccines to prevent infectious diseases in the older population: immunological challenges and future perspectives. Front Immunol. 2020:11, 117.
226. Pollard AJ, Bijker EM. A guide to vaccinology: from basic principles to new developments. Nat Rev Immunol. 2021;21(2):83–100.
227. Clem AS. Fundamentals of vaccine immunology. J Glob Infect Dis. 2011;3(1):73–8.
228. Aoshi T. Modes of action for mucosal vaccine adjuvants. Viral Immunol. 2017;30(6):463–70.
229. Gonnet J, et al. Mechanisms of innate events during skin reaction following intradermal injection of seasonal influenza vaccine. J Proteome. 2020;216:103670.
230. Migliore A, et al. Intradermal vaccination: a potential tool in the battle against the COVID-19 pandemic? Risk Manag Healthc Policy. 2021;14:2079–87.
231. Vetter V, et al. Understanding modern-day vaccines: what you need to know. Ann Med. 2018;50(2):110–20.
232. Toolbox.eupati.eu. Types of Vaccines. 2021. Available from: https://toolbox.eupati.eu/resources/types-of-vaccines/.
233. Kew OM, et al. Vaccine-derived polioviruses and the endgame strategy for global polio eradication. Annu Rev Microbiol. 2005;59:587–635.
234. D'Amico C, et al. Development of vaccine formulations: past, present, and future. Drug Deliv Transl Res. 2021;11(2):353–72.
235. cdc.gov. Adult immunisation schedule: recommendations for ages 19yrs or older. 2021.; Available from: https://www.cdc.gov/vaccines/schedules/hcp/imz/adult.html.
236. cdc.gov. About Diptheria, Tetanus and Pertussis vaccination. 2021.
237. cdc.gov. Flu & people 65yrs and over. 2021.; Available from: https://www.cdc.gov/flu/highrisk/65over.htm.
238. Singh G, et al. Recombinant zoster vaccine (Shingrix(®)): a new option for the prevention of herpes zoster and postherpetic neuralgia. Korean J Pain. 2020;33(3):201–7.
239. cdc.gov. Chicken pox vaccination: what everyone should know. 2019.; Available from: https://www.cdc.gov/vaccines/vpd/varicella/public/index.html.
240. cdc.gov. Yellow fever recommendations. 2021.; Available from: https://www.cdc.gov/yellowfever/vaccine/vaccine-recommendations.html.
241. cdc.gov. Measles, Mumps, and Rubella (MMR) vaccination. 2021.; Available from: https://www.cdc.gov/vaccines/vpd/mmr/public/index.html.
242. cdc.gov. Rabies vaccine. 2014.; Available from: https://www.cdc.gov/rabies/medical_care/vaccine.html?CDC_AA_refVal=https%3A%2F%2Fwww.cdc.gov%2Fvaccines%2Fvpd%2Frabies%2Fpublic%2Findex.html.

243. cdc.gov. Polio vaccination: what everyone should know. 2018.; Available from: https://www.cdc.gov/vaccines/vpd/polio/public/index.html.
244. cdc.gov. Typhoid vaccine information statement. 2019.; Available from: https://www.cdc.gov/vaccines/hcp/vis/vis-statements/typhoid.html.
245. cdc.gov. About Haemophilus influenzae Type b (Hib) vaccines. 2021.; Available from: https://www.cdc.gov/vaccines/vpd/hib/hcp/about-vaccine.html.
246. Amanna IJ. Balancing the efficacy and safety of vaccines in the elderly. Open Longev Sci. 2012;2012(6):64–72.
247. Fda.gov. Evaluating inclusion and exclusion criteria in clinical trials [online]. 2018.
248. cdc.gov. Immunogenicity, efficacy and effectiveness of influenza vaccines. 2021; Available from: https://www.cdc.gov/flu/professionals/acip/immunogenicity.htm.
249. Chen WH, et al. Antibody and Th1-type cell-mediated immune responses in elderly and young adults immunized with the standard or a high dose influenza vaccine. Vaccine. 2011;29(16):2865–73.
250. Young F, Marra F. A systematic review of intradermal influenza vaccines. Vaccine. 2011;29(48):8788–801.
251. Arnou R, et al. Intradermal influenza vaccine for older adults: a randomized controlled multicenter phase III study. Vaccine. 2009;27(52):7304–12.
252. Chan T-C, et al. Immunogenicity and safety of intradermal trivalent influenza vaccination in nursing home older adults: a randomized controlled trial. J Am Med Dir Assoc. 2014;15(8):607.e5–607.e12.
253. Tsang P, et al. Immunogenicity and safety of Fluzone® intradermal and high-dose influenza vaccines in older adults ≥65 years of age: a randomized, controlled, phase II trial. Vaccine. 2014;32(21):2507–17.
254. Fluzone, Fluzone High-Dose and Fluzone Intradermal. 2021, U. S. Food and Drug Administration.
255. Domnich A, et al. Effectiveness of MF59-adjuvanted seasonal influenza vaccine in the elderly: a systematic review and meta-analysis. Vaccine. 2017;35(4):513–20.
256. Puig-Barberà J. MF59®-adjuvanted seasonalinfluenza vaccine. Aging Health. 2009;5(4):475–81.
257. Caproni E, et al. MF59 and Pam3CSK4 boost adaptive responses to influenza subunit vaccine through an IFN type I-independent mechanism of action. J Immunol. 2012;188(7):3088–98.
258. Del Giudice G, et al. Aflunov®: a vaccine tailored for pre-pandemic and pandemic approaches against influenza. Expert Opin Biol Ther. 2013;13(1):121–35.
259. Della Cioppa G, et al. A dose-ranging study in older adults to compare the safety and immunogenicity profiles of MF59®-adjuvanted and non-adjuvanted seasonal influenza vaccines following intradermal and intramuscular administration. Hum Vaccin Immunother. 2014;10(6):1701–10.
260. Seo YB, et al. Comparison of the immunogenicity and safety of the conventional subunit, MF59-adjuvanted, and intradermal influenza vaccines in the elderly. Clin Vaccine Immunol. 2014;21(7):989–96.
261. Ambrose CS, Levin MJ. The rationale for quadrivalent influenza vaccines. Hum Vaccin Immunother. 2012;8(1):81–8.
262. Chang LJ, et al. Safety and immunogenicity of high-dose quadrivalent influenza vaccine in adults ≥65 years of age: a phase 3 randomized clinical trial. Vaccine. 2019;37(39):5825–34.
263. Chahine EB. High-dose inactivated influenza vaccine quadrivalent for older adults. Ann Pharmacother. 2020;55(1):89–97.
264. Dunkle LM, et al. Efficacy of recombinant influenza vaccine in adults 50 years of age or older. N Engl J Med. 2017;376(25):2427–36.
265. Ainslie KEC, Haber M, Orenstein WA. Challenges in estimating influenza vaccine effectiveness. Expert Rev Vaccines. 2019;18(6):615–28.
266. Trucchi C, et al. Influenza vaccination in the elderly: why are the overall benefits still hotly debated? J Prev Med Hyg. 2015;56(1):E37–43.

267. Demicheli V, et al. Vaccines for preventing influenza in the elderly. Cochrane Database Syst Rev. 2018;2:CD004876.
268. Khan T, et al. Tailored vaccines targeting the elderly using whole inactivated influenza vaccines bearing cytokine immunomodulators. J Interf Cytokine Res. 2014;34(2):129–39.
269. Bang Y-J, et al. Effective inactivated influenza vaccine for the elderly using a single-stranded RNA-based adjuvant. Sci Rep. 2021;11(1):11981.
270. Brosio F, et al. A novel nonlive, adjuvanted herpes zoster subunit vaccine: a report on the emerging clinical data and safety profile. Infect Drug Resist. 2018;11:1401–11.
271. Lal H, et al. Efficacy of an adjuvanted herpes zoster subunit vaccine in older adults. N Engl J Med. 2015;372(22):2087–96.
272. Cunningham AL, et al. Efficacy of the herpes zoster subunit vaccine in adults 70 years of age or older. N Engl J Med. 2016;375(11):1019–32.
273. Schwarz TF, et al. Persistence of immune response to an adjuvanted varicella-zoster virus subunit vaccine for up to year nine in older adults. Hum Vaccin Immunother. 2018;14(6):1370–7.
274. Chlibek R, et al. Long-term immunogenicity and safety of an investigational herpes zoster subunit vaccine in older adults. Vaccine. 2016;34(6):863–8.
275. López-Fauqued M, et al. Safety profile of the adjuvanted recombinant zoster vaccine in immunocompromised populations: an overview of six trials. Drug Saf. 2021;44(7):811–23.
276. cdc.gov. Shingles (Herpes Zoster) | CDC [online]. 2021.; Available from: https://www.cdc.gov/shingles/index.html.
277. Derhovanessian E, Pawelec G. Vaccination in the elderly. Microb Biotechnol. 2012;5(2):226–32.
278. Dorrington M, Bowdish D. Immunosenescence and novel vaccination strategies for the elderly. Front Immunol. 2013;4:171.
279. Carroll TD, et al. Efficacy of influenza vaccination of elderly rhesus macaques is dramatically improved by addition of a cationic lipid/DNA adjuvant. J Infect Dis. 2013;209(1):24–33.
280. Leng J, et al. Efficacy of a vaccine that links viral epitopes to flagellin in protecting aged mice from influenza viral infection. Vaccine. 2011;29(45):8147–55.
281. Lim JS, et al. Flagellin-dependent TLR5/caveolin-1 as a promising immune activator in immunosenescence. Aging Cell. 2015;14(5):907–15.
282. Bulut O, et al. Overcoming immune dysfunction in the elderly: trained immunity as a novel approach. Int Immunol. 2020;32(12):741–53.
283. Hodgins B, et al. A single intramuscular dose of a plant-made virus-like particle vaccine elicits a balanced humoral and cellular response and protects Young and aged mice from influenza H1N1 virus challenge despite a modest/absent humoral response. Clin Vaccine Immunol. 2017;24(12):e00273-17.
284. Hodgins B, et al. Prime-pull vaccination with a plant-derived virus-like particle influenza vaccine elicits a broad immune response and protects aged mice from death and frailty after challenge. Immun Ageing. 2019;16(1):27.

Index

M. Orlu, F. Liu (eds.), *Pharmaceutical Formulations for Older Patients*, AAPS Advances in the Pharmaceutical Sciences Series 51,
https://doi.org/10.1007/978-3-031-35811-1

MIX
Papier aus verantwortungsvollen Quellen
Paper from responsible sources
FSC® C105338

If you have any concerns about our products,
you can contact us on
ProductSafety@springernature.com

In case Publisher is established outside the EU,
the EU authorized representative is:
Springer Nature Customer Service Center GmbH
Europaplatz 3, 69115 Heidelberg, Germany

Printed by Libri Plureos GmbH
in Hamburg, Germany